**THE LIBRARY
ST. MARY'S COLLEGE OF MARYLAND
ST. MARY'S CITY, MARYLAND 20686**

D1564536

Theories of the Earth and Universe

THEORIES OF THE EARTH AND UNIVERSE

A History of Dogma in the Earth Sciences

S. Warren Carey

Stanford University Press   Stanford, California

Stanford University Press
Stanford, California
© 1988 by the Board of Trustees of the
Leland Stanford Junior University
Printed in the United States of America
Last figure below indicates year of this printing:
98 97 96 95 94 93 92 91 90 89

CIP data appear at the end of the book

To my students

Preface

THE SEED OF THIS BOOK was sown by an Italian colleague who suggested that I trace the evolution of man's conception of Earth and the universe from the Stone Age to beyond Einstein, that I should address the book to the educated layman rather than to technical specialists, and that the tone should be autobiographical in the sense of tracing the beginnings and maturation of my own unorthodox philosophy of the earth and universe.

The Stone Age specification was easily satisfied, because in the early 1930's fate had placed me alone in the interior of New Guinea among primitives whose neolithic culture was no further advanced than that of our own ancestors millennia before Ur and Babylon.

The saga of my philosophy is much like Goethe's *Faust*. In his twenties, Goethe adopted the centuries-old Faustus tapestry and, to the resentment of contemporary European literati, interwove his own *mädchen* fabric from vibrant threads of his personal pains and passions. During the next fifty years, he reworked it over and again with his own evolving philosophy. Similarly, the roots of this book began with my adoption of Wegener's treatise just as he was perishing in a Greenland blizzard. I pulled out many threads and interwove major themes peculiarly my own, culminating in my Doctor of Science thesis written in 1936, wherein I first nailed my colors to the mast. During the fifty years that followed, the warp remained but I have perennially reworked the woof, resolving fuzzy areas and threading new themes as I was called on for presidential addresses, commemorative lectures, keynote speeches, and books. As with Goethe's *Faust*, each version grew directly out of its precursors. Therefore I should here review briefly some turning points along the path that ends (for the time being) with this book.

Some bellwether ideas were first expressed in my doctoral thesis, such as megashearing on the thousand-kilometer scale, a wholly heretic concept then (which led to the first recognition of the Tethyan torsion, treated here in Chapter 21), as well as rhombochasms (which really only paraphrased Wegener) and oroclines (softly, very softly, lest examiners lose patience with my heresy).

After the Second World War closed my New Guinea period, my debut as chief government geologist of Tasmania called for a maiden speech to the Royal Society of Tasmania in 1945, in which I called my creed. This stimulated the professor of physics in the University of Tasmania to invite the professors of mathematics, chemistry, biology, and engineering, the science curriculum adviser for high schools, and the director of the Tasmanian Museum to meet me in his home for a day-long Socratic inquisition. This had a good deal to do with the decision to found a geology department in the University of Tasmania, and subsequently, after due public advertisement, to appoint me as the foundation professor.

Two decades earlier, the American Association of Petroleum Geologists had convened a continental drift symposium in 1926 in New York in which the consensus had discredited continental drift as physically impossible. So during my early years as a professor I pondered deeply the physics of the crystalline flow of glaciers, of rock salt, and of the rocks of the earth's crust, the forces available in the earth to make them flow, and the effect of the scale of size and time on these phenomena. From these meditations came the rheid concept, the geotectonic scale concept, the mechanics of glacier flow, and the basic principles of folding. Gradually these insights became the foundations of my developing philosophy of global tectonics and led me to abandon the generally accepted canons of compressional tectonics. As such, they make up Chapters 15 to 19 of this book.

Meanwhile I began a counterattack on the stated reasons for the rejection of Wegener's theory of continental drift. First, Sir Harold Jeffreys, the mogul of contemporary geophysics, had long asserted that the fit across the South Atlantic, whence the idea of continental drift had sprouted, was a bad misfit, and everyone had believed him. I knew Jeffreys was wrong and could prove it, and in 1953 when Dr. G. M. Lees, my former chief in the Anglo-Persian Oil Company and then president of the Geological Society of London, repeated Jeffreys's claim in his presidential address, I confronted him. Lees saw that my rebuttal was published, and as a result this objection to continental drift has never been repeated since. This matter is discussed in Chapter 8.

Next I turned to Jeffreys's objection that it was physically impossible for continents to move about the globe, and in 1953 I sent a paper to the *Journal of Geophysical Research*, proposing a mechanism essentially the same as that adopted two decades later in the so-called "new global tectonics." Because of the prevailing dogma that the Wegener theory was a fantasy, the paper was rejected by the referees as naïve.

I had realized in the mid-1930's that if Wegener was right, any mountain belts that were actively forming during the time of Wegener's continental movements would be stretched or bent and would record the nature of the movements; I also knew that such a concept was too exotic for acceptance then. However, by the early 1950's, I felt that the time for attack had come, and I submitted the orocline concept to the Geological Society of Australia. The referees, three of the then most eminent geologists in Australia, turned it down, but the Royal Society of Tasmania published it for me in 1955. Of course it is now universally known to be valid, since all the rotations I proposed have since been confirmed by paleomagnetic measurements (listed in Chapter 9).

The next milestone was the international Continental Drift Symposium, which I convened in Hobart early in 1956. During this symposium I first realized that the Wegnerian model was not possible without gross expansion of the earth. In the symposium I explained and illustrated the separation of continents by rift valleys progressively widening into oceans by the repeated insertion of paired slices of mantle rock—the process that two decades later became the cornerstone of the "new global tectonics." The symposium volume was reprinted twice, but after that I considered it out of date and refused another printing. It has now become a classic, and there is pressure on me to reissue it.

Chester Longwell, then Chairman of Yale University's geology department, who attended this symposium as principal guest, invited me to Yale as visiting professor for a year (an invitation I was not able to take up until 1959). That led to my warm friendship with Harry Hess at Princeton and a grand tour of mobilistic evangelism through North America, recounted in Chapter 9.

Next followed a series of updates in various special lectures. My new ideas on folding (Chapter 16) formed the theme of the Honorary Anniversary Address to the Alberta Association of Petroleum Geologists, which I gave in Calgary in 1960. My review of the scale of tectonic phenomena (Chapter 15) was prepared when the Geological Society of India asked me to write a major article for their new journal. The asymmetry of the earth was the theme of my presidential address to the geology section of the Australian and New Zealand Association for the Advancement of Science (ANZAAS) in 1962. The cambium-like growth-strips parallel to the mid-ocean spreading ridges were featured in my Holland Memorial Oration in Calcutta in 1964. My Stanley Memorial Lecture, "Orthodoxy, Heresy, and Discovery," in Papua in 1966, is the source for parts of Chapter 23. A prognosis of future discovery was the theme of the Occasional Address at the annual meeting of the Royal Australasian College of Physicians in Hobart in 1967, and later the same year my Clarke Memorial Address to the Royal Society of New South Wales focused on the evolution of the southwestern Pacific. My 1970 presidential address to ANZAAS in Port Moresby updated my theory of Earth expansion, identifying the center of maximum continental dispersion near the Falkland Islands and the center of minimum dispersion in eastern Siberia (Chapter 21).

In the early 1970's I started to draw all these threads together in collaboration with the Elsevier Scientific Publishing Company of Amsterdam, first in 1975 as a 38-page summary in its *Earth Science Reviews*, then in 1976 by a comprehensive book, *The Expanding Earth*, which also incorporated my basic work on structural geology and tectonics generally. I also discussed the earth as a planet, including both differential displacements within and dynamic interactions with the sun and moon.

By this time, the global tectonic revolution in North America had routed all opposition to the gross dispersion of continents and had reached what I had been teaching my students in the early 1950's, but I was disgusted that the "new global tectonics" had gone only halfway. It still assumed axiomatically, notwithstanding the patent rapid growth of new oceanic crust, that the size of the earth had remained essentially constant. Hence it had to go back to the mechanism I had adopted in the 1930's and 1940's of swallowing great areas of crust down the ocean trenches, but which, after 20 years of working with it, I had found by 1956 to be unworkable on a global scale. Hence my Elsevier book set out to quash this subduction myth (see Chapters 12 and 13 herein).

Meanwhile another fundamental trend began to dominate my thinking. We all believe that the laws of nature are universal, exactly that. In other words, any phenomena we recognize on Earth must be compatible with those of the whole universe. I had concluded that not

Preface

only had the volume of the earth greatly increased, but so had its mass. What about the moon, the other planets, the sun, the other stars, the galaxy, indeed the whole universe?

In my Elsevier book, I recognized these as proper questions and suspected that the solution would prove to be cosmological. I suggested that everything in the universe canceled to zero—positive and negative charges, momenta, and mass and energy—as inseparable opposites like two sides of a coin. Indeed, how else could a universe come into being? If there had been a state of zero before the beginning, everything would have to add to zero after the beginning. I devoted my 1977 Johnston Memorial Address to the Royal Society of Tasmania to this theme, developed it further at the 1983 international Symposium on Earth Expansion, which I convened at the University of Sydney early in 1981, and crystallized it in the final chapters of this book. This has involved uniting Newton's and Hubble's laws into a duplex empiricism, and exploding the "Big Bang."

This armchair philosophy of an aging scientist does not warrant a formal bibliography, but I have listed below the succession of my own papers through which this book has rooted, matured, and come to harvest. Each of these has its own catalog of references; indeed, the Elsevier book has 820 of them.

1938   Tectonic evolution of New Guinea and Melanesia. D.Sc. Thesis, University of Sydney.

1945   Tasmania's place in the geological structure of the world, Address to the Royal Society of Tasmania, May 15, 1945.

1954   The rheid concept in geotectonics, *Journal of the Geological Society of Australia*, v. 1, pp. 67–117.

1955   Wegener's South America–Africa assembly, fit or misfit? *Geological Magazine*, v. 43, no. 3, pp. 196–200.

The orocline concept in geotectonics, *Proceedings of the Royal Society of Tasmania*, v. 89, pp. 255–88.

1958   The tectonic approach to continental drift, in S. W. Carey, ed., *Continental Drift: A Symposium*, pp. 177–363. Hobart: University of Tasmania.

1959   The tectonic approach to the origin of the Indian Ocean, in *Third Pan–Indian Ocean Science Congress, Madagascar, Proceedings*, pp. 171–228.

1961   Glacial marine sedimentation, in *First International Symposium on Arctic Geology*, pp. 903–32. University of Toronto Press. (With Naseeruddin Ahmed.)

1962   Folding, *Journal of the Alberta Society of Petroleum Geologists*, v. 10, no. 2, pp. 95–144. (Honorary Anniversary Address.)

Scale of geotectonic phenomena, *Journal of the Geological Society of India*, v. 3, pp. 97–105.

1963 The asymmetry of the Earth, *Australian Journal of Science*, v. 25, pp. 479–88. (Presidential Address.)

1964 Tectonic relations of India and Australia, Holland Memorial Oration to the Mining, Geological, and Metallurgical Society of India.

1967 2000 A.D.—Prognosis, *Medical Journal of Australia*, June 24, 1967, pp. 1235–42. (Occasional Address.)

1970 Australia, New Guinea and Melanesia in the current revolution in concepts of the evolution of the Earth, *Search*, v. 1, no. 5, pp. 178–89. (Presidential Address.)

1972 The face of the Earth, *Australian Natural History*, v. 17, no. 8, pp. 254–57.

1975 The subduction myth, *Proceedings of the Southeast Asia Petroleum Exploration Society, Singapore*, v. 2, pp. 41–69.

The tectonic evolution of Southeast Asia, *Indonesian Petroleum Congress, Jakarta, Proceedings*, pp. 1–31. (Invited lecture.)

The expanding Earth: An essay review, *Earth Science Reviews*, v. 11, no. 2, pp. 105–43.

Palaeomagnetism and Earth expansion, *Chayanica Geologica* (Calcutta), v. 1, no. 2, pp. 152–95. (Invited paper.)

1976 *The Expanding Earth*. Amsterdam: Elsevier.

1978 A philosophy of the Earth and Universe, *Proceedings of the Royal Society of Tasmania*, v. 112, pp. 5–19. (Johnston Memorial Address.)

1979 Genesis of the Himalayan system from Turkey to Burma, in *Himalayan Geology Seminar* (New Delhi), sec. IIA, pp. 401–16. (Geological Survey of India, Miscellaneous Publication No. 41.) (Invited lecture.)

1983 Evolution of beliefs on the nature and origin of the Earth, in S. Warren Carey, ed., *The Expanding Earth—A Symposium, University of Sydney, February, 1981*, pp. 3–7.

Tethys and her forebears, ibid., pp. 169–87.

Earth expansion and the null Universe, ibid., pp. 365–74.

The necessity for Earth expansion, ibid., 376–96.

1985 Geotectonic setting of Australasia, Principal Address to the Second South-Eastern Australia Oil Exploration Symposium, Melbourne, November 1985.

1986 Tethys and her forebears, in *International Symposium: Shallow Tethys 2, Wagga Wagga, Australia, September 1986*. Rotterdam: Balkema. (Opening Address.)

Diapiric krikogenesis, in *The Origin of Arcs* (Developments in Geotectonics series). Amsterdam: Elsevier. (Keynote Address.)

La Terra in espansione. Rome: Laterza. (In Italian.)

Preface

The figures in this book have been drawn from all of these publications (which explains some of the variations of style). A vista of draftsmen have been of great assistance to me, but over recent years I should particularly acknowledge Mrs. June Pongratz, whose skills and patience are deeply appreciated.

My entry to this narrative was as a geologist. But, even as a raw graduate working in the Werrie Basin of New South Wales, questions emerged not hitherto asked, which I felt obliged to answer. During my New Guinea years, geologic problems were rooted in other sciences, but convention frowned on transgression into other fields. Science had become so specialized that each branch must be left to its oracles. Generalists like Aristotle, Leonardo, Hooke, and Darwin belonged to past ages. Let the tailor stick to his needle!

The passing decades crystallized three observations. First, specialists in other fields were preoccupied with their own problems, and not inclined to ponder questions asked from without. Second, when they did answer, they did so cursorily, with scant attention to the geological evidence. Third, each scientist had a responsibility to pursue the implications of his conclusions to finality, wherever they might lead in other disciplines.

So pursuit of the earth impelled me into the universe and the cosmos, where my indoctrination was thin. Therefore I am bound to have made errors, but so be it. The very errors may elicit the truth. On the other hand, I may have been spared false axioms inherent in the established creed.

S.W.C.

Contents

PART ONE. RETROSPECT

Chapter 1. Philosophy in the Stone Age   3
     The Dawn of Science, 5   Primitive Gods, 6   Man, the Ultimate Conception, 7   Revelation Versus Observation, 9   The Enigmatic Universe, 12

Chapter 2. Planet Earth   14
     Earth, Center of the Universe, 16   The Central Sun, 20   The Shape of the Earth, 22   Earth, a Magnet, 27

Chapter 3. Fossils   33
     Myths and Superstitions, 33   The Renaissance in Italy, 36   The Man Who Witnessed the Flood, 38   Baron Cuvier, 39   Concepts of Evolution, 41

Chapter 4. Neptunists and Plutonists   45
     Werner and Hutton, 46   Werner's Neptunism, 48   The Challenge of Plutonism, 51   Basalt: Sediment or Lava? 52   The Granite Controversy, 53   Plutonism Wins Out, 56   Catastrophism and Uniformitarianism, 57   Aspects of Uniformitarianism, 60

Chapter 5. The Ice Age   63

Chapter 6. The Age of the Earth   70
     Lord Kelvin, 71   Sophist Appeasers, 73   Courage of Convictions, 74   Radioactivity to the Rescue, 75   Radioactivity Clocks, 77   The Geological Time Scale, 78

Chapter 7. Numeracy in Geology   80

PART TWO. MOBILE CONTINENTS

Chapter 8. Continental Drift 89
The South Atlantic Fit, 89   Wegener's Predecessors, 90
Wegener, 92   The AAPG Symposium, 97   Du Toit, 97
Flow and Convection in Solids, 100   Decades in Contempt, 101

Chapter 9. Sowing the Seeds of Revolution 105
Paleomagnetism, 105   Oroclines, 113   Ocean-Floor Spreading, 114   The Hobart Symposium, 117   American Evangelism, 118

Chapter 10. The Kuhnian Revolution 120
Paleomagnetic Polarity Reversals, 122   Plate Tectonics, a Shotgun Wedding, 126   Transform Faults, 129   The Revolution to End Revolutions, 132

PART THREE. THE EXPANDING EARTH

Chapter 11. Development of the Expanding-Earth Concept 137
The Gravitational Constant, 141   Evidence from Mining Geology, 142   Volume of Seawater, 142   My Conversion to Expansion, 143   Heezen and Wilson, 143   A Pulsating Earth, 145   Revival in the 1960's, 145

Chapter 12. The Earth Is Expanding 150
The Arctic Paradox, 150   The Paradox of Paleopole Overshoot, 151   The Pacific Paradox, 153   The Pacific-Perimeter Paradox, 156   The Gape Artifact, 158   The India Enigma, 159   Missing Archean Crust, 162   Missing Ophiolites and Flysch, 164   Cartographic Precision, 164   NASA Geodetic Measurements, 167   The Earth *Has* Expanded, 172

Chapter 13. The Subduction Myth 174
The Africa Enigma, 174   The Peru–Chile Trench Anomaly, 176   The Kermadec Trench Anomaly, 178   Himalaya and Tethys, 178   The Myth of the Iapetus Ocean, 180   The Zodiac Fan Anomaly, 184   Why No Residual Ocean Floor? 186   Subduction *Is* a Myth, 187

Chapter 14. Criticisms of Earth Expansion 190
Surface Gravity, 190   The Volume of Seawater, 191   Paleo-

Contents   xvii

magnetism, 192   Growth Lines of Fossil Corals, 195   The Blinkers of Dogma, 197   Other Planets, 199

PART FOUR. VERTICAL OROGENESIS

Chapter 15. *Gravity Rules the Earth*   205
Vertical or Horizontal? 206   Isostasy, 208   The Significance of Scale, 209

Chapter 16. *Folding*   217
Similar and Concentric Folding, 217   Similar Folding Implies Flow, Not Static Compression, 219   Concentric Folding Implies Décollement at Depth, 222

Chapter 17. *Diapirs*   225
Salt Domes, 225   Gneiss Domes, 232   Glacier Analog, 234   Tectonic Diapirs, 235

Chapter 18. *A Simple Model of an Orogen*   237
Folding and Thrusting, 239   The Myth of Alpine Foreshortening, 240   The Orogenic Root, 241   The Appalachians, 245   The Alps, 247   The Himalaya, 248

Chapter 19. *The Benioff Zone*   251

PART FIVE. TECTONICS OF THE WHOLE EARTH

Chapter 20. *Global Extension*   261
The Hierarchy of Expansion, 261   Relative Movement of Primary Continental Prisms, 266   Mantle-Welded Continents, 269

Chapter 21. *Global Torsions*   271
Tethyan Torsion, 278   Transverse Extension Across Tethys, 282   The Cause of Tethyan Torsion, 285   A Dextral Conjugate to the Tethyan Torsion, 286   Global Expression of the Conjugate Torsion, 289   Integration of Spreading Ridges and Torsions, 302

Chapter 22. *Evolution of the Lithosphere*   306
What Is the Tethys? 306   Earlier Analogs of the Tethys, 307   Birth of the Pacific, 313   Summary, 321

PART SIX. A PHILOSOPHY OF THE UNIVERSE

Chapter 23. The Earth and Cosmology   325
    Hubble's Law and the "Big Bang," 329   Olbers' Paradox, 332
    Newton Attraction and Hubble Repulsion, 334   The Null
    Universe, 338   The Cosmological Principle, 342   Elton's
    Planetary Evolution Process, 347

Chapter 24. Philosophical Speculations   351
    Antimatter and Black Holes, 351   Gravity Waves, 352
    Dimensional Equivalence of Mass and Energy, 353   Cosmic
    Rotation, 355   Mass, Substance, Energy, and Mind, 356
    Fogs of Notation, 357   Are We Alone? 360

Epilogue   363

Glossary   369

Index   391

PART ONE
RETROSPECT

CHAPTER 1

Philosophy in the Stone Age

FIFTY YEARS AGO, I was a geologist working deep in the interior of New Guinea. Weeks would pass without my seeing any other white man, and when I did see one he would be another of our exploration company. Months elapsed between mail from the world beyond. I had no radio, thus no communication or news or entertainment. My companions—my carriers and the local tribes—were Neolithic Papuans who had never seen a wheel, nor any metal before ours. As communication deepened, I learned to understand them, to love and respect them. I found honesty rather than deceit, and more goodwill than malice, more kindness than greed, more loyalty than treachery, more courage than cowardice. I found compassion too and, yes, brutality, yet none worse than the Inquisition in the name of Christianity, nor jihad in the name of Islam. Indeed, they were very like our so-superior selves, but in general a kinder race than us, and, notwithstanding their lack of the "civilized" amenities, happier.

But what of mind and intellect? Karl von Zittel, president of the Bavarian Academy of Sciences, and perhaps the greatest systematic paleontologist ever, had written in 1899: "The wide chasm between the childish Saga of Creation handed down by the Bushmen, Australians, Eskimos and Negroes, and the grand poetic conceptions of the Aryan-Germanic races of Europe, conveys to us the immense difference at that time in the condition of culture and intellectual capacity of these peoples." Still more inferior would be my Papuans. But they taught me what they knew of nature, and how all came to be so. They knew intimately the prolific jungle life, and its potential yield of foods, fibers, fabrics, tools, vessels, drums, weapons, adornment, medicines and potions, herbs and spices, and construction materials for houses, rafts, and canoes. They hunted with impunity the slow-witted crocodile, which flees sudden confrontations, but woe betide them if that fearsome reptile lay in wait for them at fords or watering places. Natural brine springs were valued because culinary salt was savored, and hostile tribes sequestered them from the sea. Tough stones were sought in river gravels for grinding axe and club heads—such a laborious task that, once made, they were highly prized, inherited, traded, or seized in combat. Neither peat nor coal was recognized as fuel—why should either be, with firewood so abundant? Only the more advanced tribes selected clays and fired crude pots (shaped by hand, not spun); they could not make glass, nor any metal, the first rung on the ladder of technology. They dug ochres for pigments. They knew where shale contained fossil shells, which they burned for lime to blend with their betelnut or to bleach their brown hair. They guessed that somehow these shells belonged to the sea, but had no inkling that the sea had been there 7 million years ago.

They measured the passage of the day, not only by the Sun but by the jungle sounds of birds and insects. They counted time as so many suns, so many moons, and by the position of the "Year-man" (the star cluster we call Orion, who pursues the "Year-woman"—the Pleiades), who rose later each night until after 13 moons he was back where he had started. The 13 was counted and communicated on the digits of their hands and feet.

A dozen or so kilometers would find a different dialect. There was no writing, no ideographic symbols (except taboo and warning signs), although miming was understood and effective. Some information could be transmitted from village to village by the rhythmic pounding of the great wooden drum, and at least one mountain tribe (the Arisili) had developed a remarkable system of communication via bugle-like conch shells, yodel, or whistle. This was not by code but rather by miming the rhythm and cadence of the message. Music was already essential to them—rhythm, duration of notes, and two-part harmony of tenor and baritone (especially the Arisili). They had percussion instruments (the wooden *garamuts* and hand-held *kundus* with lizard-skin diaphragms) and bamboo pipes, each cut to length for specific notes.

They had rules of land tenure and property—every tree and palm in the jungle had its owner and every patch of forest its legitimate hunters and recognized access for their slash-burn-sow-harvest-abandon style of gardens. They had formal kinship and taboos, moral codes and sanctions. Many tribes had not realized that a single coitus could suffice for pregnancy but believed that a baby had to be repeatedly "made" over a sufficient period. Of course there were many supernatural beliefs and superstitions; there were sprites and demons, good and evil.

Over the years, I came to realize that Professor von Zittel's judgment was jaundiced; he had never seen a Papuan, still less lived with them as his only companions for months at a time. The Papuans' intellect, their capacity to observe and think, was not less than ours. The difference lay not in our inborn capacities but in our legacy of knowledge, of beliefs and inferences inherited step by step through the millennia by our ancestors, winnowed and re-winnowed, refined and re-crystallized, toward the quintessence of our science. By contrast, the repertory of knowledge and dogma of my Papuan companions was as embryonic as that of our own Neolithic ancestors must have been 6000 years ago, with only the most primitive pots and before any metals, glass, or writing. Yet in solving problems, involving matters they already understood, their minds were certainly as nimble and efficient as mine, and I listened and learned from their lips and relied on their judgment. Their dogma of accepted truth, just like that of our own ancestors, contained much that we now believe to be false. But I have no doubt that our own orthodox dogma still contains falsities within the self-evident axioms we believe we know to be true. Our most intractable beliefs are those we took with our mother's milk.

The Dawn of Science

Let us then, in retrospect, retrace from the earliest times the step-by-step recognition and rejection of false axioms among our own ancestral beliefs about the earth and universe. Like me, you may be surprised to find that the rate of recognition of false axioms (and adoption of bold new ones) has accelerated through the millennia, the centuries, and the decades, right down to our own lifetime. Only the naïve would believe that at last our dogma is pure. The most likely site for error is in the most fundamental of our beliefs.

How and when did science start? Anthropologists debate the order of development and the relative "humanizing" contributions of our descent from the trees, our upright gait, enlargement of the brain, and an effective social conscience, the harbinger of morality. Recognition of cause and consequence, and anticipation of consequence from cause, clearly preceded the anthropoids, echoing from vertebrates far back, but such recognition, although essential to science, remained irrelevant without communication to other individuals. Dogma began with communication. Science was born with the first telling to another of the how or the why of something. Repetition, amendment, and codification of that communication was the beginning of dogma, to be passed on thereafter as knowledge and belief, ossifying further with each retelling. Before the written word, only narrative and ritual transmitted beliefs through the generations.

Primitive Gods

So much remained utterly beyond the comprehension of my Papuan friends that superior beings had to be invoked to explain them. Storm and tempest, thunder and lightning, volcanoes and earthquakes, seasons, rivers and floods, the seas—all were attributed to gods. Even the more sophisticated natives, faced with the incomprehensible possessions of white men in Rabaul and Port Moresby—great ships, airplanes, guns, motor trucks, telescopes, cameras, gramophones, radio conversations with people far away over the sea, canned meat, miracle medicines, anesthesia—knew that such impossible things could only be the work of gods. Their pagan priests convinced them and their tribes that, with proper incantations and rites, they could expect that great ships would come laden with such cargoes for them, and that even those cargoes now arriving and taken by the white men were rightly theirs. Every few years, to this day, a new epidemic of such "cargo cults" erupts.

Since the dawn of prehistory everyone who has contemplated the stars has wondered how, whence, why, and whither? Most have shunted the problem into the too-hard basket. Their savants, denied such escape, invoked multiple gods, single gods, matriarchs, or pantheism. Most assumed a special initial state: the Babylonian battle of Marduk and Tiamat the goddess of the sea, the Egyptian egg, the Hindu tortoise bearing the elephants that supported the world, the Polynesian air-god Tangaloa, or the pregnant chaos of Milton and the Garden of Eden. Before the Aryan invasion of the eastern Mediterranean about 2000 B.C., the Neolithic people there sacrificed to a supreme fecund matriarchal goddess and her three nubile nymphs, who copulated for pleasure with no recognition of any male function in fertilization, which was attributed to the winds or rivers.

Hesiod's poem *Theogony*, in the eighth century B.C., relates the genealogy of the gods from Cronus to Zeus, and thence the history of the creation of the earth, hell, ocean, night, sun, and moon. Democritus of the fifth century B.C. was the first on record to conceive a self-created universe from the random occurrence of atoms (but whence his atoms?). Today's cosmologists postulate a "Big Bang" when the whole mass of the universe was created by God at a point, which then exploded to expand at colossal speed to form the universe as we know it. But modern Russian materialists divorce morality from theology and dismiss all gods as superstitions we should have outgrown.

With our own ancestors, two kinds of gods emerged: the meddling god of the Zoroastrians and of Plato, who interfered in human affairs as and when he chose, and the aloof god of Descartes, who created matter and motion, then left the universe to develop according to the laws he ordained, and which he himself obeyed. For the former, the physical and spiritual realms interact. Heaven and hell are real places. For the latter, they have no physical existence, below our feet or in the sky, and soul has no mass, charge, electromagnetic field, or any other physically detectable property whatever. God of Judaism, Islam, and Christianity is clearly of the interfering category, for otherwise there could be no miracles, nor exorcism, nor prayers for amelioration of human affairs, and practices such as the blessing of fishing fleets would be irrelevant.

Let me cite an analogy from mathematics. In the algebra of complex numbers, where an equation contains terms both "real" and "imaginary" (those containing the square root of minus one, there being no number whose square is -1), such an equation can be replaced validly by two independent equations, one of which contains all the "real" terms and the other all the "imaginary" terms. Similarly, with complex propositions involving both material and spiritual, these immiscibles should yield segregated propositions, each complete in itself. Surprisingly, some mathematical propositions containing only "imaginary" terms have significant practical value in electromagnetic theory; likewise it does not follow necessarily that spiritual propositions, segregated from physical reality, thereby lack relevance or beneficence.

Some gods were conceived in the form of animals, real or imaginary, but supreme gods were created in man's own image, since nothing superior could be conceived. Before me as I write is a bronze miniature of the great statue of Zeus casting his thunderbolt. He is the supreme perfection of an Athenian athlete.

Man, the Ultimate Conception

The egoism of man, who created his god in his own image, then reversed the honors to have his god create man in *his* image, has pervaded his dogma and philosophy. Organic evolution ends with man.

His world was the center, not only of the solar system but of the whole universe, rather than a minor satellite of a very ordinary sun among billions of others in an ordinary galaxy among billions of others in a universe without identified limits in time or space. He has believed and taught that oceans are fixed and eternal, and has taken it for granted that the size and mass of his earth have been much as now since inception, and that the universal inheritance of matter and energy was determined at creation, to remain constant forevermore.

In contrast to the concept that the physical and biological environment of *now* was the norm of the past, paleontologists retrospect the ascent of life from simple molecules, nearly 4 billion years ago, to the oldest known proteinoid microspheres (in the 3.8-billion-year-old Isua Quartzite in Greenland), through primitive organisms that had not yet evolved a true cell nucleus (the prokaryotes of 2.8 billion years ago), to eukaryotes with fully developed cell nuclei (1.7 billion years ago), to metazoa, which had developed specialized tissues and organs, thence through the invertebrate phyla to vertebrates, and so on to the emergence of life on land some 400 million years ago, to the reign of reptiles that climaxed some 200 million years ago, thence through early mammals to the primates, and finally to man. Embryologists see each one of us individually retrace in nine months this 4-billion-year sequence from blood chemicals to single-celled ova, thence through fetal stages similar to the whole evolutionary path, to human birth. When Robert Chambers stated in 1844 that ontogeny (development of the individual) recapitulates phylogeny (development of the race), his book created a sensation. This principle was emphasized by the German biologist Ernst Heinrich Haeckel (1834–1919).

But let us set aside our conceit, and turn from our retrospect of evolution to ponder the future eons, and to contemplate what species, genera, families, orders, classes, and phyla of beings may follow us. Could other vertebrates eventually supersede us, perhaps descendants of the so-clever dolphins? Could even our vertebrate lineage be superseded by progeny from some other phylum, perhaps from the cephalopods (octopuses do have very large brains for their size!) or arthropods? Arthropods were the dominant phylum 600 million years ago, and today outnumber all the rest combined. Reptiles, not our ancestral mammals, were the dominant class 200 million years ago. What group will lead 100 million years, 1000 million years hence?

Revelation Versus Observation

In medieval Europe, all clergy and nearly all the population believed in the literal truth of the scriptures, absolutely. Any suggestion to the contrary was heresy, inspired by Satan and deserving of severe punishment and eradication.

For a thousand years the biblical dogma, that the scriptures recorded the literal truth, straitjacketed geology and astronomy. Observation and deduction were redundant, because the truth was written for all to read. During the Dark Ages the flat Earth was revived, with vaulted dome above the firmament that separated the waters, and then, above that, heaven. In the fifth century A.D., Father Lucius Lactantius denounced notions of a spherical Earth as absurd heresy, notwithstanding that Pythagoras had proved that the earth is spherical a thousand years before him. In the sixth century, the ecclesiastical dogma in Europe was as naïve as the tenets of ancient Babylon. Although Eratosthenes of Alexandria, in the third century B.C., had produced a map of the then known world with latitudes and crude longitudes, Cosmas, also of Alexandria, in 540 made it flat. His *Topographicus Christiana* depicted Christendom as a flat rectangle (Fig. 1), with indentations for the Mediterranean, Caspian, and Red seas and

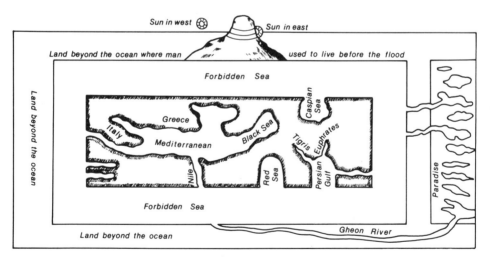

Fig. 1. World map by Cosmas, 540 A.D., in *Topographicus Christiana*. The words have been translated from the Latin and Greek. The Gheon River from Paradise feeds the Nile. Pythagoras had demonstrated the spherical earth a thousand years earlier.

the Persian Gulf, opening into the forbidden seas, also rectangular, surrounding the central land. Beyond the forbidden sea along the north side was *terra ultra oceanum ubi ante diluvium habitabant homines* (the land beyond the ocean where men used to live before the flood). There, too, was a very high mountain, behind which the sun passed at night, high up in summer so that nights were shorter, and lower down in winter. Cosmas (also latinized as "Indicopleustrus"—Indian navigator) should have known better, for he had sailed the Mediterranean, the Red Sea, the Persian Gulf, and south to the tropics at Zanzibar before becoming a monk recluse in a Sinai monastery, where the scriptural dogma erased his own experiences.

Much of the Greek progress had been lost in the destruction of the great library of Alexandria, first by Julius Caesar in 48 B.C. and finally by burning by the Arabs after the collapse of the Roman empire in 642 A.D. (400,000 precious volumes, including the works of Aristarchus, Eratosthenes, and Ptolemy, perished). Here again absolute belief in "the book" (that is what *al-Quran* means) prevailed. The Arab general Amru's instruction from Caliph Omar was: "The contents of those books are in conformity with the Quran or they are not. If they are, the Quran is sufficient without them; if they are not, they are pernicious. Let them, therefore, be destroyed."

Ironically, some of the Greek scholarship was preserved in Islamic courts along the old silk route. Some ruling caliphs patronized as status symbols various local and foreign scholars, whose Arabic translations are our only source of many Greek manuscripts. The Arabs also made some advances of their own in mathematics and astronomy, less so in biology, not at all in geology. Papermaking, invented in China, was transmitted by the Arabs throughout the vast Islamic empire to Morocco and Spain, and thence to Western Europe. For seven centuries the Arabs, virtually alone, nurtured the spirit of inquiry. But they also wantonly destroyed. Erudition and culture were fashionable, but whole libraries of irreplaceable manuscripts were sacked and burned by rival warlords.

Aryabhton, in India in the fifth century A.D., taught that Earth turned daily on her axis, although still the center of the system. Ibn-Sina (980–1037), better known as Avicenna, who was born and lived in Tadzhikistan, translated much of Aristotle into Arabic and believed that slow erosion by running water had carved valleys out of the mountains, and that land and sea had changed places many times through an everlasting history of the earth, and that fossils recorded these events. As these ideas conflicted with the literal Koran, Moslem clergy persecuted him and burned his papers. A contemporary, the Uzbek philosopher Ahmad al-Biruni (972–1048), likewise interpreted the fish and shell fossils found far inland as proof of the repeated flooding of the land by the sea.

Finally, Pope Sylvester II, a French Benedictine who had studied Greek, mathematics, and astronomy before his enthronement in 999 and knew the works of Plato and Eratosthenes, reestablished the spherical Earth, though it remained motionless at the center of the universe. (Even Brahe six centuries later would deny the earth's rotation because a cannonball fired east did not go farther than one fired west.) Nevertheless his influence had turned the tide, so that 1000 A.D. was the nadir of the Dark Ages.

Unfortunately, however, the renaissance brought enslavement to another dogma—the dead hand of Aristotle—and the next five centuries saw a tug-of-war between dogma and observation, between revelation and deduction. The Inquisition, for four centuries a powerful arm of the Roman Catholic Church, brought the harassment and slaughter of a succession of savants. In the thirteenth century, Roger Bacon (ca. 1214–94), *Doctor admirabilis*, a philosopher and scientist and only coincidentally a Franciscan friar, was convicted of heresy and witchcraft through the jealousy of his less erudite brothers and imprisoned for ten years. In jail he wrote his three great works, copies of which he was ordered to send to Rome to Pope Clement IV, whose predecessor had forbidden him to read them during the time he had been papal legate in England. After ten years of freedom, Bacon was jailed for another ten years, this time for segregating science and theology.

In the sixteenth century, the Catholic Church sternly denounced Nicolaus Copernicus's heresy that Earth and all the planets revolved around the sun, and added his work to the list of banned books, where it remained for two centuries. Martin Luther, though the quintessential Protestant, also damned Copernicus because, he said, only a fool would advance a thesis in direct contradiction to the Holy Scriptures. Giordano Bruno (1548–1600), a natural philosopher of remarkable geological vision, was condemned by the Inquisition and burned at the stake in Rome for his suggestion that there were other worlds, and for his propagation of the Copernican system. Galileo narrowly avoided the same fate in 1633 by recanting. Not until 1984 was Galileo's damnation for heresy revoked by the Church.

In the middle of the last century an acrimonious and trenchant controversy erupted between Tayler Lewis, professor of oriental languages and lecturer on biblical and classical literature at Union College in New York state, and James Dwight Dana, professor of natural history at Yale, and probably the most celebrated geologist in the United States. Dana was an ultraconservative Christian who believed the biblical doctrines absolutely and maintained that all the works of science endeavored to confirm the scriptures. Lewis considered even this use of science to be impious; he asserted that the *only* permissible study of the origin and nature of the earth was the study and interpretation of the written word, which he did in Latin, Greek, and Hebrew. For him, interpretation of the scriptures by science was presumptuous and contemptible blasphemy.

Nearly all the leading scientists of Christendom and Islam, right up to the present, have affirmed the axiomatic belief, inherited from Judaism, of a single God. In the eighteenth century, Jean André de Luc, of Geneva, coiner of the name "geology," devoted his life to the study of natural phenomena in order to prove the validity of the Holy Scriptures. This initial premise has excluded contrary hypotheses and allowed retention of concepts impossible without divine intervention, either at the beginning or along the way. Albert Einstein's cosmos begins with the divine creation of matter, notwithstanding that such creation is contrary to the conservation laws at the very foundation of physics. In this, rightly or wrongly, we do as did our most primitive ancestors—invoke the supernatural when we reach the limit of our own observation and deduction.

The Enigmatic Universe

Who should probe the fundamental enigma of the origin and destiny of the universe? A cosmologist? Astrophysicist? Mathematician? Biologist? Theologian? Perhaps even a geologist, who at least has attempted to unravel the history of the earth. Alexis Carroll in his *Man the Unknown* wrote that to an anatomist, man is a system of bones and muscles; to a physiologist he is a sack of organs, to a biochemist a symphony of enzymes and chemical complexes, to a psychologist a mind, and to a priest a soul. Man is all these, and more. Each discipline sees but an abstraction of the totality; so with the universe. There has also been an oriental bias to an imaginative mysticism, whereas the Golden Age of Greece was more abstractly philosophical, the Roman age more pragmatic, and post-Renaissance Europe empirical. But throughout, it is dogma that dogs us—our heritage of belief, our creed, be it a doctrine of orthodox science or of religion. The wise but untaught may see new truths invisible to the savants.

I used to tell my students that I would do my best to teach them what I believed to be true, but I had no doubt that some of what I told them would turn out to be wrong; unfortunately I did not know which parts. Every professor before me had later found he had been wrong in some of the things he had conscientiously taught, and I should not expect to fare better. So my charge to them was not "believe what I say," but "disbelieve *if you can.*" In my own independent way, I had followed the path of René Descartes three centuries before. A bishop is charged with transmitting inviolate the scriptures entrusted to him, although he should check and recheck that no alteration or translation error has occurred since the source; a professor is likewise charged with the transmission to his students of the dogma he has received, but he must persistently seek errors in its source, its premises, and its logic, and urge his students to do the same.

CHAPTER 2

Planet Earth

SAVANTS AND PRIESTS of the earliest cultures, through the Babylonians, Egyptians, Hebrews, and early Greeks, *knew* that their Earth was flat beneath the vault of the heavens, where coursed the Sun-god, the Moon, the five planet-gods, and the stars. They knew also that beyond the heavens lay the spirit domain, and the realm of fire. To think otherwise was absurd. If the earth had another side beneath our feet, rain would have to fall upward, water would not stay in lakes, and people would have to walk upside down. The earth was still; we would feel the motion if it were not. Sun came up every morning in the east—indeed, that is the only meaning of "east"—passed overhead, and went down in the west, with much speculation about how he got back for the next morning. Moon and stars made similar transits. With the dawning of the Golden Age of Greek philosophy, Pythagoras, in the sixth century B.C., deduced that the earth was spherical, as did his pupil Parmenides a decade younger. But validity has never shielded denial of accepted dogma from scorn, and Lactidurus in the fourth century B.C. still ridiculed the idea of people walking upside down on a spherical Earth.

Nevertheless Aristotle (384–322 B.C.), Plato's pupil and founder of the Lyceum in Athens, who contemplated rocks, minerals, metals, earthquakes, rivers, and springs, insisted that Earth must be spherical, because ship's hulls went below the horizon while masts were still visible, and approaching ships sighted mountain peaks first and lower lands later; on a calm day at sea the horizon could be seen to be an arc, higher at the center; Earth's shadow on the moon during eclipses showed her spherical shape; the rising and setting of the sun, moon, and stars differed for more northerly localities, and the pole star rose higher from the horizon farther north. From all this Aristotle also deduced that the earth-sphere must be relatively small compared with the distances of the planets: "The size of the Earth is nothing, absolutely nothing, compared with the whole heavens. The mass of the Sun must be far greater than the mass of our globe, and the distance of the fixed stars from us is much greater than that of the Sun."

Even so, Zeno (340–265 B.C.), the founder of Stoic philosophy, remained unconvinced that the earth was spherical. Finally, aesthetic harmony and symmetry, so important to Pythagoras and fast becoming a dogma with the Greeks, implied the harmony of a spherical Earth at the center of the celestial spheres. And so the obvious axiom, that the earth is flat, was abandoned, but for the wrong reason. Herein Pythagoras and Aristotle contrast "the two cultures," arts and science, contemplation and observation.

Quite independently of the Greeks, Chinese scholars, at least by the first century A.D., believed the earth to be spherical. They may have learned this from the Sanskrit Aryans, who knew this much earlier.

A century later, Eratosthenes (276–194 B.C.), librarian of Alexandria and pioneer geographer, with a level of luck that does not always grace good reasoning, measured the circumference of the spherical Earth to within 2 percent of the correct value. He had observed that the sun fully illuminated the bottom of a vertical well at Syrene (now Aswan) at noon on the day of the summer solstice, whereas at Alexandria the sun was one-fiftieth of a circle from the zenith on that date (Fig. 2). Camels, whose daily stage was 100 stadia, took 50 days to go between these cities, so the circumference of the earth must be a quarter of a million stadia. Like the difference between American and British gallons, the stadion meant different measures to different people: the stadion used by travelers was 157 modern meters, that of Greek officialdom was 185 m, and the Royal Egyptian stadion was 210 m. As Eratosthenes was dealing with the daily travel of camels, he would have used the first, and this was confirmed by the elder Pliny (23–79 A.D.), whose writings are the only remaining source of much ancient knowledge. This would be 39,250 km, only 2 percent below 40,008 km, the present polar circumference of the earth (the journey was along the meridian).

To realize that the relative length of solar shadows carried such fundamental information was as brilliant as the flash of sunlight from the bottom of the well. I am sure that other yet undiscovered truths lie latent in our most everyday experience, awaiting the flash of genius to see them.

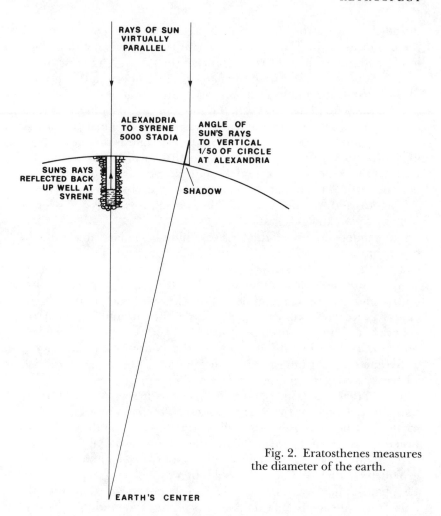

Fig. 2. Eratosthenes measures the diameter of the earth.

Earth, Center of the Universe

Philosophers of northern India, some 2000 years before Pythagoras, had taught that the sun was at the center and that it holds the earth in its power, and that the earth also has a similar power of attraction. From *Science Age* (New Delhi), I quote from J. Arunachalan's translation from the Sanskrit of the Rig-Veda (the earliest literary work in any Indo-Aryan language):

In the prescribed daily prayers to the Sun (*sandhya vandanum*) we find . . . *Soura mandala madhyastham Sambam.* (The Sun is at the center of the solar system.) The word *mandala* means curved, referring perhaps to the curved path of the planets at the center of which the Sun is located. . . . The students ask, "What is the nature of the entity that holds the Earth?" The teacher answers, "Risha Vatsa holds the view that the Earth is held in space by the Sun." In the *sandhya vandhana* we find the phrase: *Mitro dadhara pritivi.* (The Sun holds the Earth.)

The early Indians had identified correctly the relative distances of the known planets from the Sun, and they knew that Moon was nearer to Earth than was Sun. The Vedas also recorded that the equinoxes came a little earlier each year, and gave the rate. This "precession of the equinoxes," which was rediscovered centuries later by Hipparchus, is caused by the attraction of the sun and moon for the tilted earth's equatorial bulge, which makes the earth's axis slowly turn like a spinning top. Considering the links between the Greek language and the Sanskrit, it is surprising that the Greeks were ignorant of the ancient Brahman teaching. Or was it that they thought they knew better?

The spherical Earth adopted by Aristotle was still motionless at the center of the universe, which revolved daily about her, borne on transparent crystal spheres. As the sun moved through the stars once a year, and the moon once a month, three independent crystal spheres were necessary. The five planets also moved independently of the stars and of each other, so each needed its own crystal sphere. To complicate things further, even as far back as Babylonian times Mar-Istar had observed that Jupiter sometimes looped backwards with respect to the fixed stars, and soon Mars and Saturn were known to do likewise. Here emerged the great enigma of the next two millennia, "Plato's problem": to account for the apparent motions of the heavenly bodies while preserving their underlying perfection.

Another of Plato's students, Eudoxus of Cnidus (407–355 B.C.), the inventor of the sundial, gave an elegant mathematical solution to the enigma by bearing the stars, planets, Sun, and Moon on 27 coupled transparent spheres around a stationary Earth. This solution became the basis of dogma for a thousand years. Callipus removed some residual discrepancies, such as the summer period between equinoxes being longer than the winter period, but this required the adding of seven more spheres (the discrepancy being due to the unrecognized ellipticity of Earth's orbit), so the five planets and Sun and Moon had to be corrected individually. Now there were 34 spheres. Aristotle

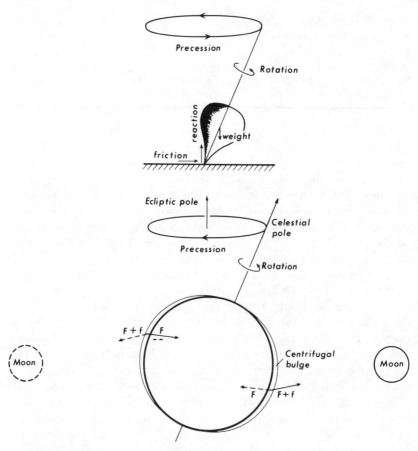

Fig. 3. Precession of Earth caused by the differential attraction of the moon (and to a lesser extent by the sun) for the near and far sides of Earth's equatorial bulge. The couple formed by the weight and the reaction is clockwise for the top, counterclockwise for the earth; hence the precession is in the opposite direction.

added still more to make 56, including a divine outermost sphere, his *primum mobile*, which drove all the rest and produced day and night.

Even 2000 years later, Girolamo Fracastoro (1483–1553), a physician, poet, and professor of philosophy at Padua, resuscitated the Eudoxus model with 79(!) crystal spheres, 8 bearing the stars and planets (equivalent to the octave of Pythagoras), 6 for the daily rotation and precession, 6 each for Sun and Moon, 10 for Saturn's motions, 11 for Jupiter, 9 for Mars, 11 each for Venus and Mercury, plus the outer

Planet Earth

primum mobile. He did this so as to revert to having all rotations centered on man's Earth (*Homocentria*, he titled his work), eliminating the epicyclic motions Hipparchus had meanwhile introduced to match the apparent looping motion of the planets caused by their heliocentric, not geocentric, motion.

Hipparchus (160–120 B.C.), who had, among other notable achievements, discovered the precession of the equinoxes caused by the 25,800-year toplike gyration of the Earth's axis (Fig. 3), dispensed with the crystal spheres, whose growing number had become too cumbersome. Instead he conceived the celestial bodies as floating in Aristotle's ether: the pristine weightless, transparent substance of the heavens. He solved Plato's problem of planetary motions by having each planet

Fig. 4. This sixteenth-century clock in Exeter Cathedral, the gift of Bishop Courtenay (1478–87), is still working. The earth is the sphere at the center. The moon revolves around her, turning on her own axis with the illuminated side always facing the sun, so that the earth always sees the correct moon phase. The sun, indicated by the *fleur-de-lis*, also revolves around the central earth, its stem pointing to the number of days since the last new moon, and its tip pointing to the hour or part hour in the outer circle, daytime hours if the sun is above the horizon. The fixed stars, indicated in the corners, are beyond the sun. The motto means, "They [the hours] perish and are reckoned against us." This clock in the holy cathedral, faithfully recording the visible behavior of all the heavenly bodies, was for the worshipers tangible proof of the ecclesiastic doctrine.

move in a small circular orbit, the center of which itself moves in a circle around Earth. Having no concept of gravitational attraction, Hipparchus could not explain why the celestial bodies should be constrained thus, but his scheme did accurately predict their motions. (Indeed, when Newton so much later stated the law of universal gravitation, he took the problem only one step back by showing that the force he postulated would cause the observed results; he could not explain why such a force should exist at all. Einstein's field concept does no more to explain such action-at-a-distance.)

The famous astronomer Claudius Ptolemaeus (126–61 A.D.) of Alexandria, commonly known as Ptolemy, compiled and integrated the earlier works of the Arabs and the Greeks, particularly of Hipparchus. Because very little of the work he summarized has survived, Ptolemy is probably accorded greater status than he merits. Ptolemy's system follows closely that of Hipparchus, but although for him all motions are circular and all speeds are uniform, he made the centers of the primary circles slightly eccentric from Earth. His reconstruction was as perfect mathematically as was possible while holding to geocentricity and circularity. For 15 centuries many tried to improve on it, but none succeeded. Moreover, it had the blessing of the Christian church, for it maintained the dogmatic prescription of a stationary Earth under the vault of heaven, at the center of the universe, with man as the ultimate creation (see Fig. 4).

The Central Sun

So the Ptolemaic system reigned supreme until the sixteenth century, accepted and believed by all. True, other suggestions had been made, but in earlier times they had been scorned and ridiculed, and after the rise of Christianity they became blasphemous heresies. In the fourth century B.C., Heraclides, a pupil of Plato and contemporary of Aristotle, seems to have been the first to observe that if Earth rotated once a day, this would explain the main motions of the sun, moon, planets, and stars. He also explained the motions and variations in brightness of Venus and Mercury by suggesting that they revolved around the sun, which along with the moon, planets, and stars revolved around Earth. This would explain why Venus sometimes heralded the sun as the morning star (Phosphorus) and at other times followed its descent as the evening star (Hesperus), and further, why it seemed much brighter and closer at times or much fainter and farther away. These were brilliant observations, and significant steps toward heliocentricity. But the contemporary establishment rejected such heresy because the asymmetry of his system was indigestible. The Pythagorean vision of universal harmony and symmetry was then the ruling dogma.

A century and a half later, Aristarchus of Samos (220–144 B.C.), a predecessor of Eratosthenes as librarian at Alexandria, declared that all the planets, including Earth, revolve around the sun, that Earth rotates once a day, and that all the celestial motions are thus simply explained without the multiplicity of imaginary crystal spheres. A brilliant mathematician, Aristarchus deduced that the sun was at least 18 times farther away than the moon and was very much larger than Earth. He was probably the first to estimate the diameter of Earth, and Eratosthenes may have got the idea from him. Had the works of Aristarchus survived, he would probably now rank as one of the greatest thinkers ever. But alas, Aristarchus, whom we now know to have been wholly correct, was totally rejected, even by Archimedes, and persecuted by Ptolemy Philometor, and, apart from a short essay on the distances of the sun and moon, none of his works was preserved. Archimedes was a generation older, and more physicist and engineer than astronomer. Besides, everyone knew that man and his Earth were the center of the universe, knowledge that became firmly fixed by the rise of Christianity (Fig. 4). Aristarchus's discovery was lost for seventeen centuries.

When at long last the heliocentric system was resurrected, it happened within the folds of the Church itself when Nicolaus Copernicus (1473–1543), Canon of Frauenburg, more astronomer than priest, rather timidly and apologetically restated the heliocentric system. Copernicus realized the relativity of motion and saw that a heliocentric system could give identical apparent motions as seen from Earth as did the Ptolemaic system. He also recognized that the rotation and revolution of Earth could together explain many enigmatic features of the celestial motions. His monograph was circulated only privately for three decades, and although he received covert favorable comments, it was only published cautiously in 1543, the author virtually on his deathbed, and then not in his own country.

Tycho Brahe (1546–1601), a nobleman at the apex of the contemporary astronomy establishment, had impeccable grounds for rejecting the heliocentric system, because, if Copernicus were correct, Venus and Mercury would each show phases like the moon, and the stars should show parallax (that is, their apparent position would change a little when observed from the extreme positions of the earth's orbit). None such had ever been observed by any competent astronomer. Brahe's reasoning was sound, but although the predicted phases and parallax do indeed occur, as is now well known, contemporary instruments were not sensitive enough to detect them, because astronomic distances are so much greater than had been imagined. (Stellar parallax was not observed until 1838.) Herein lies a lesson! How often has sound logic, processing false premises, denied the truth?

Ironically, it was Brahe's own precise observations of planetary positions that ultimately confirmed the Copernican notion that Earth orbits the sun. Brahe's data enabled Johannes Kepler, his student, to announce after eight years of labor that planetary orbits were ellipses with the sun at one focus, and that a constant area from focus to orbit is swept in each unit of time. Thus collapsed Pythagoras's harmonic symmetry of circular orbits around a central Earth, and so did Earth's divinely special status. Kepler's laws, in turn (rather than a falling apple!), provided the foundation for Newton's law of gravitation.

Surprisingly perhaps, this demotion of the earth from center of the universe to mere solar satellite did not trigger any reexamination of Earth herself.

The Shape of the Earth

For two millennia after Eratosthenes' ingenious estimate of its radius, Earth remained a sphere, at least to the wise (although flat-earth cults persist to this day!). In China, almost another millennium later, the T'ang dynasty consolidated the warring factions after four centuries of decadence and created the need for more accurate maps to cover its vast empire. Thus the great mathematician and astronomer I-Hsing, in the years 723–26, measured sun shadows along 2500 km and thus calculated the radius of the earth and the length of a degree by Eratosthenes' method.

In Europe, after the lapse of nearly another millennium, maps and navigating instruments improved, but more exact measurements of the length of a degree were needed to meet the rising international competition in world exploration and trade. Quite apart from their colonial aspirations, the French were particularly anxious to do this exactly, because their meter's standard length was to be one ten-millionth of the 90 degrees from the equator to the pole. Although they went to great pains to achieve this, the standard meter bar of platino-iridium made for this purpose is a bit short because of irregularities in the shape of the earth.

In 1669 Jean Picard measured the distance from Amiens to Malvoisine, France, by baseline and triangulation. Giovanni Domenico Cassini (1625–1712), brought from Bologna to Paris by the French Academy to head the Paris Observatory, extended this arc northward to Dunkirk and south to the Pyrenees. He found that the length of a degree north of Paris was 267 m shorter than one south of Paris. On the face of it this meant that the earth curved more sharply in the north, thus that the earth must be prolate (an ellipsoid with its long axis through the poles).

Between Picard's survey and Cassini's extension of it, Jean Richer (1630–96), a Paris astronomer and clockmaker who had been commissioned to build a clock for an observatory in Cayenne, French Guiana, found in 1672 that his pendulum clock, which had kept correct time in Paris, lost 2 minutes 28 seconds a day in Cayenne, and was correct again when returned to Paris. We now know that this meant that the force of gravity was less in Cayenne than in Paris, implying that the earth is oblate (an ellipsoid with its *short* axis through the poles).

Unfortunately the significance of the slower pendulum swing was not then generally understood. Galileo had discovered in 1583 that the time of swing was proportional to the square root of its length, irrespective of how wide the swing was. The Dutch mathematician and astronomer, Christian Huygens (1629–95), had discovered that the time of swing depended also on the force of gravity, which he reported in a cryptic anagram in the fashion of the time, and did not publish the actual relationship until 1673. Sir Isaac Newton (1642–1727) had told Edmund Halley (of comet fame) in a private letter that he had discovered this in 1671. In 1687 in his *Principia Mathematica*, Newton stated that gravity was reduced by centrifugal force, which being greatest at the equator would make the earth oblate. Otherwise, he added, the ocean waters would run to the equator, leaving all high latitudes bare. He also quoted the analogy with Jupiter, which was observed to be flattened at the poles.

However, Jacques Cassini (1677–1756), who had succeeded his father in 1710 as director of the Paris Observatory, insisted that the length of a degree had been proved by measurement to decrease toward the poles, and that therefore the earth must be prolate (even though he had observed the polar flattening of Jupiter himself). Thus sparked off a controversy, destined to rage for half a century, between the French empirical egg and the English theoretical orange, an argument fanned by endemic jealousy between the English Royal Society and the French Academy, and heightened by filial undertones as Cassini son followed Cassini father for four generations as director of the Paris Observatory.

In 1735, to prove the English wrong, the French Academy sent an expedition to northern Peru (now Ecuador) under Charles-Marie de la Condamine (1701–74) and Pierre Bouguer (1698–1758), and sent another expedition to Finnish Lapland the next year under Louis Moreau de Maupertuis (1698–1759) to measure and compare the lengths of equatorial and Arctic degrees. The Finnish expedition went smoothly and was completed in 14 months. The Peru expedition, suffering the hardships of rugged mountainous terrain, tropical rain forests, and uncooperative Spanish officials, took nine years to complete. It turned out that a meridional degree in Peru was 900 m shorter than a degree in Lapland, proving that the earth is oblate. Hence Voltaire's sarcastic quip to la Condamine:

> Vous avez trouvé par de longs ennuis
> Ce que Newton trouva sans sortir de chez lui.
> [You have found with great labor and care
> What Newton found without leaving his chair.]

Such had been Maupertuis's certainty that he would humble the English conceit that he *knew* there must be some hidden error in his data or calculations. For two months, he remained silent at his base, but two more months of thorough rechecking, coupled with the painful fact that his pendulum (which had been set in Paris to beat precise seconds) consistently gained 59 seconds per day instead of beating more slowly as he had expected, left no doubt. As Thomas Henry Huxley has said, the greatest tragedy in science is the slaying of a beautiful theory by an ugly fact. Maupertuis too suffered Voltaire's barb: "Congratulations! You have flattened the poles—and the Cassinis."

In his 1743 book, *La Figure de la Terre*, the distinguished French mathematician Alexis-Claude Clairaut (1713–65), who had accompanied Maupertuis to Lapland, rigorously computed the oblate ellipsoidal figure as an equilibrium between centripetal gravitation and centrifugal rotational force.

What then was wrong with Cassini's original measurements that a degree north of Paris was 267 m shorter than a degree south of Paris? (They had indeed been carefully made!) But the difference sought, even between Dunkirk and the Pyrenees, was scarcely larger than the instrumental errors, and Cassini did not know about allowance for the gravitational attraction of the Pyrenees or for their low-density roots, which distorted the level and vertical and made the degree south of

Paris appear longer than normal for that latitude. Bouguer encountered similar anomalies caused by the attraction of the Andes, and going on to investigate this problem established the theory of isostasy (see Chapter 15). He corrected his observed measurements of gravity for the elevation of the stations (gravitational attraction decreases with increasing distance from the center of the earth), and also subtracted the attraction of the mountains according to Newton's law. Surprisingly, these "Bouguer values" were found to be lower than expected as the elevation increased, thereby showing that mountains generally are supported by less dense material below. Such anomalies, still called Bouguer anomalies, yield important information about density distribution in the earth's crust.

For those with a mathematical mind (others may skip this explanation), the following is the modern version of the spheroid that fits best the shape of the earth. It is expressed in the form of surface gravity, because where gravity is higher, the radius is shorter, in reciprocal relation:

$$\gamma = c_1 + c_2\sin^2\phi + c_3\sin^3\phi + c_4\sin^2 2\phi + c_5\sin(\lambda - c_6)\cos\phi$$

In this equation, γ is the theoretical gravity at latitude ϕ and longitude λ, and the constant c_1 is the actual mean value of gravity at the equator. So if we rubbed out all the other terms, the earth would be a sphere, with the same radius and gravity everywhere. In the second term, c_2 is the centrifugal acceleration at the equator; $\sin^2\phi$ is zero at the equator and increases to 1 at the poles. This means that effective gravity is greater at the poles than at the equator by the amount of the centrifugal acceleration; the resulting model is an oblate ellipsoid, with the equatorial diameter some 55 km longer than the polar diameter.

The third term corrects for the fact that there is more continental crust (which rises higher than the oceanic crust) in the northern hemisphere than in the southern hemisphere, which makes the earth a bit peach-shaped, with the pointed tip at the north pole and the flatter side at the south pole. (This is usually called "pear-shaped," but this is a poor analogy because pears are prolate, whereas the earth is oblate.) The constant c_3 is half the difference in the polar radii; $\sin^3\phi$ is zero at the equator, positive in the northern hemisphere, and negative in the southern hemisphere and thus achieves the statistical peach shape. The correction is not large, the north-pole radius being nearly 50 m longer than the south-pole radius. With these corrections, this Earth model is not an ellipsoid but a rotational spheroid, symmetrical about the polar axis.

The fourth term corrects for the fact that the density of the earth is not uniform but increases steeply inwards, with a marked jump in density about halfway down. The effect of this factor is to cause a slight midlatitude depression in both hemispheres. $\sin^2 2\phi$ is zero at the equator and at the poles, increasing to 1 at 45° north and south. This correction amounts to only about 5 m in the middle latitudes.

The last term corrects for the fact that the equator is not exactly circular but bulges out a bit near Sri Lanka (gravity less), and is a bit depressed near the Solomon Islands and from Iceland to Cape Town (gravity more), which makes the equator very roughly elliptical. Two additional constants correct for these observations: c_5 is the statistical difference in gravity between the high-gravity meridians and the low-gravity meridians, and c_6 is the longitude of the Sri Lanka gravity low; $\sin(\lambda - c_6)$ then varies from zero at the low-gravity meridian to 1 at the high-gravity meridian, and $\cos\phi$ in this term means that this correction is full along the equator, declining to zero at the poles. The effect of this term is to make the equator somewhat elliptical; the resulting Earth figure is still an oblate spheroid, but no longer a rotational spheroid.

Of course this is still not the real shape of the earth, only the nearest we can get to it with a mathematical expression, which is needed by geodesists to calculate the expected force of gravity at any given place and the curvature of the earth there. Each of the constants c_1 to c_6 is the value that gives the least statistical error from a large number of field measurements, so that this spheroid has the least statistical error between calculated gravity and observed gravity anywhere on the earth's surface. The effects of the continents must be smoothed out over the southern hemisphere against the northern hemisphere, and over the land hemisphere against the smoothed water hemisphere. Also, as Cassini in Europe, Bouguer in South America, and later Everest in India found to their sorrow, mountains distort the field, as do regions of denser or lighter rock. In fact the earth is not spherical, nor ellipsoidal, nor even the best theoretical spheroid: global sea level is a "geoid," which simply means "earth-like."

The difference between the gravity actually observed and the gravity calculated for that place from a formula like the one above is called a "gravity anomaly." Although many people do so, it is quite meaningless to speak of "*the* gravity anomaly" at a place, because there are many possible values of gravity anomaly there according to the theoretical figure assumed, rock densities assumed, and the way correction is made for the height of the station above sea level. During the last two decades the bumpy shape of the sea level surface (that is, the geoid) has been mapped with 2-meter contours by the U.S. space agency from sophisticated analysis of the aberrations of satellite orbits. If the earth were a perfect sphere and there were no other disturbing forces, the orbits would be perfect ellipses, but every departure from sphericity shows up as an aberration of the orbits.

Earth, a Magnet

Magnetism has been known in Europe since the time of Thales of Miletus (?640–546 B.C.), and it was discussed by Plato and later writers. The name comes from the Turkish town Magnes (now Manissa), because natural magnets (lodestones) were found there. "Lode" comes from an old Teutonic root meaning to lead or show the way. Miners still follow a lead or lode to the orebody. The mysterious powers of a lodestone, particularly its ability to pass on this power to an iron wire stroked with it, gave rise to myths and fears, and lodestone spoons were used as lie detectors and for fortunetelling. Lodestones consist of the mineral magnetite (iron sesquioxide, Fe_3O_4), which is strongly magnetic, but most minerals containing iron are magnetic, even if only weakly. Such magnetism remains fixed in direction in the mineral unless it is redirected by a stronger magnetic field or wiped out by heat, and it has been found to endure for a thousand million years or more. This is the basis of paleomagnetism, of which more in Chapters 10 and 12.

The Chinese knew about lodestones at least a thousand years ago, perhaps even three thousand. Although their first recorded magnetic compass dates from 1080 A.D., they had probably made a lodestone spoon compass as early as the second century B.C. I-Hsing measured the magnetic declination (the local angle between magnetic north and true north) in 750 A.D. during the T'ang dynasty, a blossoming epoch of Chinese history. Through the next eleven centuries the Chinese recorded the slow variation in the declination.

The compass was known in Europe in the twelfth century, and may have come there from China, but there was a conceptual difference: in Europe the magnet was regarded as pointing north, whereas in China it was said to point south. The Chinese had also accumulated knowledge of the *variation* of declination, which was not known in Europe until the end of the sixteenth century. The magnet's power to point to the north was believed to be due to the same *primum mobile* that moved Ptolemy's epicycles. In 1269, Peter de Mericourt, in his book *Epistola de Magnete*, described his experiments with a magnetized sphere to represent the earth, but the Ptolemaic dogma still ruled the establishment, and his seeds fell on stony ground.

The fact that a magnetic needle, if free to do so, rests horizontally at the magnetic equator and stands vertically over the magnetic poles had been reported in 1544 by Georg Hartmann, vicar of St. Sebaldus in Nürnberg. In England a London maker of nautical instruments, Robert Norman, measured this magnetic dip in 1576 and wrote a book on magnets in 1581.

Like de Mericourt three centuries earlier, the English physician William Gilbert (1544–1603) also realized that the earth itself was a great magnet, and in his benchmark book *De Magnete* (1600) he rejected the alleged link with the planets and stars: "As regards this *primum mobile* with its contrary and most rapid career, where are the bodies that propel it? . . . And what mad force lies beyond the *primum mobile*? . . . The agent force abides in bodies themselves, not in space, not in the interspaces." Gilbert entertained Queen Elizabeth with a small magnetic needle and a globe containing an internal bar magnet, which modeled faithfully the behavior of a compass moved about the earth. Gilbert hoped to enable navigators to determine their latitude by the magnetic dip and their longitude by the declination (longitude was very difficult before reliable chronometers became available), but neither of these goals was achieved.

When variation of the declination first appeared, it was disbelieved. A retired naval officer, William Borough (1536–99), had measured the declination in the garden of his London home in 1580. Forty-two years later Edmund Gunter, professor of astronomy at Gresham College, measured the declination in the same place, but because he found it to be 5° 25′ less than Borough had reported, he rejected Borough's results as unreliable. (Later statistical analysis of both sets of readings showed the mean deviation for Borough to be 4 minutes and Gunter's 10 minutes of arc.) Another 12 years later, Henry Gellibrand, who had succeeded Gunter at Gresham, repeated the measurement and found that the declination had decreased by 2° more, with a mean deviation of only 4 minutes, and the reality of the variation of declination was firmly established.

Edmund Halley (1656–1742), the famous astronomer who first recognized the 76-year period of the spectacular comet named after him, made extensive studies of the declination, and in 1701 published his *General Chart of the Variation of the Compass*, on which there was a poem by Halley (in Latin of course) praising the unknown inventor of the magnetic compass. Strangely, Halley mapped only the declination and ignored the magnetic dip.

The perennial jealousy between the French Academy and the English Royal Society embittered debates in this field also. The great French philosopher and mathematician René Descartes (1596–1650) had attributed the variations of the compass to random deposits of ironstone and lodestone under the surface of the earth, and Halley accordingly set out to "reconcile the observations to some general rule" instead of "causes altogether uncertain," thus to "put a stop to all further contemplation." Halley concluded that "the whole globe of the Earth is one great magnet, having four magnetic poles, or points of attraction, near each pole of the equator two, and that, in those parts of the world which lye near adjacent to any one of these magnetical poles, the needle is governed thereby, the nearest pole being always predominant over the more remote." Halley went on to study the variation of the declination from records made since the compass came to Europe from China, and estimated that the declination moved westward with a period of "700 years or thereabouts," but not precisely about the earth's poles, because the variations do not follow the parallels of latitude.

For two centuries after the pioneering books of Gilbert and Halley, scarcely any advance was made in the understanding of the earth's magnetism, but the technology of the mariner's compass, and the accuracy of magnetic charts of declination and its variation, improved rapidly in the heroic age of exploration and navigation inspired by the intense colonization competition of the European powers. So highly was the compilation of such magnetic data then regarded that the British Association for the Advancement of Science named geomagnetism as its paramount object and pressured the British government in 1839 to mount a major program: "The object, it is true, is to perfect a theory, but it is a theory pregnant with practical applications of the highest importance. The laws of magnetism, under any circumstances to a great maritime power, are every day acquiring additional interest by reason of the introduction of iron vessels into navigation."

The British government responded by commissioning several expeditions, and established magnetic observatories at St. Helena, Toronto, Cape Town, and Hobart. Valuable data resulted that were pooled with other data from the East India Company and the French and Germans to yield reliable charts of the world distribution of magnetic field intensity, dip, and declination. Secular variation was described and measured, as well as more rapid fluctuations clearly originating from the sun, such as daily cycles and intense "magnetic storms" related to sunspots. But little progress was made in understanding what it all meant. Indeed, Halley's remarkable insight into the real nature of the earth's magnetism had been centuries ahead of his contemporaries. He had written:

> The external parts of the globe may well be reckoned as the shell, and the internal as a *nucleus* or inner globe included within ours, with a fluid medium between . . . for if this exterior shell of Earth be a magnet having its poles at a distance from the poles of diurnal rotation; and if the internal *nucleus* be likewise a magnet having its poles in two other places distant also from the axis. . . .

He suggested that the outer shell and inner nucleus turned at slightly different rates to explain the westward drift of features of the earth's magnetic field.

When Sir Edgeworth David first reached the south magnetic pole 70 years ago, it was located near Mount Erebus in Antarctica. Now it is more than a thousand kilometers farther west at about the same latitude. That is, the magnetic poles are not fixed in position on the earth's surface, and the magnetic axis is inclined to the rotation axis and gyrates about it. The fluid core of the earth 2900 km below the surface is now known to be where the main magnetic field is located, equivalent to a bar magnet with its axis tilted more than 11 degrees from the earth's rotation poles. We can be sure of this, because if the magnetic poles were at a shallow depth, the magnetic dip at the surface would fall off from vertical above the magnetic poles to much flatter angles in quite a short distance. The rate of change of dip away from the magnetic pole leaves no doubt that the poles are located near the top of the fluid core.

Systematic charting of isopors (lines of equal rate of change of declination) indicates that the fluid core circulates much the same way as the atmosphere does, with rising convection cells resembling the "lows" on the weather map and descending cells resembling the "highs"; this circulation of electrically conducting fluid produces the magnetic field, like a self-excited dynamo. Just as the cells of high and low pressure on the weather map drift eastward a couple of hundred kilometers per day, so the isoporic cells in the core are found to drift westward by a bit more than 20 degrees per century, enough to drift right around the earth in about 1600 years. But the drift is far from uniform, and because there are other variables than those covered by the above account, Halley's estimate of "700 years or thereabouts" was very good for the data available to him. This means that the earth does not rotate as a solid body, for the mantle is slowly overrunning the core, and the atmosphere is slowly overrunning the surface.

As mentioned earlier, the earth's axis precesses like a spinning top owing to the attraction of the sun and moon on the earth's equatorial bulge (Fig. 3). The different ellipticities of the core and mantle imply that the core, if free to do so, would precess in some 34,000 years as against 25,800 years for the whole earth. The mantle continuously drags the core along to keep up with its own precession. Hence the core suffers a torque tending to displace its axis of rotation, and the mantle suffers a similar torque from the reaction of the core. This may be the cause of the precession of the magnetic axis (that is, the core axis) in about 1600 years with an amplitude of about 11°; the mantle axis precesses in the opposite direction with an amplitude of about half a degree, because of its much greater moment of inertia.

Halley's concept of an outer magnetized shell has also been confirmed. Rudolf Wolf (1816–93), a Zürich physicist, successfully imitated the earth's magnetic field by placing a magnet inside a globe and gluing iron filings over the ocean. The reason his model worked is that ferromagnetism (that is, the induction of a magnetic field in iron) drops to zero at a temperature of a few hundred degrees called the Curie point. This temperature is reached within the earth at a depth of a couple of hundred kilometers. Therefore it is only this outer shell that is magnetized by the magnetic field in the core. But in that shell, the rocks under the oceans have much greater magnetic susceptibility than those under the continents.

By the 1940's understanding of the earth's magnetism had stabilized to a point where the essentials were known. True, work was proceeding on the effects of activity on the sun, and there was serious discussion on the cause of the earth's field; this seemed to result from convective circulation in the electrically conducting fluid core, which is so large that the motion could compress the magnetic field more rapidly than the field could decay (a study called magnetohydrodynamics). It seemed that all that was left to do was tidy up the loose ends. But, as has happened so often in science, such complacency is the forerunner of convulsion. After World War II, the history of the earth's magnetism was found to have been recorded in the rocks through thousands of millions of years, history that contained many surprises. More astonishing, the north and south magnetic poles had switched places, not once, but again and again, and were certain to do so in the future, perhaps soon, and the floors of the oceans held the continuous story of these somersaults, like a giant tape recorder. The consequences of these discoveries were so profound that discussion of them will be taken up in a later chapter, "Sowing the Seeds of Revolution."

CHAPTER 3

Fossils

FOSSILS ARE TRACES of former living things that have been preserved by natural processes in geologic deposits. Fossils may be the actual bones, shells, or tissues, or the imprints or impressions retained by other material, including hollow molds whence the original material has been removed, or natural castings made up of some other mineral that has filled such a mold. Tracks where an animal has crawled or walked or burrowed, and petrified excreta (called coprolites, and valued as gemstones because their phosphatic content often forms turquoise), are also fossils.

Before the nineteenth century, "fossils" included anything dug out of rocks (Latin *fossilis*, something dug up). Hence minerals were fossils, and the organic curiosities found in rocks were part of the study of mineralogy. But when it was realized in the nineteenth century how enormously important such things were, not only for the reconstruction of the history of life but also for establishing the relative ages of the rocks that contain them, paleontology expanded as a major science, independent from mineralogy. The single modern criterion for fossils is that they record the former presence of living organisms.

Myths and Superstitions

Such objects, of course, aroused curiosity from the earliest times, especially when things just like seashells were found far from the sea, even high in the mountains, or things just like bones of existing animals, sometimes of huge size, but somewhat different, were found in cliffs. Some became mystic fetishes, treasured by primitive priests for their special powers and kept in sacred places, or buried with the dead. Myths arose that the shell banks high in the hills had been stopping places of travelers who had brought their sea food with them, and that bones, some larger than those of known animals, indicated haunts of former giants. With the rise of god-antigod faiths, fossils could have been sown there by the devil to tempt the faith of believers. This view was taught in Oxford University as recently as the eighteenth century.

Early Greeks believed that amber ("lynx-stones"), which came down the trade route from the Baltic, were the petrified urine of lynxes, which explained to them why complete insects of many kinds were found in amber. Here is an instructive example of how science progresses. The frequency of insects in amber, perfect even to the finest hairs, was an enigma. Amber must have been a gentle fluid when it enclosed the insects. The color suggested urine. Blood hardens, so why not urine? Amber came from the north, the home of the arctic lynx. This solution was believed for centuries, although it involved a false assumption. Much of the fabric of science, even to this day, includes initial threads that happen to be false. But as we grow up with them woven within the tapestry of our creed, we do not suspect them.

Belemnites, common fossils in western Europe, come from a relative of squids, but the hard part that is found as a fossil looks like a petrified cigar (Fig. 5). These were called thunderstones, from the belief that they were formed when a thunderbolt penetrated rocks; they were evil omens. Even in seventeenth-century England, competent biologists like Edward Lhwyd, a Welshman who prepared the excellent catalog of all the fossils in the Ashmolean Museum at Oxford University, had no idea what kind of animal the thunderstones came from. Belemnite fossils do somewhat resemble true fulgurites (from Latin

Fig. 5. "Tonguestone" (shark's tooth) and "thunderstone" (guard of belemnite rostrum). Steno showed that fossils like these were not created inside rocks but came there while the rocks were still soft mud on the seafloor.

fulgur, lightning), which are formed when lightning strikes through a thin layer of dry sand to the highly conducting wet sand below, forming a finger-sized tube of fused silica.

Fossil shark teeth up to 12 cm long were also fairly common, because sharks are very primitive fish that do not have bony jaws, and the teeth are shed (Fig. 5). They were not recognized as teeth until the seventeenth century but were called tonguestones, as they were believed to be the ossified tongues of reptiles or dragons, sown by evil spirits.

An early but long-lived notion was that fossils were seeds or embryos that had failed to mature, a notion that accorded with ideas of the spontaneous generation of living things in rocks. Indeed, corroborated tales of miners and quarrymen finding live frogs or toads in newly broken rock far underground recur to the present day.

From very early times, some savants, for instance Xenophanes of Colophon (540–480 B.C.) and Herodotus (?484–425 B.C.), concluded that fossils were indeed remains of previously living things, and the question to be asked was not what they were, but how they got there. Xenophanes reasoned that seashells far inland were the consequence of past submergence below the sea. Herodotus proclaimed that Egypt was the gift of the Nile when he recognized that the Nile delta had been built out by that great river's annual floods, and by estimating each year's new contribution he realized the vastness of time. The whole of lower Egypt had been under the sea, and that was how fossils had got there. This conclusion was endorsed by Aristotle in his *Meteorology* and confirmed a couple of centuries later by Eratosthenes, who found abundant seashells 3000 stadia (some 600 km) inland. Strabo (63 B.C.–25 A.D.), the Greek geographer and historian, fully agreed, and among many examples of movement of the land relative to the sea he cited Spina (near Ravenna), formerly a seaport but in his time 90 stadia (about 18 km) inland.

However, the insight of Xenophanes and Herodotus and others of like mind did not end the matter, and argument and affirmations about the nature and significance of fossils ranged through the next 2000 years in Europe, wherever thinking aloud was not fraught with danger. Meanwhile in China, Lo Han in the fourth century A.D., Li Tao-Yuan in the sixth, Yen Chen-Chang in the eighth, Shen Kua in the eleventh, and Chu-Hsi in the twelfth century recognized the true nature of fossils and discussed their petrifaction, and so did Mohammed Kazwini in Arabia in the thirteenth century.

The Renaissance in Italy

Finally the Renaissance dawned in Italy. The Platonic Academy sprouted in Florence under the patronage of Cosimo de' Medici, followed by the Padua Academy in 1520, then the Academy of Natural Sciences in Naples in 1560, and the Academy dei Lincei in Rome in 1570. The Italian lead was followed during the next two centuries by a northward progression: the Académie Française in 1633, the Académie des Sciences in Paris in 1656, the Royal Society of London founded by Charles II in 1660 from informal precursors a decade earlier in Oxford, the Akademie der Wissenschaften in Berlin in 1700, the Royal Society of Sciences in Uppsala in 1725, and the St. Petersburg Academy founded by the Empress Catherine, also in 1725. The birth of these academies, the ascendence of Latin as the learned *lingua franca*, the loosening of the scriptural straitjacket, the disintegration of feudal society, and the rise of individual capitalism were cognate phases of a fundamental evolution throughout Christendom. But the Italians remained geology bellwethers until the rising tide of regionalism and political strife of the nineteenth century.

The precursor of this intellectual renaissance in Italy was Gerard of Cremona (ca. 1114–87), who had translated into Latin many of the Arab manuscripts at Toledo that had regurgitated early Greek and Roman doctrines. Then followed Boccaccio, Leonardo da Vinci, and Fracastoro.

The Florentine poet Giovanni Boccaccio (1313–75), who was also a Greek scholar, wrote that fossil shells in the Tuscany hills had formerly lived in the sea, which at some past time had been there—this notwithstanding his close association with the Popes in Avignon and Rome. A century or so later, Leonardo da Vinci (1452–1519), who ignored dogma and worked things out for himself, recognized that fossil shells were the remains of former marine animals and wrote so in his notebook, though not publicly (he was physician to Pope Paul III).

Girolamo Fracastoro (1483–1553), a Verona physician and philosopher 30 years younger than Leonardo and strongly influenced by him, took the same view—a view, however, like Leonardo's, that was not published until after his death. Geronimo Cardano (1501–76), professor of mathematics in Milan and physician and astrologer of Verona, stated in 1552 in his *De Varietate Rerum* (On various matters) that fossil shells indicated that the sea had once submerged the local hills. Bernard Palissy (1510–81), the Huguenot potter famous for his fired enamels, whose life was also at risk from the Inquisition, came to the same conclusion; he also added the significant observation that fossil species differed consistently in essential details from comparable modern species. Extinctions prior to Noah's flood were incompatible with the biblical account, and even the deluge itself must have caused no extinctions, because the scripture is specific that Noah took two of every creature into the ark.

Thus, contemporary publications sequestered fossils completely from real life. No other view would have passed the censors, rigidly controlled by the clergy. Michel Mercati in 1574 meticulously described and accurately figured the Vatican fossil collection, but stated that they were only stones that had acquired their shapes under the influence of the stars. The German physician Andreas Libavius wrote that fossils grew in rocks from seed and lapidifying juices (*succus lapidius*). This was no more irrational than the contemporary belief that maggots arose spontaneously in rotting meat and rats from old rags. Others treated fossils as sports of nature, and many (such as Olivi of Cremona in 1584) regarded fossils as a category of their own (*sui generis*), not related to real creatures but no doubt also created by God for reasons of his own. Even late in the eighteenth century, Elie Bertrand, a Swiss naturalist, suggested that fossils had been placed there directly by the Creator, "with the design of displaying thereby the harmony of His work, and the agreement of the productions of the sea with those of the land."

Nevertheless, increasing numbers of scientists believed privately that fossils were remains of real animals that had lived at the places where they were now found, and as the Christian dogma had already been proved wrong in its insistence that a stationary Earth was the center of our planetary system, the authority of the Church generally in such physical matters began to crumble. The acceptance of fossils as former living creatures hardened and became more open, as professed by Conrad Gesner (1565), Bernard Palissy (1580), Imperato (1599), Fabio Colonna (1616), Ceruti and Chiocco (1622), and many others. But caution and discretion were still necessary lest the inquisitors strike.

Robert Hooke (1635–1703), for many years secretary of the Royal Society, is remembered for his contributions to physics and microscopy, and for his extraordinary inventions, including the double-barreled air pump (which enabled his master, Robert Boyle, to establish Boyle's law), the spirit level, areometer, marine barometer, the balance spring for watches, the anchor escapement for clocks, and a sea gauge. Few realize, however, that his contributions to geology were profound and precocious but were lost in obscurity, mainly because of his bitter feud with Newton (who had turned his back on geology). That feud robbed Hooke of due recognition as one of the most brilliant and original thinkers ever. He discovered the control of the external form of crystals by their internal structure. He was the first to recognize fossils as documents of history, recording not only the paleogeography and climates of the past, but the sequence of events. He discussed the processes of petrifaction, and also concluded that many of the fossils he collected were the remains of creatures long extinct. As a young man he had published an important monograph on earthquakes, and he suggested that the uplift and submergence of land was caused by them. Although Hooke's geologic papers were completed with all necessary figures, he did not publish them, apparently because of the bitter battle and persecution by the theologians that would have ensued. They were published posthumously in 1705 but were ignored and forgotten for 60 years. They must have been read by Hutton (see the next chapter), although he made no acknowledgment of Hooke's priority.

Finally came the work of the brilliant young Danish surgeon and naturalist Niels Stenson (1638–86), better known simply as Steno, whose prodromus (a sort of outline or abstract) published in 1669 established the principles of sedimentation (that strata now tilted or contorted had begun as horizontal beds deposited in water), developed the law of superposition (that strata deposited on top of other strata were therefore later, and fractures and mineral veins cutting across them were later still), and demonstrated that animals buried in such strata became fossils. His studies of the comparative anatomy of both fossil and living forms was so detailed and precise that there was no room for doubt. His drawings side-by-side of fossil tonguestones and teeth taken from a dead shark established their identity. Because most new converts were devout Christians, their easiest conclusion was to explain all such marine fossils on land as evidence of Noah's flood, a solution that was to dominate for more than a century.

The Man Who Witnessed the Flood

This solution brought its own new problems. The flood should scarcely have affected marine life, yet the most abundant fossils were of sea creatures. If the biblical flood was really the explanation, there should have been people among the fossils, lots of them. Indeed, Dr. Johan Jacob Scheuchzer (1672–1733) of Zürich, a physician and a devout subscriber to the flood theory, after reading John Woodward's 1695 "Essay Towards a Natural History of the Earth" reported an abundance of human remains: vertebrae and whole skeletons, including his most famous exhibit, which he called *Homo diluvii testis* (the man who witnessed the flood). But his vertebrae turned out to be from fish and lacked the hole for the spinal cord. Petrus Camper examined the "human" skeletons in 1787 and reported that they were lizards, but they were finally identified as salamanders. Scheuchzer's vertebrae specimens are still preserved in the University of Zürich and now known to be some 200 million years old, and one of his "human" salamander skeletons (5 million years old) is preserved in the Teyler Museum in Haarlem.

Detailed examination showed that whereas fossils were generally similar to known creatures, they differed in detail, as the bones of a horse would from those of an ass, but in the case of most fossils, no living creatures were known to match them exactly. An English anatomist, Martin Lister (1639–1712), was so impressed by this anomaly that he reverted to the older view of *sui generis*, that fossils were things in a different category of their own, because the scriptures were clear that no living creature had been omitted from the ark.

Baron Cuvier

Onto this stage a century later came Baron Georges Léopold Chrétien Frédéric Dagobert Cuvier (1769–1832), who from his laboratory at the natural history museum at the Paris botanic gardens was to become the greatest comparative anatomist ever to devote his life to the study of fossil vertebrates. It was he who precisely identified Scheuchzer's fossil "human sinner" as a salamander, which he named *Andrias scheuchzeri*.

Cuvier was a confirmed Lutheran; indeed, in 1822 his services to the faith were acknowledged by his appointment as grand master of the faculties of Protestant theology in the University of Paris. Nevertheless he was most caustic about Scheuchzer's biblical blindfold: "Nothing less than total blindness on the scientific level can explain how a man of Scheuchzer's rank, a man who was a physician and must have seen human skeletons, could embrace such a gross self-deception. For this fragment, which he propagated so sententiously, and which has been sustained for so long on the prestige of his word, cannot withstand the most cursory examination." It has always been so. Faith blinds trained eyes and deludes sound minds.

Notwithstanding religious doctrine, Cuvier was a strict empiricist. He observed critically and thoroughly, and reasoned from what he saw. He compared equivalent bones in all groups, living and fossil, irrespective of changing function, thus setting systematic paleontology on a sound foundation. Cuvier was primarily a zoologist, but his colleague, Alexandre Brongniart, was primarily a geologist. Together they established the fact that successive strata yield distinct suites of fossils, which identified particular strata throughout the region, thus founding stratigraphic paleontology, which provided the Rosetta Stone to earth history. Cuvier established the succession of faunas back through the Tertiary epochs of the Paris basin to the Cretaceous chalk. He proved that giant sloths and ancestral elephants had lived in Europe in the not-so-distant past, but such species had become extinct, although there was no provision for such extinctions in *Genesis*.

Contemporaneously, William Smith made the same discovery in England, that particular strata bore their characteristic fossil suites. He applied this knowledge in the siting of canals and the tracing of coal measures, and went on to prepare the first geological maps of Britain. As with most significant discoveries in science, precursors can be found. Giovanni Arduino of Verona had published stratigraphic maps in 1740, and more than a century earlier Robert Hooke had suggested that fossils could be used as chronological indices, as coins could in archeology.

Cuvier had no difficulty in finding evidence of a catastrophic deluge that had obliterated contemporary living things—rather, that this had certainly happened many times, with long intervals of intervening peace during which a new fauna and flora replaced the earlier ones, perhaps immigrants from a distant region not destroyed by the last catastrophe. Similar animals filled each ecological niche, but they were different, perhaps somewhat superior, species. Although older faunas were more primitive than later ones, Cuvier did not attribute this to evolutionary descent but regarded each as a new creation, perhaps successive experiments by the Creator. The permanent uniqueness of each creation was deadlocked in Cuvier's faith. Following his censure of Scheuchzer he was thus hoist with his own petard. His rigid rejection of inherited environmental adaptation, and absolute repudiation of any suggestion that living species could be descendants of extinct fossil ones, soured his friendship with his former mentor, the evolutionist Jean Baptiste Pierre de Monet de Lamarck, who had sponsored his initial appointment and rapid rise.

Concepts of Evolution

To understand Cuvier's reasoning, consider the succession of models of Ford automobiles through eight decades of this century. We can trace the evolution of the cars overall, or of any single function, such as carburetion, ignition, lubrication, transmission, noise suppression, shock absorption, braking, lighting, signaling, body aerodynamics, exhaust pollution, weatherproofing, and so on and on through the multiplicity of details, gross and trivial. Certainly each model is a derivative of its predecessors with minor changes here and major steps there, each motivated by environmental advantage or performance competition in the marketplace. But we know with certainty that no genetic link connects the models, *except in the minds of the designers.* That is how Baron Cuvier would have envisaged the evolution of the horse, from goat-sized *Eohippus* prancing with five toes on the ground some 50 million years ago to a modern Clydesdale striding majestically stiltlike on a single toe, each successive new species being a separate new divine creation. A species is a species and forevermore shall be so!

Lamarck's interpretation of the same succession of horses would be that each individual suffers minor anatomical and physiological changes during its lifetime, conditioned by the constraints and opportunities of its environment, which are then passed on to its progeny. A particular bone or muscle or organ may be strengthened, or weakened, or required to perform a somewhat different function. Although changes in any one generation might be minor, cumulatively over many generations they produced major changes, such as the evolution from doglike land animals to walruses and seals. The concept of a giraffe doe stretching for higher branches above the reach of her competitors, and thereby passing on to her gestating calf a trend toward a longer neck, was created by Lamarck's satirical critics.

Charles Darwin's interpretation of this succession was that within every species there is a significant spread of variation. Although Darwin did not discuss humans, his concept is clear when applied to ourselves. In some 5 billion humans in the single species *Homo sapiens*, no two are identical (except genetic twins). If a particular characteristic (be it physical or physiological robustness, aggressiveness, kindness, or whatever) increases the probability of that person's successful mating, while another characteristic has the opposite effect, the percentage of people having the first characteristic will increase through the generations, whereas the percentage with the second characteristic will progressively decline. When we invent a new insecticide, a tiny fraction of the target insect population will not be affected by it. After extensive use of the insecticide, that immune fraction will progressively increase until the whole population is immune, and will rapidly increase in numbers until limited by other environmental constraints. If a small fraction of a percentage point of the human population happens to be immune to high-energy radiation, that moiety might survive a nuclear holocaust, and subsequently the entire human population would be immune.

Still another explanation was offered by Linnaeus (Carl von Linné, 1707–78, born in Sweden), a devout Christian originally intended for the ministry. He became sorely troubled by the kinship he repeatedly observed between plant species. But to remain consistent with his faith, he conceived an initial creation followed by extensive hybridizations to produce and maintain the multiplicity of related species, and to rationalize their descent from fossil ancestors. He did not realize that interspecific hybrids were usually sterile and that crosses did not occur between genera.

In Cuvier's concept, there is no genetic connection whatever between the successive species, and each species is distinct and immutable, but direct divine interference was essential for the creation of each species that has ever lived. In the Lamarck concept, changes induced in an individual by environmental constraints may be inherited by its offspring, and species thus change progressively through the generations; descendants from a common ancestor many generations back may become so different that they can no longer interbreed, like pumas and tigers. Lamarckism involves no divine intervention. In Darwin's concept, characteristics acquired by an individual through environmental pressures are not transmitted to offspring; rather, any character that happens to be advantageous in the particular environment progressively forms an increasing proportion of the population; any characters no longer useful, like the legs of a whale, dwindle in proportion to the total population and ultimately become vestigial. Darwinian evolution does not involve divine intervention, but does not explain the initial variation either. Linnaean evolution involves unique initial creation followed by unlimited hybridization through the generations.

Cuvier's evolution is catastrophic; the whole fauna is wiped out and replaced by a new fauna, which in due course suffers the same fate. Lamarck's and Darwin's evolution models are continuous and imperceptibly slow. Darwin said *Natura non facit saltum* (Nature does not jump), which accorded well with the field experience that spawned his theory and with experimental evidence that random mutations constantly occur in genes of organisms, which yield observable changes that are thereafter inherited by following generations.

However, Darwin and his disciples were left with the problem of "missing links"—large discontinuities within apparent evolutionary lineages difficult to explain with small mutations. The Jurassic *Archaeopteryx*, the first known bird (about the size of a pigeon, fully feathered but with toothed reptilian jaws, long tail and a skeleton more like that of a small dinosaur than that of any bird, and wing bones not yet fused) is the lonely link between ancestral reptiles and descendent birds, but the jumps from reptile to *Archaeopteryx* and thence to birds are uncomfortably great. It could be argued that this only reflects the gaps in the fossil record, which could scarcely be claimed to be one-thousandth complete, if even that. The only specimens of *Archaeopteryx* come from the Bavarian Solenhofen lithographic limestone, which is quite exceptional in its preservation of fine details. Several million years separate *Archaeopteryx* and the next fossil bird. This argument has usually carried the day; nevertheless, the number and rank level of such missing links are at least embarrassing.

Moreover, fossil evidence generally suggests rather strongly that there have been bursts of "explosive" evolution, when several diverse taxonomic orders appear at about the same time, possibly from a common ancestor. For example, eight orders—the pterosaurs (flying reptiles including the pterodactyls, with the wing membrane supported on the greatly elongated fourth finger), ornithischian herbivorous dinosaurs (with birdlike pelvises), saurischian carnivorous dinosaurs (with lizardlike pelvises), the sauropod herbivorous dinosaurs (including the giant *Diplodocus* and *Brontosaurus*), crocodilians (crocodiles and their relatives), birds, snakes, and lizards—seem to have originated as distinct lines from primitive reptile ancestors (thecodonts, with tooth sockets) 200 million years ago, and remained separate thereafter. That thecodont ancestor seems to have sprung in a burst 50 million years earlier that yielded the turtles, the plesiosaurs (the highly successful group of carnivorous marine reptiles), rhynchocephalians (literally "snout-headed" because of their beaklike skulls, of which the New Zealand tuatara is the sole survivor), pelycosaurs (a wholly extinct group of synapsid reptiles), and therapsids (leading to mammals). The mammals themselves seem to have sprung from an evolutionary explosion when most of the existing orders appeared about 200 million years ago.

Persistent missing links and evidence of bursts of explosive evolution troubled George Gaylord Simpson, patriarch of twentieth century zoology. Although he strongly supported Darwin, he became convinced that additional processes, yet to be discovered, were implied, and he introduced the concepts of microevolution (Darwinian), macroevolution, and megaevolution (perhaps even Cuvieran).

For my part, I agree with Simpson, but far from being a problem, such explosive bursts should be anticipated. A definite fraction of the carbon photosynthesized by plants and thence incorporated into the food chains is radioactive carbon (^{14}C), which is continuously produced by cosmic radiation of atmospheric nitrogen, and ultimately incorporated in all living cells. A very small but definite fraction of the carbon in genes must therefore be radiocarbon, the spontaneous reversion of which to nitrogen (stable ^{14}N) must alter that gene, often lethally, but rarely yielding a new viable gene. (Carbon and nitrogen are the crucial atoms in amino acids of gene tissue.) The effect of this spontaneous alteration would vary greatly over the hundreds of millions of carbon atoms in a gene, as would the probability of rare crucial concurrent substitutions. These could trigger a burst of evolution by making viable many of the more frequent mutations that could not be effective without that single ultra-rare key.

Quite apart from such questions of explosive organic evolution, controversy raged during the eighteenth and nineteenth centuries between those who favored slow gradual evolutionary processes and those who believed that only episodic catastrophic processes could achieve in the time allowed by the scriptures the long and complex series of events recorded in the rocks. The next chapter takes up this debate.

CHAPTER 4

Neptunists and Plutonists

THE EVOLUTION OF SCIENCE in general, and that of geology in particular, has proceeded in bursts. Primitive pantheism, when all mysteries were attributed to gods, gave way to the golden age of Greek philosophers and the more practical Romans. With the fall of the Roman empire, the Dark Ages regressed for seven centuries, until the renaissance of scholarship in Italy in the twelfth century revived the Greek style of contemplative, "armchair" science. For another seven centuries, scriptures (of Aristotle as much as Moses) were self-evident truths and the foundation of knowledge. The late eighteenth century saw a rising tide of empirical pragmatism. Earth herself, with her rocks, mountains, and seas, slowly replaced ancient writings as sources of data to be observed and contemplated.

Of course, from the earliest times there had been iconoclastic loners who observed and contemplated rather than believed inherited lore, but like Aristarchus of Samos, who anticipated Copernicus by two millennia, their wisdom drowned in the tide of dogma. Standing tall and clear-sighted among the pioneers of the eighteenth century was Mikhail Vasilievich Lomonosov (1711–65), who recognized the principle of actualism (the uniformitarianism later developed by Hutton and Lyell), the crucial role of the earth's internal heat, the rise and fall of the lands, and the great age of the earth. He recognized that folding, rupturing, and uplift of strata were imperceptibly slow processes, in contrast to the catastrophic revolutions postulated by Cuvier and Hutton who came later. Why were Lomonosov's ideas not immediately adopted? His work was too far ahead of his contemporaries. Small steps are acclaimed, but large leaps are scorned. A century passed before Europe reached Lomonosov's level of understanding of geological phenomena.

The golden age of geology of the late eighteenth and early nineteenth centuries, like the golden age of Greek philosophy two millennia earlier, was inspired not by peripatetic professors (whose conservative dogmas inhibited new concepts) but by leisured independents like Buffon, Telliamed, von Buch, d'Aubuisson, Hutton, Hall, von Humboldt, Murchison, Lyell, Darwin, and Wallace, whose financial freedom allowed wide travel, not only geographically but through flights of fancy, so that exotic blooms did not wither on the vine. With such stimulation, the literate public became more interested in geology than at any time before or since. Adding spice to the times was a vigorous debate between two schools of thought under two great figures.

Werner and Hutton

Abraham Gottlieb Werner (1749–1817) was the pre-eminent geological synthesizer, teacher, and inspiration of students of his time. Among his many notable disciples were Goethe (who became a keen geologist through Werner's influence), von Humboldt, von Buch, and von Schlotheim in Germany; Reuss in Bohemia; Jameson in Scotland; Kirwan in Ireland; Greenough in England; d'Aubuisson, de la Métherie, Cuvier, and Brongniart in France; and Maclure and Eaton in America. These and their leader were the Neptunists; they held that the visible earth was born of water. Fundamentally opposed to them were the Plutonists (theorists of subterranean fire), led by James Hutton of Edinburgh; among his followers in Scotland were Hall and Playfair, with Lyell in England, Breislak and Fortis in Italy, Desmarest in France, and Fichtel in Hungary. The Neptunism–Plutonism controversy predated both Werner and Hutton but was crystallized, publicized, and critically focused by them. Interwoven with but independent of it was the contest of uniformitarianism versus catastrophism, to be discussed next.

Although Werner pontificated *ex cathedra* a philosophy of the entire earth, his field experience was very circumscribed, limited to the hills of his native Saxony as far as the Erz Mountains bordering Bohemia. He was, however, from his childhood days deeply interested in mines and mining. Devoted to systematization and classification, Werner studied comprehensively all that had been written by his predecessors. He was a supreme orator, but avoided writing.

James Hutton (1726–96), son of an Edinburgh merchant and city treasurer, was like Werner a lifelong bachelor. After a widely ranging education and a short medical practice in London, he gave up medicine to join his boyhood friend John Davie in Edinburgh to manufacture sal ammoniac from soot. When he inherited a farm in Berwickshire, he became a scientific farmer and found time to pursue his beloved geology with extensive observations in the field. He now read rocks rather than books. Many decades ahead of his contemporaries, Hutton recognized that rocks recorded the history of the earth, and he strove to master their language by seeking symptoms in natural exposures. In 1785 he read his revolutionary paper, "Theory of the Earth," before the Royal Society of Edinburgh. His essays were rigorous but also dull, prolix, and uninspiring. But his contagious enthusiasm infected two formidable disciples, Hall and Playfair.

Sir James Hall (1762–1831), the leisured Baronet of Dunglass, at first skeptical of his neighbor's heterodox ideas, nevertheless roamed the glens with him and was converted. He then carried out many well-planned and quite original experiments to test and prove Hutton's theories on the folding of strata, the crystallization of basalts from molten magma, and the thermal metamorphism of limestone to marble. Triggered thus by Hutton, Hall became the father of experimental geology.

John Playfair (1748–1819), a minister of the church and brilliant professor of mathematics and philosophy in the University of Edinburgh, became a devoted geologist under Hutton's guidance and enthusiasm. His concise writing was as bright as Hutton's was dull, and he became the medium whereby Hutton's work stirred the community at large. Among Hutton's other important friends were Dr. Joseph Black, the famous chemist, and James Watt, of steam-engine fame, but it was the Hutton–Hall–Playfair synergistic triad—Hutton, observer and thinker, Hall, critic and experimentalist, and Playfair, mathematician and communicator—that combined to establish modern geology.

Werner and Hutton polarized two divergent concepts of Earth evolution—Werner, that the waters of the deep and deluges flooding the lands had formed and molded all the world's rocks by precipitation, crystallization, erosion; Hutton, that heat from the deep interior, expressed through volcanoes, earthquakes, and vertical movement of the lands, was the root cause. These processes, which had been treated from the earliest times in terms of the primitive elements of water and fire, were now dubbed Neptunism and Plutonism.

Werner's Neptunism

Building on a foundation laid a century earlier by Steno and Benoît de Maillet (1655–1728; he wrote under his reversed name Telliamed), Werner integrated into a comprehensive and universal theory of the earth several contemporary concepts. These included the Italian stratigraphic model of Primary, Secondary, Tertiary, and Quaternary mountains, first proposed by Lazzaro Moro (1687–1740), and developed by the Venice professor of mineralogy, Giovanni Arduino; French ideas from Comte de Buffon's encyclopedic classic, *Les Epoques de la Nature*; the nine successive sedimentary rock units recognized by George Christian Füchsel (1722–73) in the Harz and Thuringian Mountains; the ideas of the Berlin lecturer of mineralogy and mining Johann Gottlieb Lehmann (1719–67); and the work of the Swedish mineralogist Torbern Olof Bergman (1735–84), *Physical Description of the Globe*. Thus, Werner's philosophy was an organized synthesis of the contemporary concepts of geology and mineralogy, except that he excluded the Italian observations of land rising from the sea. To Werner, land was stable, but oceans came and went. His stimulating lectures attracted scholars from as far away as America, but his only publication was a brief 28-page summary: "Kurze Klassifikation und Beschreibung der verschiedenen Gebirgsarten" (Brief classification of various rock categories; here *Gebirgsarten* is used in a special Wernerian mining sense).

Werner's synthesis, all-embracing and a distinct advance on any preceding concept, was case-hardened by his disciples, with the authority and aura of the master, to the status of a creed, to be propagated and defended to the ends of the civilized world. Such was the hypnotic persuasiveness of Werner's personality. The seeds of a narcotic dogma had been sown. The parallel with plate-tectonics dogma of the 1970's is stark.

Werner's school in Freiberg became pre-eminent in Europe, not only in fundamental theory and mineralogy, but in the application of geology to mining and engineering. The neighborhood of Freiberg, extending through Saxony to Bohemia, had been the focus of mining technology for two centuries before Werner. In mineralogy, prospecting, and mining, Werner relied heavily on Georg Bauer (1494–1555) of Joachimsthal, better known under his adopted Latin name, Georgius Agricola. The international leadership of the Bergakademie Freiberg was due at least as much to Werner's inspiring personality as to its being the womb of Neptunism. Neptunism did not survive Werner's death, but his inspiration lived on for many decades.

In brief summary, the Neptunists held that the earth was an inert passive body with pronounced surface relief that was originally deeply submerged under a universal calm ocean. From this primeval sea, primitive crystalline precipitates deposited a layer of granite, which encrusted the highest elevations as well as the lowest plains, to become the foundation of all later rocks. In time, while the waters receded somewhat, crystalline precipitation continued, forming foliated rocks like gneiss, mica schist, serpentine, and slate, at decreasing altitudes as the waters receded. A partial return of the waters, now with more turbulence, resulted in mechanical erosion of some of the first-formed rocks to yield layered rocks, sometimes overlapping each other or lying on eroded surfaces of earlier rocks (that is, unconformable). Life had already appeared, as evidenced by fossils in many of these strata. Further recession of the seas closed this Primitive (*Uranfänglich*) period.

The second period, named *Floetz* (layer) because of the dominance in it of horizontally bedded sedimentary rocks, was a time of markedly fluctuating sea level characterized by violent tempests, with much erosion of primitive rocks, and irregular deposition of the debris at the lower levels, producing hummocky and hilly terrains when finally abandoned by the seas. Calm interludes between the storms left some crystalline precipitates of limestone, gypsum, and rock salt. Twelve successive formations of the Floetz (largely those of Lehmann and Füchsel) cover what is now known to range from the Permian to the Tertiary Periods, including many zones rich in fossils. (For a list of geological periods, see p. 79.) A late return of the sea led to the precipitation of crystalline basalt above some Floetz strata. Werner emphasized the flexibility of his prescriptions to accommodate the regional variability of these strata. Later he introduced a transition period (*Uebergangsgebirge*) between the Primitive period and the Floetz to better accommodate some field observations of limestone, graywackes, and diabase (of the late Paleozoic).

After the final retreat of the seas the earth entered the Volcanic and Alluvial (*Aufgeschwempte*) periods, which overlapped in part, because examples of both still occur. As Volcanic Werner included contemporary lavas, pumice, and volcanic ash (which hardened to tuff) but excluded basalt and porphyry, which he regarded as Floetz or Primitive. He also recognized "pseudo-volcanic" rocks such as some cherts, hornstones, and jaspers, which had been baked and altered by volcanic heat.

Werner's Neptunist model explained quite well the outcrops in the mountains and lowlands of his native Silesia, Bohemia, and Saxony, and it attracted wide approbation, for it integrated for the first time in a single global theory such a wide range of disparate facts and theories. All the mountain ranges of Europe did have cores of crystalline rocks, overlain by dipping strata, then by flatter lying strata with increasing proportions of fragmental formations and less of the "chemical precipitates" like limestone and gypsum. Indeed, its ability to explain so many different phenomena was a defense used for many decades by the Neptunists, who implied that such a universal theory must be right and that the apparent anomalies must eventually yield before such an all-embracing hypothesis.

On the excuse of indifferent health, Werner did not travel much. But as his disciples (von Humboldt, d'Aubuisson, von Buch, and others) traveled abroad, discrepancies accumulated. Although they therefore gradually moved away from his Neptunism, none of them ever lost respect and reverence for their inspiring mentor. Indeed, if Neptunism were erased from the record, Werner would still rank as one of the greatest geologists of his time by virtue of his stimulation of so many others. To quote Sir Archibald Geikie (1897):

No teacher of geological science either before or since has approached Werner in the extent of his personal influence, or in the breadth of his contemporary fame. . . . But never in the history of science did a stranger hallucination arise than that of Werner and his school, when they supposed themselves to discard theory and build on a foundation of accurately ascertained fact. Never was a system devised in which theory was more rampant; theory, too, unsupported by observation, and, as we now know, utterly erroneous. From beginning to end of Werner's method and its applications, assumptions were made for which there was no ground, and these assumptions were treated as demonstrable facts. The very point to be proved was taken for granted, and the geognosts, who boasted of their avoidance of speculation, were in reality among the most hopelessly speculative of all the generations that had tried to solve the problems of the theory of the Earth.

Advance the date by 80 years, and Geikie could have been writing about the current subduction school, which to fit a theory creates 300 million square kilometers of ocean crust (which never existed except in their minds) to vanish it again down the trenches! Subduction is as universally taught today and as blindly believed as was Werner's Neptunism throughout western Europe during the eighteenth century. But, of course, today is different—subduction *is* true, as Werner's *geist* stalks the world anew.

The Challenge of Plutonism

Meanwhile the Plutonists' Earth, far from being passive and inert, was dynamic, with internal heat governing her behavior. Granite, by far the most abundant of the crystalline rocks, was not an aqueous precipitate but had crystallized from the molten state, and instead of being the oldest came later than the gneisses and slates, which it had intruded. This was also true for porphyry. Basalt (the "blue metal" of quarrymen) was not an aqueous precipitate either but a lava, which had poured out over the surface and flowed many kilometers before solidifying. (Indeed, the name "basalt" goes back to Pliny for lavas vented from Vesuvius and was revived in the sixteenth century by Agricola.) Hutton and his Plutonists were also "neptunists" to the extent that they recognized that sandstone and shale originated from sediment eroded from the land and accumulated as beds on the seafloor interbedded with limestone derived from seashells. Hutton believed that heat hardened them into rock (just as heat makes bricks from clay) and eventually to slate, schist, marble, and gneiss.

Scipio Breislak (1748–1826), who had observed the dislocations and uplift associated with volcanoes in Italy, thought that uplift and subsidence of active land in a passive sea was more probable than oceans waxing and waning over inert land. Where, Breislak asked, if the Neptunists were right, did the great volumes of global ocean go to at times of low level—and repeat such coming and going so many times? Besides, to hold in solution or suspension all the solid material of the crust would require many times the volume of ocean that the Neptunists had conceived.

Nowadays, we smile at their naïveté in believing such things. But in our own times all the authorities on the Proterozoic banded iron-formations, from which comes nearly all the world's iron ore, seriously believe that these enormous deposits are true Wernerian precipitates from seawater, which could not possibly have held so much iron in solution, nor switched on and off from silica to iron precipitation through the millions of fine laminations characteristic of these rocks. (I treated this problem in my 1976 book *The Expanding Earth*.)

Basalt: Sediment or Lava?

The Achilles' heel of Neptunism was basalt. Granite, too, was crucial but is still enigmatic even today. Basalt is common in Saxony and Bohemia, far removed from any visible volcanoes, commonly as flat layers, sometimes capping the hills, elsewhere interlayered with the strata of the Floetz. To Werner it was just another chemical precipitate indicating a temporary return of a high sea level. Although he had never seen a volcano, he was aware that identical rock was found as a volcanic lava, but that presented no difficulty because the volcano itself was merely the vent from burning coal seams (such as he had seen in a Bohemian coalfield), which had melted Floetz rocks, including basalt. Certainly, in the limited region known to Werner, wherever there was basalt there were coal seams farther down. Werner cited the outcrops on Scheibenberg Hill in Saxony showing a perfect transition from Floetz sandstone and mudstone into basalt. The whole sequence, he said, had clearly been deposited in water. A characteristic feature of his precipitated basalt was that on crystallization it cracked into beautifully symmetrical hexagonal columns (called columnar jointing; see Fig. 6), which distinguished his common basalts from the rare basalt lavas at contemporary volcanoes (so he taught).

Sadly for Werner, his basalt concept was soon to be disproved by his own devoted students von Humboldt, von Buch, d'Aubuisson, and Voigt. Leopold von Buch (1774–1853) for some years after leaving Freiberg remained a confirmed disciple of Werner, including specifically the Neptunist origin of basalt. He was shaken in Italy where he saw feldspar porphyry as a volcanic product (a primitive precipitate according to Werner). He then visited the *extinct* volcanoes in the Auvergne district of the French Massif Central, which had been identified as volcanoes in 1751 by Jean Etienne Guettard (1715–86), and

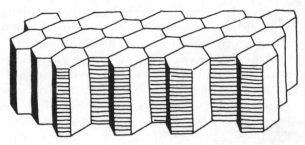

Fig. 6. Columnar jointing in basalt, caused by shrinkage during cooling.

where basalts had long been known as lavas through the meticulous work of Nicolas Desmarest (1725–1815). Desmarest and Guettard were the first to recognize old mountains as extinct volcanoes in their varied stages of denudation. Desmarest also found that genuine basalt lavas *did* have columnar jointing. Von Buch found that some such basalts were lying on granite at least 400 m thick, which according to the Werner theory had to be the primitive foundation, below which there could not be subterranean burning coal seams. Volcanic heat must come from the deeper interior, as the Plutonists claimed!

Jean François d'Aubuisson de Voisins (1760–1819) completed a treatise on the basalts of Saxony that was in accord with the Wernerian philosophy. He too went next to Auvergne, where he made the same observations and reached the same conclusions as von Buch had before him. Basaltic lavas had originated from a heat source below the granite. Moreover, he wrote, these basalts were identical in mineralogical, textural, and field characters with the basalts of Saxony, and he could not but conclude "that there has been an entire identity in formation and origin."

Johann Karl Wilhelm Voigt (1752–1821) studied law, but under Werner's magnetic influence he gave up law for geology and became Councilor of Mines at Ilmenau in Thuringia. However, Werner's warm friendship was shattered in 1779 when, after a critical examination of the contact at the base of the basalt at Werner's prize exhibit on Scheibenberg Hill, Voigt concluded that the apparent transition was simply due to the fact that the basalt became very fine grained toward the base, as known volcanic flows had been observed to do, and the mudstone in contact with it had been baked to a dark fine-grained hornstone that closely resembled the adjacent basalt.

The Granite Controversy

To all who had thought about it, granite seemed to be the primitive foundation on which all other rocks had been laid; to suggest otherwise was indeed an absurdity. However, Hutton had concluded as early as the mid-1760's that both basalt and granite were igneous rocks, which had ascended from the deep subterranean furnaces of the earth as molten magmas and crystallized. Granite, he concluded, had congealed in huge irregular bodies at some depth below the surface and had become exposed only after long periods of erosion. Basalt had flowed out on to the surface as lavas, but Hutton also found large bodies of similar rock that had been intruded in the molten state between layers of strata and were now laid bare by erosion as the whinstone hills of Northumberland and the Midland Valley of Scotland.

Blind faith in Werner's doctrines became a creed that persisted for decades after its basic tenets had been discredited. Because they *knew* the rules established positively by the master, his apostles did not need to observe and describe, but only identify where the outcrop before them fitted into the Werner system. W. H. Fitton described the reaction of one of them to an intrusive contact of trap (basalt) into sandstone:

It was arranged that the party should go to Salisbury Crags, to show Dr. Richardson a junction of the sandstone with the trap, which was regarded as an instructive example of that class of facts. After reaching the spot, Sir James [Hall] pointed out the great disturbance that had taken place at the junction, and particularly called the attention of the doctor to a piece of sandstone which had been whirled up during the convulsion and enclosed in the trap. When Sir James had finished his lecture, the doctor did not attempt to explain the facts before him on any principle of his own, nor did he recur to the shallow evasion of regarding the enclosed sandstone as contemporaneous with the trap; but he burst out into the strongest expressions of contemptuous surprise that a theory of the Earth should be founded on such small and trivial appearances! He had been accustomed, he said, to look at Nature in her grandest aspects, and to trace her hand in the gigantic cliffs of the Irish coast; and he could not conceive how opinions thus formed could be shaken by such minute irregularities as those which had been shown to him.

Religious dogma has frequently been blamed for retarding progress in geology, but it is not religion as such but faith in a dogma that is the culprit, and the Werner dogma was just as culpable.

Hutton found initial proof that granite had crystallized from a molten state in the Portsoy granite of northeast Scotland. This is a "graphic" granite, so called because it consists of re-entrant wedge-shaped quartz embedded in feldspar, resembling runes or cuneiform writing. It was impossible for such a texture to result from accumulation of aqueous precipitates, as required by the Neptunists.

Hutton next set out to prove that the granite had intruded the pre-existing or "country" rocks as an ascending invasive magma and had not resulted from the fusion of parts of the country rocks. When he visited an area in the Grampian Mountains where a large body of granite was surrounded by mica schists, he was delighted to find many wide veins (or dikes as they would now be called) of the red granite cutting across the grain of the schist. Werner had explained crosscutting veins as precipitates that had filled cracks from above. This could scarcely be applied to granite, which he had claimed to be the primary foundation precipitate, laid down before such later primary rocks as gneiss, schist, and serpentine. Although in his later lectures Werner had spoken of the first, second, and third granite, even this escape failed when dikes were found to grow narrow and terminate in an upward direction.

Whereas Hutton had disproved the Neptunist theory of granite genesis and shown that what was now granite had been at least in part an invasive magma, his observations had not disposed of the possibility that granite might be derived in large part from the conversion of the pre-existing rocks. Granite bodies are so very large that the question arises, what occupied that space before their intrusion; and the country rocks may be truncated across their grain for such distances that we must ask, where now are the missing parts of those truncated rocks? Indeed, these questions have been revived almost cyclically throughout the two centuries since. Sir Archibald Geikie (1834–1924), in his 1882 presidential address to the Geological Society of London, observed that the question of the transformation of country rocks into granite had again come up for debate. Early in this century only the magmatic version was seriously taught, but during the 1930's and 1940's granitization of sediments was again strongly supported and widely debated.

Only in this decade have these questions been finally answered. John Elliston has for many years pointed to the uncertainty of the common assumption that an intrusive rock is necessarily "igneous." In a brilliant series of benchmark papers in *Earth Science Reviews* beginning in 1984 he first discussed exhaustively orbicular granites, crystalline rocks with the same minerals as granite (and some other combinations) arranged concentrically, onionlike, to form a coherent mass of orbicules. Using advanced knowledge of the surface chemistry of siliceous sols and gels, Elliston demonstrated that the orbicules were formed from mobilized sediments, that the fluid mass was intruded in the precrystalline state, and that the higher temperatures were attained during crystallization after intrusion. In his second paper, he examined with similar thoroughness a specific category of granites, the rapakivi granites, and demonstrated that they too had a similar genesis and history. Thus the rapakivi granites are both intrusive *and* derived from sediments, but were never molten. Finally, Elliston applied this new knowledge to granites generally. Nevertheless, it is well established that some granite can result as the last stage of differentiation of a basaltic magma; hence, as he commented, there are granites and granites. A couple of decades ago, at an early stage of his enlightenment, Elliston was in the predicament von Buch had been in when he first studied the basalt flows in Auvergne. As Geikie told it:

> He could not yet break entirely the Wernerian bonds that held him to the beliefs he had imbibed at Freiberg. He could not bring himself to admit that all that his master had taught him as to the origin of basalt, all that he had himself so carefully noted down from his extended journeys in Germany, was radically wrong. He, no doubt, felt that it was not merely a question of the mode of origin of a single kind of stone. The whole doctrine of the chemical precipitation of the rocks of the earth's crust was at stake. If he surrendered it at one point, where was he to stop?

And so it was with Elliston. If he really believed what seemed to him to be true for the feldspars of his porphyroids, where could he stop before the granites themselves were involved? Because Elliston has made a big step from orthodox dogma, it may be a decade or so before his work is fully recognized.

Ever since Hutton, petrologists have thought of granite as having crystallized from a molten magma, even though there has been debate whether that magma came up from the depths, from melting of the roots of the geosyncline, or from transformation in place. The principles of chemistry still taught by petrologists are those of solutions and macromolecules, whereas in the sediments of a geosyncline these principles are dwarfed and superseded by the surface chemistry of minute particles of colloidal size. A one-meter cube of rock has a surface area of 6 square meters, but a cube of wet clayey sediment whose particles are one nanometer has a surface area of 60 million square meters, and the surface energy is so great that ion adsorption and desorption and hydration and dehydration processes such as syneresis, thixotropy, and rheopexy dominate the behavior. During the last two decades, physical chemists have understood this field, but this knowledge has not crossed into the field of petrology, except with Elliston. The symphony of sol-gel and gel-sol transitions, in harmony with stress perturbations, has profound consequences through petrology and the generation of ore deposits. The training of petrologists must now go right back to the assumptions of the freshman year.

Plutonism Wins Out

The final ascendency of Plutonism over Neptunism may be dated from the formation of the Scottish Geological Society in 1834 by disciples of Hutton to replace the Natural History Society, which had been founded by Robert Jameson (1774–1854), professor of geology in Edinburgh University. For 50 years Jameson blinkered his students with the Werner faith and retarded for decades the elucidation of the geology of Scotland. Clearly, Jameson's personality had failed to attract perceptive students.

Retrospect leaves no doubt that Werner's dogmas were grossly wrong, and many a modern teacher would cite him as a buffoon. The reality is not so black and white; Werner's doctrines did have a progression of Earth history, which Hutton lacked. Werner still stands tall in the vista of great synthesizers and great teachers. The test of a professor is, where are his students? Werner's inspiration yielded the richest harvest of great geologists ever. After all, the "facts" students learn from their professor are transitory, but the inspiration he inseminates outlasts their lives.

By contrast, Hutton's greatness lay not in his inspiration of students (although the bubbling enthusiasm of his personality fascinated his friends) but in his recognition of new principles, which time has substantiated: a vast age for the Earth is needed to accommodate repeated cycles of sedimentation, mountain-building, slow erosion back to a peneplain, and final submergence to start the cycle again; "the present is the key to the past" (later called uniformitarianism); the internal heat of the earth provides the motive power for mountain-building and the conversion of sediments to rocks; basalts had been molten lavas, and dolerites and granites were intrusive rocks that had crystallized at high temperatures.

Hutton's personal life, too, inspired his vision: Earth was a heat-driven machine, like Watt's steam engine; Earth was an organism, with cycles of wasting and renewal (Hutton's doctoral thesis was on the circulation of the blood); Earth, like his own good farm, continuously regenerated the essential fodders for life; Earth was a chemical factory, with its stock of elements ringing the changes.

Catastrophism and Uniformitarianism

Catastrophism and uniformitarianism were catchwords coined sarcastically in retrospect by William Whewell (1794–1866), Cambridge professor of mineralogy, when he reviewed Lyell's treatise, "Principles of Geology, Being an Attempt to Explain the Former Changes of the Earth's Surface by Causes Now in Operation," which appeared in three volumes from 1830 to 1832. However, the debate between

Plato's catastrophism, Aristotle's steady state, and the Stoic's cyclic cosmos has ancient roots. It was still the fundamental issue between Werner (and others like Cuvier) on the one hand and Hutton, followed by Lyell, on the other.

Catastrophism, which had been the ruling dogma over the preceding century and a half, assumed that the significant processes of Earth history involved processes of much greater magnitude than any now known. Like most broad intellectual tendencies, catastrophism had variants. Some invoked divine intervention, others only natural agents, but of a kind not currently operating. Cuvier, in particular, believed that the scale of the discontinuities between his series of fossil faunas was of a higher order than any existing processes could produce. The time difficulty was severe, because the earth was known from the Bible to be only 5000 years old. Some strictly divided the past into the human, or postdiluvial period, when current processes operated, and the antediluvial time, when quite different conditions prevailed. Others, especially English geologist-clergymen like Buckland, Conybeare, Sedgwick, and Fleming, tied this classification to the holy scriptures.

Uniformitarian ideas go back to Herodotus in the fifth century B.C. and continued intermittently to Robert Hooke in the seventeenth century. Comte de Buffon in the mid-eighteenth century wrote: "In order to judge what has happened, or even what will happen, one need only examine what is happening. . . . Events which occur every day, movements which succeed each other and repeat themselves without interruption, constant and constantly reiterated operations, these are our causes and our reasons." But the concept was crystallized by Hutton in 1785 in abstract and formally in his 1795 book, *Theory of the Earth*. Uniformitarianism claimed that the currently observable processes of erosion, sedimentation, vulcanism, and changes of sea level could be, and in fact had been, responsible for all the events of Earth history, and that this implied almost unimaginable duration for the evolution of the earth. "The present is the key to the past" (coined by Lyell) and "we find no vestige of a beginning—no prospect of an end" (coined by Hutton) were the key concepts. Uniformitarianism also had variants. In western Europe (Füchsel, for example) and Russia (Lomonosov), it meant the continued operation of current processes, or "actualism," not quite the same as the projection through the distant past of the known laws of nature, which had long been accepted. Some, such as Hutton, Playfair, and Hall, saw extremely long quiescent periods of erosion during which ancient mountain systems were worn away to peneplains, separated by short violent revolutions when strata were folded up into mountains and seas were driven from rising lands. This was a blend of uniformitarianism and catastrophism; indeed, Hutton used the word "catastrophe" for his revolutions.

Cuvier, a brilliant thinker and the father of comparative anatomy, was quite unable to reconcile the minuscule changes he saw in current processes with the enormity of the changes apparent in the succession of strata. Uniformity of process had to be punctuated by violent catastrophe of a wholly different order of magnitude. His blindfold was that he had not realized the vastness on human scales of geological time. Note the parallel with Brahe, who had no conception of the vastness in human terms of distances to the stars.

Charles Lyell (1797–1875) was born of a wealthy father and thus was able to give up law to study geology, which he did so successfully and with such captivation of the public that he became a national figure and was created a baronet. It was he who recognized that not only the slow denudation stage but also the "impulsive" stages of volcanism, mountain-building, and the uplift and subsidence of continental areas could also be brought about by slow, currently operating processes. He replaced all impulsive revolutionary events by low-key gradualism, and by skillful pleading and patient reiteration, eventually convinced his critics that this was so. Lyell's effective exposition of uniformitarianism was enthusiastically adopted in Britain, but acceptance was slower in Germany, where the lingering influence of von Buch and von Humboldt was strong, and in France, where the Cuvier tradition was maintained by Elie de Beaumont (1798–1874) and Alcide d'Orbigny (1802–57). Parochial dogma dies hard. But even in Britain, William Thomson, Lord Kelvin, always ready to debunk geologists, commented that Lyell's uniformitarianism was in essence perpetual motion, and therefore impossible. But this was an overstatement, because, although Hutton could "find no vestige of a beginning," both he and Lyell were Christians who accepted initial divine creation, although it lay beyond the horizon of their rocks.

Among both groups, most recognized recurrent cycles of erosion and sedimentation and of uplift and mountain-building, but to many (like Hutton and Lyell) there was no overriding trend, progression, or directionality, whereas others claimed a steady decline in the intensity of activity, whether this be due to the heat loss of a cooling planet or to some other cause. But none suggested the opposite trend of an exponential increase in tectonic activity, which I will later argue to be really the case. Clearly progression and directionality in the history of the earth are real and significant—for example the progression of life from prokaryotes to hominids, which stands as an empirical fact, irrespective of whether the Cuvier, Lamarck, Darwin, or Linnaeus model is chosen to explain it.

To interpret all ancient phenomena in terms of present processes is fallacious. During the first three-quarters of Earth history there was no land vegetation, so wind was the main transporter of rock fragments, and as the atmosphere had little oxygen, weathering processes were different. In these and several other constraints, the present is only an imperfect key to the past.

The unique event is still another category of catastrophe, outside this controversy, but one that has gained prominence in recent decades. The termination of the Mesozoic Era nearly 70 million years ago, accompanied by mass extinctions and geochemical anomalies, has been attributed to the impact of a large asteroid. Sir Edward Bullard once commented that "with a system as complicated as the earth, almost anything can happen occasionally." P. E. Gretener has added that the chance of throwing six on all eight dice in a single throw is one in 1.5 million, but in 5 million throws the probability of all eight sixes turning up at least once is 95 percent—almost a certainty. So in 4 billion years of Earth history, as Bullard said, "the impossible becomes possible, the possible becomes probable, and the probable virtually certain."

During my student days I was deeply impressed by a benchmark paper by Joseph Barrell (1869–1919) written not long before he died, "Rhythms and the Measurement of Geological Time." It emphasized the importance of rhythms and episodic pulses in the sedimentary record, which is punctuated by repeated gaps that Barrell called diastems. I witnessed this personally in September 1935, when I was near the epicenter of an exceptionally violent earthquake in the Torricelli Mountains of New Guinea (it was magnitude 7.9 with shallow focus). The consequential changes to the rivers and to sedimentation on the piedmont were greater than would result from centuries of normal processes. Professor Derek Ager of Swansea has aptly compared sedimentation to the life of a soldier—long periods of boredom interrupted by moments of terror!

Aspects of Uniformitarianism

In summary, the principle of uniformitarianism stands as the cornerstone of geology, of other retrospective sciences, and, to some degree, of all science. However, uniformitarianism wears a coat of many colors, which scientists view through their individual color filters, and we have not yet agreed on a common palette. We can easily name ten aspects of uniformitarianism:

1. The Universe is rational, without arbitrary divine interference. Most if not all scientists up to the early twentieth century would limit this to the time since an initial creation.

2. The laws of nature are constant through space and time. Kepler asserted this in 1618.

3. The history of the earth can be compiled from the sequential relation of strata, unconformities, fossil faunas, vulcanism, and so forth (the law of superposition).

4. All geological phenomena and processes have been the same as those currently observable. Hutton and Cuvier excluded the orogenic revolutions, but Lyell applied this principle universally.

5. Gradualism: The great changes of the past were accomplished by the summation of very small changes acting over a very long time. The monotonously slow removal of mountains by weathering, erosion, and river transport was easily demonstrated, but an orogenic revolution was less obvious, as it was commonly represented only by an unconformity, with a change of fauna in the rocks above it. Lyell's observation of successive uplifts in Italy following earthquakes or vulcanism convinced him that new mountain systems were also the result of addition of small currently observable phenomena.

6. Individual species are fixed and immutable. This was believed by Cuvier, and by Lyell until reluctantly converted by Darwin. Few if any scientists accept it today.

7. Catastrophism: The past history of the earth included events many times more violent than any observed during human history. This was believed by Cuvier and Hutton, among many others, but denied by Lyell.

8. Unique events had occurred during the history of the earth, not necessarily catastrophic, but they resulted in a change in the course of history. Such could be an initial divine creation, the birth of the moon from the earth or its capture from an earlier independent orbit, a collision with a large asteroid, a natural nuclear explosion (such as the one identified in the Precambrian rocks of Gabon, where it appears that a critical mass of radioactive elements, perhaps even transuranium elements, occurred), the beginning of life, the breakup of Pangaea if this be interpreted as unique, the change from a carbon-dioxide atmosphere to one containing free oxygen, or others yet to be contemplated.

9. Cycles: Certainly there are many natural cycles that affect geological process, such as the earth's daily rotation, the monthly lunar cycle, the annual revolution, the nutation cycle of the earth's axis (18.7 years), the motion of the sun about the center of gravity of the solar system to balance the attraction of the planets (178 years), the precession of the earth's axis (25,700 years), the motion of the solar system around the galaxy (about 250 million years), the relativistic advance of the perihelion of the earth's orbit (600 million years), and probably others. Several Russians (including Peter Kropotkin) and many European geologists have emphasized the importance of pulsations of various periods. Such cyclicity is not inconsistent with any of the other phenomena.

10. Directionalism: Most have assumed a progressive increase in entropy (roughly, disorder) from an energy-rich beginning toward a final extinction of free energy in accordance with the second law of thermodynamics. In contrast, biological evolution implies a progressive decrease in entropy. Most scientists accept as axiomatic that the Earth–Moon–Sun is a closed thermodynamic system with unchanging total of mass-energy. Geological activity is assumed to be running down from decline of radioactivity and outflow of primal heat (Conybeare and Kelvin). Several cosmologists now contend that the gravitational constant has diminished with time. Lyell maintained a steady state, with no progressive change. With the exception of Charles Schuchert of Yale, almost no one has advocated that the rate of geological activity has increased with time. In the final chapter I will argue for this and will argue that mass and energy both increase with time but remain constant by mutually canceling to zero.

CHAPTER 5

The Ice Age

THE ICE AGE is a household term today, practically a cultural icon, but until two centuries ago, ice was not considered a significant geological agent in erosion or transport. Then proponents of three dogmas in turn battled for acceptance by the establishment: the diluvial, drift, and continental ice-age theories. A universal cataclysm was slowly replaced by a continuing process still observable in action today, catastrophism versus uniformitarianism under new banners. Progress only came by throwing off dogma.

The diluvial concept, in the sense that great floods had ravaged the lands and left behind marine shells, goes back at least to Anaximander in the sixth century B.C. and to Xenophanes a little later, and flourished in Christendom as various anomalies were attributed to the biblical inundation. But Diluvialism in the strict sense is of more recent vintage.

Early in the last century, naturalists throughout Europe were puzzled by curious superficial deposits that seemed to call for a unique kind of explanation. Strange blocks up to tens of tonnes in weight and meters in diameter, unlike any rocks in their vicinity, occurred on hillsides and even hilltops, sometimes perched one on top of another, or standing in stream beds as though defying floods to shift them. In the lowlands, there were broad areas of "till," an impervious unstratified subsoil, which also contained similar assortments of exotic rocks, ranging randomly in size from peas to large boulders. There were rhythmically banded, sometimes contorted beds of clay, silt, and sand, also containing similar assortments of exotic pebbles, even boulders. Resting on them in river terraces were normal alluvial deposits, clearly deposited by the floods of the present rivers, some of which contained fossil plants or animals of modern type. What were these strange pre-alluvial ("diluvial") deposits, and what agency had put them there?

In England, where literal belief in *Genesis* was strongest, led by the Reverends Kidd, Buckland, Conybeare, and Sedgwick, the answer was obvious. Here was the patent proof of Noah's universal flood, which submerged Britain to depths of 1500 meters and the Alps nearly 3000 meters, whose violent currents had swept all before them, and which left debris large and small stranded wherever it finally came to rest hundreds of miles from the source. This deluge was correlated with the last of Cuvier's inundations. Christian theologians argued that although Moses's narrative did not mention earlier inundations, this did not preclude them. Many who accepted the validity of the field evidence of a general flooding nevertheless objected to its correlation with the Mosaic flood.

William Buckland (1784–1856), after graduating in theology, became reader in mineralogy at Oxford in 1813 and reader in geology in 1818. As Werner's had done in Freiberg, Buckland's charisma and excellence as a lecturer attracted large audiences of distinguished men. Buckland discussed the European cave deposits that contained bones of lions, elephants, rhinoceroses, hippopotamuses, and hyenas (but not man), which were preserved only in caves because the violence of the deluge had swept away surface remains. From study of the overlying deposits, he concluded that the deluge could not have been more than 5000 years ago (which fitted nicely with Archbishop James Ussher's calculation in 1650 that the earth was created in the year 4004 B.C. and the Flood came in 2349 B.C.). Buckland's Cambridge colleague, geology professor Adam Sedgwick (1785–1873), emphasized that here was wholly independent proof of the scriptures.

But many doubted the capacity of floodwaters, even of great magnitude, to transport such huge boulders so far from their source. Von Buch demonstrated that the exotic blocks strewn across the lowlands of Germany and Poland had come from Scandinavia. Boulders of Norwegian larvikite, an attractive and quite distinctive rock popular for facing buildings, are even found along the Yorkshire coast. As early as 1780 Ivan Lepëkhin (1740–1802), a Russian physician, had recognized that it was ice that had transported the alien blocks over the plains of northwestern Russia. In 1815 John Playfair wrote that a glacier, which conveys great boulders on its surface, is the only agent known to us capable of transporting them such a distance. Also, fossil bones of Arctic type, such as reindeer and polar birds, had been found in the south of France, implying frigid conditions in latitudes halfway to the equator. Hence it was logical to assume that it was melting icebergs that had distributed the large erratics (from Latin *erratus*, strayed) across Britain and so much of western Europe. So the diluvial theory was replaced by the glacial drift theory.

There were still specific puzzling problems. Great boulders of granite and schist on the southeast flanks of the Jura Mountains were wholly foreign to the limestone and shale of the region, but they were similar to rocks cropping out in the Alps opposite to them on the other side of the Rhône valley and the Swiss plain. Jean André de Luc suggested they had been shot nearly 100 kilometers across the valley by air pressure when land subsidence had compressed deep caverns! A famous mountaineer and naturalist, Horace Venedicte de Saussure (1740–99), observed that valley courses linked kinds of boulders with their Alpine sources opposite and invoked catastrophic floods to transport them, although this implied that the rush of water carried some boulders uphill 1000 meters to their resting places! Hutton proposed that the transport had occurred before the excavation of the Rhône valley, when the Alps were much higher and ice-bound (the high Alpine peaks like Mont Blanc were clearly eroded remnants of much more extensive terrains), and the foreign boulders on the Jura Mountains had slid downhill all the way by glacial ice transport to their present resting places.

B. F. Kuhn in 1787 had described the transport of rock debris by the Swiss Grindelwald glacier to its deposition as a moraine at its end; then he traced old moraines and realized that the glacier had formerly extended much farther down the valley. Kuhn's study of Alpine glaciers was extended by de Saussure and later by another intrepid mountaineer, F. G. Hugi.

Peasant folklore has a plausible, though often specious, explanation for all natural phenomena, but the Alpine peasants of three centuries ago were far ahead of their erudite contemporaries, having learned from their fathers that glaciers themselves had piled up the trains of boulders at the end of the ice. So it was natural that in 1815, a Swiss chamois hunter, Jean-Pierre Perraudin, recognized that the present Alpine glaciers were only shrunken remnants of their former size. Working down from their present moraines, he traced a succession of earlier moraines at lower levels far down the valley. He convinced a skeptical engineer, Ignace Venetz, who then carried the investigation further over the next decade, until he was able to announce in 1829 that an ice sheet originating in the Alps had covered the Swiss plain to a depth of more than 1000 m, spreading high up onto the Jura Mountains and over other parts of Europe, dumping Alpine rocks far and wide when the ice melted. So began a domino series, from Swiss folklore to Perraudin, who knocked down Venetz's opposition, who in turn was to do the same to Charpentier, thence to Agassiz, to Buckland, and finally the stubborn establishment.

Venetz's astonishing proposal got little support and much ridicule, but Jean de Charpentier (1786–1855), director of the salt mines at Bex in the Rhône valley southeast of Lake Geneva, adopted it after initial skepticism (indeed he had accompanied Venetz to help exorcise his friend's crackpot ideas). Charpentier proceeded to work out in detail the flow-lines of the ice by mapping the glacial scratches and grooving on the bedrock floor, by tracing the patterns of the moraines, and by following particular types of erratics back to their sources. He in turn was also generally scorned, and one of his old school friends, the brilliant young Swiss naturalist Jean Louis Rodolphe Agassiz (1807–73), accompanied him in the field to demonstrate the error of his observations, only to become himself convinced that Charpentier and Venetz before him had been right.

For the next 30 years, Agassiz led the ice-age campaign with missionary zeal in Europe, Britain, and North America. He advocated a single great ice sheet extending from the Arctic to France, and across Canada well into the United States, although later it became clear that the ice had radiated from several centers. As commonly happens to a new fundamental concept, opposition and denigration was almost universal at first. Von Humboldt, out of kindness and respect for the excellent work Agassiz had published on fossil fish, advised him to forsake these wild speculations and return to anatomy. Karl Schimper, another former school friend of Agassiz, adopted the glacial interpretation and coined the name *Eiszeit* (Ice Age).

The course of the glacial theory from the 1830's to the 1860's was like that of the theory of continental drift one century later. Buckland, who had been the leading proponent of the diluvial theory, then of the glacial drift theory, was cautious at first, but was finally convinced by the field evidence, and became a loyal supporter of Agassiz. Von Buch, now patriarchal, remained implacably opposed until he died in 1852. Murchison held fast to the drift theory, and in later years only allowed a limited role for land ice. Lyell remained unconvinced and also adhered to the drift theory until quite late, as did Darwin. George Ballas Greenough (1778–1855), who had founded the Geological Society of London and served as its first president, remained firmly opposed and scathingly sarcastic. However, although Greenough loomed large in social, political, and administrative science as well as in self-esteem, he was a mediocrity in scientific imagination who among other things had blocked the publication by the society of William Smith's pioneering geological map in favor of his own pedestrian map. William Whewell, the Cambridge professor of mineralogy, argued that an ice age would require a sharp decline in the heat flowing out from the earth's interior and its later restoration to normal, which was not physically possible. (In fact interior heat makes a negligible contribution to climate, and the internal heat flow in the glaciated polar regions is not significantly different from that of the tropics.)

Through the decades of often acrimonious debate, Agassiz persisted firmly. By the mid-1870's, the tide had turned throughout the scientific world, except for a minority of conservative diehards who in the end did just that. Continental glaciation, unthought of only a century before, had been adopted as another firm step toward understanding of our planet (Fig. 7). Oddly enough, Agassiz became a diehard against Darwin's evolution at the same time.

Establishment of the validity of the Pleistocene glaciation raised the question whether that was the first time the earth had suffered such a climatic change. In 1852 William Thomas Blandford (1832–1905) of the Geological Survey of India, in collaboration with two other geologists, his brother Henry and W. Theobald, attributed to glacial transport the 300-million-year-old Talchir boulder bed at the base of the Indian Permian strata, which they later found to be resting on glacially striated bedrock. Blandford was invited to Australia where he also confirmed PermoCarboniferous glaciation, and similar discoveries followed in South Africa and South America. Meanwhile, in 1855, Sir Andrew Ramsay (1814–71), director of the British Geological Survey, reported evidence of ancient glaciation in England. Subsequently, widespread glaciation earlier than the beginning of the Paleozoic Era was found in many countries.

Ramsay appears to have been the first to show that the Pleistocene ice age consisted of at least two glaciations with a long interglacial at least as warm as the present time. Meticulous field work in Europe and North America during the latter part of the nineteenth century and the first half of the twentieth century identified and established in detail the succession of at least four glacial stages during the last million years: the Günz glaciation, an interglacial lasting about 150,000 years, the Mindel glaciation, the "long" interglacial lasting some 300,000 years, the Riss glaciation, the last interglacial lasting about 75,000 years, then finally the Würm glaciation, which retreated about 10,000 years ago. America had a similar sequence of glaciations

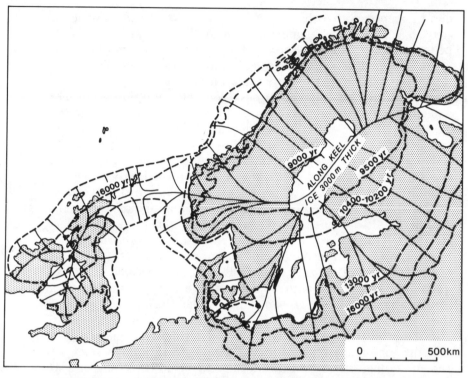

Fig. 7. Retreat of the last ice age glaciation in northwestern Europe. The outermost broken line shows the maximum extent of the ice sheet more than 20,000 years ago, and the inner broken lines mark the ice edge at later times. (After Boulton, Smith, Jones, and Newsome.)

with different names, respectively Kansan, Nebraskan, Illinoian, and Wisconsinan. So, complacently, we could look forward to 50,000 years or so before the next glaciation descends upon us and have ample time to consider how to cope.

Alas, in the mid-1950's the guillotine fell—all this turned out to be wrong. The glacial record on land was hopelessly incomplete, because advancing ice sheets could rub out the evidence of earlier glaciations, and this was not always recognizable in the peripheral deposits. But Professor Cesare Emiliani of the University of Miami found that the complete account had been kept by marine sediments. Not only had the actual temperatures been faithfully recorded by variations in the isotopes of oxygen (^{16}O and ^{18}O), but the sedimentary facies documented the status of the glaciers themselves. At first Emiliani tried to fit his readings into the glaciation pattern which all knew had been firmly established, but he soon found that the glaciations and interglacials had been shorter and much more numerous—not 4, but 20 of them! More startling, our present interglacial warmth has probably nearly run its course. While the world worries about the prospect of a "nuclear winter," we may suddenly find the seas lowering by 100 meters or more as the next glacial stage advances over Europe, Asia, and North America. As the essayist, Alpha of the Plough, has said: "The bolt comes from the blue! We take pains to guard our face, and get a thump in the small of the back!"

CHAPTER 6

The Age of the Earth

HERODOTUS CONCLUDED from his observations of the rate of extension of the Nile delta by the annual floods that the earth had a very long past. To Aristotle, the earth was eternal and everlasting. But throughout the Dark Ages, Christian dogma restricted prehistory to Moses's account.

James Ussher (1581–1656), Archbishop of Armagh, has long been the notorious butt of amusement for having declared that the earth was created on October 26 in the year 4004 B.C. at nine o'clock in the morning. In fact, this statement culminated a long series of such estimates over the previous century, going back to the sacred chronologies of Julius Africanus and Eusebius Pamphili of Caesarea (260–340 A.D.), who had calculated the date of creation at 2016 B.C. It must be remembered that the Bible was believed quite literally in medieval Europe, particularly in Britain, and the age of the earth was estimated by systematic studies of the succession of events reported in it since creation. In the ninth century the rabbinate calculated creation at 3760 B.C., which was thereafter adopted as year one counting from the beginning of the world. Hence in the Jewish calendar 1986 A.D. is 5746 A.M. (*Anno Mundi*, from the year of the world).

"Cooper's Chronicle," dated 1560, listed several such careful calculations of distinguished authorship. For example, one estimated the time from Creation to the Deluge as 1656 years, thence to Abraham 292, to the Exodus 503, to the building of the Temple 432, to Babylon 414, and to Christ 614, thus dating the Creation at 3911 B.C. Another increased the estimate of the earlier part: Creation to Deluge, 2242, Deluge to Abraham, 942, to David 941, to Babylon 485, to Christ 589, making the Creation 5191 B.C. The same chronicle lists four other calculations of the time from creation to the birth of Christ: Hebrews,

3951, Wirandula, 3491, Eusebius, 5177, Augustine, 7331, Alphonsine tables, 6952. The mean of these seven calculations is 5143 B.C. The total span of Earth and man from Creation to Armageddon was commonly accepted to be six millennial "days," four of which had expired by the birth of Christ, so there remained only a few hundred years to come.

Ussher's conclusion was the result of protracted study of sources in their original languages, including the genealogies, calendar systems, and eclipses on which precise time and date could be based. The archbishop, an authority on the classics and several Semitic languages, had amassed a remarkable library of relevant manuscripts. Any who believed the written word with absolute certainty, as he did, could only admire his erudition. Hence his calculation was widely respected and accepted long after his death.

The rise of uniformitarianism, which implied time spans of hundreds of millions of years, was strongly opposed by the church. Hutton, Playfair, Lyell, and Darwin drew larger and larger drafts from the Bank of Time. Hutton could find "no vestige of a beginning—no prospect of an end"; nothing short of endless time seemed to be the testimony of the rocks. Then the Bank of Time slammed its doors. Lord Kelvin, master physicist of his generation, armed with the formulas of that "exact" science, announced that the age of the earth could not exceed 20 million years.

Lord Kelvin

William Thomson (1824–1907), later Lord Kelvin, had been appointed professor of physics in his native Glasgow at the precocious age of 22. He was already widely renowned for his theoretical and industrial achievements in electrical transmission and telegraphy and for his formulation of the laws of thermodynamics when he confronted the geologists in 1862. He hemmed in the age of the earth on three independent grounds: the source of the sun's heat, the cooling history of the earth, and tidal friction in the earth.

Kelvin assumed that the only primary source of energy in the universe was gravitational. The sun was radiating energy at a prodigious rate, without any conceivable source of replenishment; hence the sun's past life and future life must be limited. After considering all relevant data, he concluded it improbable that the sun had illuminated the earth for 100 million years, and almost certain that it had not done so for 500 million years (an age estimated by Darwin from a study of the rate of denudation in southeast England). Said Kelvin, the sun provides the energy for all processes of erosion and denudation, so how could there be uniformitarianism when the sun's output had been diminishing through the ages? A good question. But two axioms are involved—that gravity potential is the only source of energy in the sun, and that matter and energy are fixed inheritances from the "beginning."

Independently of Kelvin, his German contemporary Baron Hermann Ludwig Ferdinand von Helmholtz (1821–94), had demonstrated that if the sun was entirely a coal fire it would burn up in 1500 years, but if its heat output came from contraction under gravity, it could last a thousand times as long, still only one and a half million years. In a lecture in 1871, he backed up Kelvin, stating that the gravitational energy from condensation could have sustained the sun's output for 22 million years and could still be relied on for many million years into the future.

Kelvin's second physical attack concerned the rate of cooling of the earth, which had already been discussed in a preliminary way by Descartes, Gottfried Wilhelm von Leibnitz (1646–1716), and Georges Louis Leclerc, Comte de Buffon (1707–86). Buffon had estimated the age of the earth to be 75,000 years. Kelvin accumulated data from mines on the temperature gradient in the earth; on average the temperature rose one degree Fahrenheit for every 100 feet of depth. Combining this with the heat conductivity of rocks gave him the rate of heat flow out of the earth. As very hot bodies lose heat at a very fast rate at first, which declines exponentially through time, even if Kelvin assumed the highest conceivable initial temperature for the earth, he could calculate how long it had been cooling to reach the present rate of heat flow. Taking extreme values for his data, the lower and upper limits were 20 and 400 million years.

Kelvin explained his third constraint on the age of the earth at a meeting of the Geological Society of Glasgow in 1868. By estimating the amount of retardation of the earth's rotation due to tidal friction, he judged that the earth became consolidated not more than 100 million years ago. This slowing of the earth's rotation means that the day is getting longer. Assuming that the rate of rotation was absolutely constant, the exact meridian at which ancient eclipses would have been seen could be calculated. But ancient records showed that where they were seen was farther to the west. From the meridian discrepancy the exact rate of retardation was known.

Most geologists wilted before the physicist's heat and tried to accommodate their theory to Kelvin's authority. It is an amusing fact that when a scientist knows the answer believed to be correct, that is the result that comes out of the data. For example, Leonhard Euler calculated in 1765 that the free period of wobble if the earth's rotation axis was disturbed would be 305 days, which remained the geophysical dogma for 130 years. But when Seth C. Chandler, a Boston accountant and amateur astronomer, found in 1892 that the earth wobbled with a period of 14 months, he was at first scorned, for after all he was only an amateur, and many distinguished professionals had failed to detect such a wobble. When his result was definitely confirmed, Simon Newcomb, the U.S. Navy astronomer, made corrections to Euler's premises and found that the proper period of free oscillation was indeed 14 months!

Sophist Appeasers

So having been bludgeoned by Kelvin into believing that the age of the earth was between 20 and 400 million years, with about 100 million years as the most probable, a procession of geologists reexamined the data and found that, of course, the age of the earth was 100 million years. First off the mark was John Phillips, professor of geology at Oxford, who compared the cumulative thickness of strata with the average depositional rate and estimated an age of 96 million years. Several others later tried this method, with variable results, but the subjectiveness of the input premises renders the method quite useless.

James Croll (1821–90), a member of the Scottish Geological Survey who later distinguished himself in the saga of the Ice Age, established his balanced neutrality (the right stance for a true scientist!) by criticism of both Kelvin and the uniformitarianists. The geological periods were obviously immensely long, but in assessing how long numerically, geologists were likely to grossly overestimate. Croll thought Kelvin's lower limit could be correct, but eclectically suggested 100 million as the most acceptable.

Next was Archibald Geikie, then director of the Scottish Geological Survey, and a friend of Kelvin. Addressing the Geological Society of Scotland in 1871, he inferred, quite subjectively, that the present rates of denudation were much slower than the average of the past, and therefore that Kelvin's age of 100 million years was compatible with the geological evidence. (Twenty years later, Geikie was to lead the dissent from Kelvin's constraints.)

T. Mellard Reade, following the common dogma in England then that continents and oceans had alternately subsided and risen, and that they were roughly of equal area, estimated that 625 million years was needed to accumulate a crust 10 miles thick. When the *Challenger* expedition found that most of the ocean floor was covered by oceanic planktonic ooze rather than terrigenous (land-derived) sediments as he had assumed, he revised his estimate with this and other modifications and concluded that 95 million years had elapsed since the beginning of the Cambrian Period. The input data were so flexible that whatever result he sought would be the one he found!

In 1893, Clarence King, first director of the U.S. Geological Survey, also entered the lists with an age of 24 million years, the time needed, he believed, for the primitive earth to cool from a temperature of 2000° C.

Samuel Haughton, professor of geology at Trinity College, Dublin, published his *Manual of Geology* in 1865, just after Kelvin's initial blast, but had not been able to take account of it. By quite sophistic reasoning, he estimated that the earth was 2300 million years old, in harmony with the prevailing uniformitarian dogma. Recoiling from Kelvin's constraints, he published a paper in *Nature* in 1878 in which he discussed the lapse of time from when polar temperatures dropped to 212° F, to allow water to condense, thence to 122° F, the coagulation temperature of albumin (to allow life to begin), thence to 48° F at the end of the Paleozoic, and to the freezing point in the middle of the Tertiary Ice Age. He went on to deduce that 153 million years had elapsed up to the Miocene. Numerical mumbo-jumbo always carries respect in science!

Professor John Joly, who followed Haughton in the Dublin chair, estimated the tonnage of sodium salts carried to the oceans by the world's rivers, divided this by the total tonnage of salts now in the oceans, and concluded that the age of the oceans was between 90 and 99 million years. He and others attempted many corrections and refinements, such as allowance for cyclic salt (salt returned to the land as seawater in marine rocks and as evaporites in salt beds), salt added to the oceans by submarine volcanism, and variation in the rate of denudation. However, the only use of this method is to get a result you expect. Recent studies have suggested that the salinity of the oceans reached its present level very early, and that thereafter input and removal have roughly balanced.

Courage of Convictions

Apart from such sophists, many geologists, although subdued under the physics yoke, nevertheless could not escape the conviction that

The Age of the Earth 75

geological time had to be much longer than their physics master would grant, and they suspected that there must be some latent flaw in the master's premises. Thomas Henry Huxley (1825–95), a strong supporter of Darwin but unable to fault Kelvin's argument, nevertheless defended uniformitarianism before the Glasgow Geological Society:

> Mathematics may be compared to a mill of exquisite workmanship, which grinds you stuff of any degree of fineness; but, nevertheless, what you get out depends upon what you put in; and as the grandest mill in the world will not extract wheat-flour from peascod, so pages of formulae will not get a definite result out of loose data.

Two years later, Lyell expressed the hope that some hitherto unknown source of energy would be discovered. Geikie, who had at first supported Kelvin, told the British Association for the Advancement of Science in 1892:

> That there must be some flaw in the physical argument I can, for my own part, hardly doubt, though I do not pretend to be able to say where it is to be found. Some assumption, it seems to me, has been made, or some consideration has been left out of sight, which will eventually be seen to vitiate the conclusions, and which when duly taken into account will allow time enough for any reasonable interpretation of the geological record.

Professor Thomas Crowther Chamberlin (1843–1928), of the University of Chicago, wrote in *Science* in 1899:

> Is present knowledge relative to the behavior of matter under such extraordinary conditions as obtain in the interior of the Sun sufficiently exhaustive to warrant the assertion that no unrecognised sources of heat reside there? What the internal constitution of the atoms may be is yet open to question. It is not improbable that they are complex organisations and seats of enormous energies. Certainly no careful chemist would affirm either that the atoms are really elementary or that there may not be locked up in them energies of the first order of magnitude. . . . Nor would they probably be prepared to affirm or deny that the extraordinary conditions which reside at the center of the Sun may not set free a portion of this energy.

Radioactivity to the Rescue

The years 1900, 1924, 1960, and 1985 were pregnant ones for geology, each presaging the birth of a revolution. In 1900 radioactivity was about to restore geological time. In 1924, translation of Wegener was about to stir geology to its foundations. In 1960, the static oceanic crust was about to acquire astonishing activity and contemporary growth. In 1985, the revolution to accelerating expansion of the earth is imminent, as NASA continues its geodetic measurements (see Chapter 12).

Antoine Henri Becquerel (1852–1908), professor of physics at the Ecole Polytechnique in Paris, established in 1896 that uranium minerals emitted spontaneously a penetrating radiation (x-rays). In 1901 Kelvin's former student Ernest (later Baron) Rutherford (1871–1937), and Frederick Soddy (1877–1956), both then at McGill University in Montreal, discovered that radioactive elements constantly released alpha particles with high amounts of energy, and Pierre Curie (1859–1906) of the Sorbonne in Paris showed that much of this energy degraded to heat. Rutherford wrote in 1904, "The time during which the Earth has been at a temperature capable of supporting the presence of animal and vegetable life may be very much longer than the estimate made by Lord Kelvin from other data."

Meanwhile, an American physicist, Bertram Borden Boltwood (1870–1927), had established in 1904 the steps whereby uranium and thorium lose electrons and alpha particles to end finally as lead, and in 1907 he suggested that by measuring the relative amounts of uranium and lead isotopes in a mineral the elapsed time since its crystallization could be calculated. Thus it was proved that the maximum age calculated by Kelvin had been at least an order of magnitude too short.

In 1905, Robert John Strutt, later Lord Rayleigh (1842–1919), found that the amount of helium accumulated in minerals from the decay of radioactive elements present in ordinary rocks was much greater than could have accumulated in the 100 million years estimated by Kelvin. Indeed, if the radioactivity present in crustal rocks persisted far into the interior, the earth, far from being a cooling body as Kelvin had assumed, would be getting hotter! The geologists now had ample time to fit in whatever processes might be indicated by the rocks, and Darwin, who had been one of Kelvin's prime targets, had ample time for the evolution of species by natural selection.

Like Zeno with the spherical Earth, Archimedes with heliocentricity, von Buch with the Ice Age, Agassiz with evolution, and later Jeffreys with continental drift, many truly great thinkers, when confronted in their later years with a new discovery that overturned their entrenched dogma, refused to acknowledge it, so Lord Kelvin never acknowledged that the geologists had been right. Indeed, he held fast to the uniqueness of unsplittable atoms and refused to believe that helium and lead could be born of pure uranium—a concept that smacked of alchemy.

The Age of the Earth

"New ideas," the great German scientist and philosopher Hermann von Helmholtz once observed, "need more time for gaining general assent the more really original they are." The high priests of science, today as ever, become as impervious to any challenge to their axioms as any of the popes of religion, and their edicts of excommunication effectively deny the heretic access to publication, appointments, promotions, research funds, and social status. Even if the revolution comes while the precocious visionary still lives, he bears yet a sinister taint, and the accolades for the great advance are worn by the *nouveau*-wise.

Radioactivity Clocks

Using several different radioactivity phenomena (such as helium accumulation, breakdown of uranium and thorium to lead, rubidium to strontium, potassium to argon, carbon-14 to nitrogen, and the isotopes of ore lead) each of which makes its particular contribution, the measurement of geological age developed rapidly, particularly through the leadership of Arthur Holmes (1890–1965), professor of geology at Durham and later Edinburgh. I well remember the impact on me in the late 1920's of Holmes's little booklet in Benn's Sixpenny Series, *The Age of the Earth*, published in 1913.

Uranium and thorium decay through several stages to stable isotopes of lead at rates unaffected by temperature, pressure, or physical or chemical state. The "half-lives" of some of these steps are many millions of years, although the statistical probability of any particular atom decaying in the next second is constant. It is like throwing a million dice, one-sixth of which may be expected to fall six. If those that fell six are removed and the remainder thrown the next day, one-sixth of these should fall six to be removed. The probability of a particular die falling six in the very next throw is always one in six, and the last dice in the game have survived a very large number of throws without ever having fallen six. It is simple to calculate how many days are needed to reduce the number of dice to half the starting number; this is called the half-life. The number removed becomes less with each daily throw, and if at any time the number of dice left is compared with the total number excluded, the number of throws can be calculated and hence the number of days the process has continued. Any mineral that contains any uranium can be used. As uranium-238, the most abundant isotope, has a half-life of 4510 million years, this method yields ages of the oldest rocks. Thorium behaves similarly.

Rubidium is a common impurity in feldspars and micas, and

rubidium-87 (which constitutes 28 percent of the natural element) has a half-life of 50 billion years, so it is most useful for measuring the ages of the oldest rocks. Potassium is even more abundant in feldspars and in feldspathoids and micas, and is commonly present in several other minerals; its radioactive isotope potassium-40 has a half-life of 1280 million years. Because extremely sensitive methods of measuring potassium and its argon product have been perfected by Jack F. Evernden and G. Brent Dalrymple at Berkeley and Ian McDougall of the Australian National University, the potassium-argon method has made possible the dating of rocks from the very oldest to the youngest geological epochs (to less than a million years old).

Carbon-14, which is continuously produced in the atmosphere from cosmic-ray impact on nitrogen, has a half-life of 5730 years and has been widely used for dating charcoal, shells, and other organic remains in archeology and back into the Ice Age. So with these four techniques, materials can be dated from any age from the beginning of the earth right up to the present.

Analysis of the relative proportions of the several lead isotopes in ore lead and in the lead in meteorites, some of which isotopes are primordial and some radiogenic from the decay of uranium and thorium, has enabled a reliable dating of the formation of the meteorites and by inference of the earth, which comes out to be 4600 million years. This compares with 3800 million years, the age of the oldest terrestrial rocks so far measured.

The Geological Time Scale

Rock radioactivity rapidly reversed the physicists' straitjacket on the age of the earth, but stranded many geologists embarrassed in the opposite direction. All had re-examined their "hourglass" methods of estimating the rate of denudation, accumulation of marine sediments, increase in the salinity of the sea, or the rate of biological evolution. Many sophists had cut back to near Kelvin's 100 million years. Most others had retreated from "no vestige of a beginning" to congregate at about 500 million years. It was Arthur Holmes of Durham, who led the world in the critical study of geological time, who bridged the gap and reconciled the evidence of geology and the new data from natural radioactivity.

Table 1, from a recent consensus of the Geological Society of America, gives the age in millions of years of the beginnings of the geological epochs, periods, eras, and eons. The birth of the earth is esti-

TABLE 1
The Span of Geological Time
(onsets of periods and epochs, millions of years before present)

Eons	Eras	Periods	Epochs	
Phanerozoic	Cenozoic	Quaternary	Holocene	0.01
			Pleistocene	1.6
		Tertiary	Pliocene	5.3
			Miocene	23.7
			Oligocene	36.6
			Eocene	57.8
			Paleocene	66.4
	Mesozoic	Cretaceous		144
		Jurassic		208
		Triassic		245
	Paleozoic	Permian		286
		Carboniferous		360
		Devonian		408
		Silurian		438
		Ordovician		505
		Cambrian		570
Proterozoic				2500
Archean				3400+

mated to be about 4600 million years ago. The ages given may be varied marginally as more data become available, and even today different laboratories adopt slightly different scales. Epochs have been defined within each of the periods before the Tertiary, but owing to uncertainties of the correlation between continents, the nomenclature varies in different regions, which would make the table too complex for our present purposes.

CHAPTER 7

Numeracy in Geology

PHYSICISTS HAVE perennially derided geology as a subjective, imprecise, descriptive subject, not really comparable with the "exact" sciences. Geologists, they scoff, run for cover at the first appearance of an integral sign. Kelvin said that geology without numbers is not science. However, Kelvin's own case highlights the Achilles' heel of physics—the vulnerability of its premises, which are never better than articles of axiomatic faith. Kelvin's elegant numbers yielded answers as false as they were final. This has been the fate of physics throughout history.

Perfection and harmony in the universe had been axiomatic to Pythagoras and the schools that stemmed from him. Stimulated by the pitch of hammers striking in a smithy, he experimented with lengths of strings, pipes, glass vessels, and trains of bells, to find for each source the harmonious intervals of the musical scale. Numbers were fundamental with Pythagoras, so, having discovered that the harmonious intervals were simple integral ratios of the length of a stretched string, inspiration flashed that such ratios must be the relative distances of the celestial bodies to produce the harmony of the spheres (a melody that only Pythagoras and immaculate maidens could hear, although beloved by poets for millennia thereafter). Moon was the nearest and the lowest tone, and the stars the highest and most distant, with Sun and the five known planets at the intermediate intervals, the whole range from Earth to stars being one octave (Fig. 8). The relative distances of all these bodies were therefore precisely known. Science had become exact! Pythagoras's estimate of one such celestial interval, the distance to the moon, was said to be 126,000 stadia (about 23,000 km, about one-twentieth of the actual distance), but it is not recorded how he arrived at this, because Pythagoras himself left no writings whatever. Another four centuries passed before Hip-

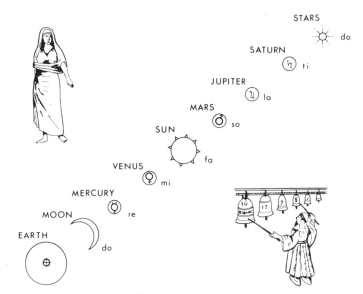

Fig. 8. Only Pythagoras and immaculate maidens could hear the harmony of the spheres.

parchus of Nicea (160–120 B.C.) estimated the distance to the moon by comparing the radius of the earth's eclipse shadow on the moon with the earth's radius, which had been measured by Eratosthenes (Fig. 2). Under the influence of Pythagoras, universal harmony became an axiom, a faith. Pythagoras really believed he heard the melody of the spheres. Its falsity notwithstanding, the harmony of the spheres became accepted dogma for the next 2000 years. Shakespeare's courtiers speak of it in terms that indicate this as part of the general knowledge of gentlemen. But alas, the precise numbers of Pythagoras were a delusion.

Archbishop Ussher's "exact" numbers, processed with scholarship and erudition unmatched before or since, gave him precisely the date of creation of the earth; but his answer was as wrong as it could possibly be. Impeccable logic and precise computation produced rubbish from a false premise.

Harold Jeffreys, the most competent mathematical geophysicist ever, who insisted throughout his long life that the continents were fixed entities on the face of the globe, pooh-poohed geologists who argued otherwise. It was not his numerical skills that let him down, but his intuitive creed.

The giants of physics, from Newton through to Einstein, knew with certainty the validity of initial creation, whereby all the atoms of the universe came into being where nothing had been before, contrary though this be to their laws of conservation. If this premise be false, can any conclusion based on it, even with the best mathematics, be accepted as necessarily valid? This question is examined in later chapters.

Huxley's caustic comment that the most elegant mathematics won't yield wheat-flour from peascod raises the general question of the status of numeracy in geology. First let us recognize intrinsic differences between scientific fields.

A mathematician may be a good one while knowing nothing whatever of any other science. A physicist may succeed knowing naught but mathematics and physics. But every action in chemistry roots in physics, and thence in mathematics. Biology involves complex chemistry, and hence physics and mathematics, together with its own intrinsic variables. Geology founds on all four.

Physics may formulate a problem theoretically or experimentally, allowing only one parameter to vary. But the ascent of the ladder of complexity to geology involves so many factors, and these of such gross scale, that neither mathematical nor experimental isolation of variables is conceivable; so the geologist must rely more and more on empiricism and qualitative eclectic judgment, a sophisticated synonym for common sense.

Many of the basic principles of physics may be tested experimentally in the laboratory. But in geology the scales of size and time are so very large that experimental tests are quite out of the question. Dynamically similar models can be made, but no one can be sure whether other terms may be involved that are below the level of detection on the scale of possible observation but surpass all other terms when size or duration is extended. Such effects of scale are discussed in Part Four and again in the final chapter.

Geologists and geophysicists approach truth from opposite sides. The geophysicist represents some aspect of the earth by a theoretical mathematical model, which restricts the number of variables and quantifiable parameters. The conclusion may be valid for the model but may have less or no relevance to the real earth, although the geophysicist may firmly believe it, especially if it is printed out by an impartial computer. By contrast, the geologist takes the real earth, warts and all, and reasons qualitatively and inductively. Hazard for the geophysicist lies in the deficiencies of the model. The geologist's Achilles' heel is in the validity of intuition. Neither can be sure that all relevant laws of nature have already been discovered, or that some threshold of discontinuity of behavior has not been crossed. The quantitative and empirical approaches are both needed, and each can help the other in proposing new avenues of theory and experiment.

Comparable competence in mathematics or geology implies equivalent intelligence, but difference in kind. Mathematics demands pure logic from stated premises. Geology differs from most other sciences in three ways. First, as illustrated above, geology requires a capacity for eclectic thinking with intuitive weighting of multiple variables. Second, geology involves retrospective sequential thinking, as crystallized in the law of superposition (first enunciated in 1669 by Steno)—that the sequence of strata, their deformation and fracturing, and their injection by veins, implies a definite historical sequence of events. Third, geology requires an ability to visualize in three dimensions—to go from a random cross-section in a microscope slide to optical and crystal axes of the mineral; from outcrops on a geological map to the three-dimensional structures of folds, faults, unconformities, and intrusive contacts; from stereoscopic aerial photos to configuration underground; from borehole intersections to the form of an orebody; from the random surface trace of an eroded fossil to a mental picture of the whole animal; from projections of lineations or cleavage poles to visualization of the stress that made them; from the convolutions of multiply folded features to the individual deformations they imply; and so on.

For the reasons outlined above, quantification in geology came late. Pythagoras's numerical distances of the heavenly bodies from Earth rested on a false faith. Four centuries later, Eratosthenes fared much better when he measured the diameter of the earth with valid reasoning based on sound observation. Newton correctly deduced that the earth was oblate because centrifugal force due to rotation would progressively counteract gravity from poles toward the equator, from which Newton calculated the earth's ellipticity. A century later, Henry Cavendish calculated the weight of the earth; he put a gold ball at each end of a bar suspended from a fiber at its middle, then measured the torsion in the fiber when a heavy mass was brought near one of the balls. As he knew the weight of the attracting masses, by applying Newton's gravitation law to the attraction force measured by the torsion in the fiber he was able to calculate the gravitation constant in Newton's law. From that it was a simple matter to calculate how much the earth weighed, again using Newton's law. But physicists claim all these and other Earth measurements as physics, not geology.

Physics was numerate from birth. Quantification of chemistry came some time later, and of biology much more recently. Numerical geophysics slowly developed from Osmund Fisher, Lord Kelvin, and Harold Jeffreys to a veritable phalanx of significant recent workers, overlapping with the explosion in applied geophysics financed by the mineral industry since the late 1920's. Commencing with Arduino two centuries ago, the qualitative relative order of geological events was gradually crystallized, but numerical ages for the geological periods were only defined within this century. Petrology also became quantitative, extending gradually through geochemistry more generally. The last two decades have brought an explosive expansion in numeracy throughout geology. This was already in train, but it was enormously accelerated by the sudden availability of powerful computers capable of analyzing and processing voluminous and complex data.

Although inevitable, and in the long run destined to advance geology rapidly, numeracy is not an unmixed blessing, and is apt to become a blind fetish. A rising generation may emerge of lab-bound expert number-jugglers and twisters of dials on black boxes, applying programs whose ramifications they do not really understand, in ever-narrowing fields of specialization, smugly believing their printouts, and scornful of methods of their wiser predecessors. I have witnessed Little Jack Horners proudly accumulating stacks of computer printout—potentially rich ore—but without refining any bullion. Huxley's garbage in, garbage out warning is ever more relevant. Constraints of cognate fields may be overlooked. Traps lurk in the assumptions and shortcuts involved in the software.

False conclusions arising from one or other of these hazards are numerous: Kelvin's omission of a hitherto unknown factor is one I have already mentioned. Others I will treat include Simpson's invalid use of statistics (Chapter 8); McElhinny's belief in the output from an invalid program (Chapter 14); interpretation of the blueschists in terms of an irrelevant experiment (Chapter 14).

Populations of animal fossils in particular localities—fossil faunas—have been used for more than a century to measure kinship between regions. Faunal relationships imply migration or isolation between areas; thus the growing body of evidence leads to the definition of faunal provinces and realms. These in turn impose constraints on reconstructions of paleogeography. Much work in recent years has scrutinized the validity of faunal lists and the systematic bias that can be introduced by the regionality of the identifying paleontologist. Rigorous statistical methods, applied after dropping cosmopolitan fossil species and scaling down doubtful identifications, yield hard numbers to measure kinship. With this kind of thoroughness does Dr. Clive Burrett calculate his "Provinciality Index" by dividing the number of common genera by twice the number of non-common genera. After duly recognizing difficulties with making correlations at short time intervals and ensuring that the habitats compared are truly comparable, Burrett lists his numbers and reviews current tectonic hypotheses. Like Ussher's and Kelvin's before him, Burrett's erudition and care are impeccable; the difficulty lies in the premises.

J. Hoover Mackin, formerly of the University of Texas, has written:

> When mechanical processes *replace* reasoning processes, and when a number *replaces* understanding as the objective, danger enters. . . .
>
> The very act of making measurements, in a fixed pattern, provides a solid sense of accomplishment. If the measurements are complicated, involving unusual techniques and apparatus and a special jargon, they give the investigator a good feeling of belonging to an elite group, and of pushing back the frontiers. Presentation of the results is simplified by the use of mathematical shorthand, and even though nine out of ten interested geologists do not read that shorthand with ease, the author can be sure that seven out of ten will at least be impressed. It is an advantage or disadvantage of mathematical shorthand, depending on the point of view, that things can be said in equations, impressively, even arrogantly, which are so nonsensical that they would embarrass even the author if spelled out in words.

Long before the beginning of electronic computers, Thomas Chamberlin, founder of Harvard's *Journal of Geology*, wrote: "The fascinating impressiveness of rigorous mathematical analysis, with its atmosphere of precision and elegance, should not blind us to the defects of the premises that condition the whole process. There is, perhaps, no beguilement more insidious and dangerous than an elaborate and elegant mathematical process built upon unfortified premises."

Should geology then remain qualitative? Of course not. Because of the wide spectrum of interdependent variables inherent in it, geology should profit more than most other sciences from vast computing capacity. But broad training across the board and thorough experience in observing real rocks in the field is even more necessary than before. Even logistic aids like the jeep and helicopter have their drawbacks: before their advent, geologists walked to their next destination and stubbed their toes on outcrops yielding unexpected information contrary to their current assumptions.

PART TWO
MOBILE CONTINENTS

CHAPTER 8

Continental Drift

MENTAL MODELS bias our thinking, and "continental *drift*" hobbled Wegener's concept in the English-reading world from the outset. Wegener's word was *Verschiebung*, which was correctly translated by Skerl as "displacement." "Drift" was substituted by detractors, and as they were the majority, the term gained currency; the theory, saddled with that name, was successfully slanted toward fantasy.

The South Atlantic Fit

Sir Francis Bacon (1561–1628) appears to have been the first to record the similarity of shape of the opposing African and South American coasts, in Aphorism 27 of Book II of his *Novum Organum* (1620). Some have said that he had suggested that this implied Earth expansion with the opening of the Atlantic, but I cannot read that into his text:

Verum his missis, etiam in ipsa configuratione mundi in majoribus non sunt negligendae instantis conformes; veluti Africa, et regio Peruviana cum continente se porrigenti usque ad Fretum Magellanicum. Utraque enim regio habet similes isthmos et similia promontoria, quod non temere accidit. Item Novus et Vetus Orbis; in eo quod utrique orbes versus septentriones lati sunt exporrecti, versus austrum autem angusti et acuminati.

[In fact, neglecting the former observation, we must not overlook the similarities of large masses in the configuration of the earth; for instance, Africa and the Peruvian region with its landmass extending down to the Straits of Magellan. Both areas have similar isthmuses and similar promontories, a fact not due to mere accident. Again, the New and Old World resemble each other in the fact that both worlds are wide toward the north, while toward the south they taper to sharp points.]

Certainly *exporrecti* (from *exporrigere*, to stretch out) can be translated as "expanded," but here it is used in a descriptive sense only.

Not much later, R. P. François Placet expressed similar ideas in the third edition of his book (which appeared in Paris in 1688): *Où il est montré que devant le déluge l'Amérique n'était pas séparée des autres parties du monde* (Wherein it is shown that, before the Flood, America was not separated from the other parts of the world).

During the next two centuries, many were struck by the matching shapes across the Atlantic and offered various guesses as to what they might mean—Buffon, about 1780; von Humboldt, about 1800; Young, in 1810; Richard Owen, 1857; William Lowthian Green, 1857; Antonio Snider, 1858; Henry Wettstein, 1880; the Reverend Osmund Fisher, 1882; C. B. Warring, 1887; W. H. Pickering, 1907; R. Mantovani, 1909; Frank Bursley Taylor, 1910; Howard Baker, 1911; and of course Alfred Wegener in 1912, and probably several others who have escaped my notice.

Wegener's Predecessors

Several early writers considered that the Americas had formerly been continuous land with Europe and Africa but had been separated by the subsidence of the mythical land Atlantis.

The first suggestions that the continents had actually drifted apart seem to have been made in 1857 independently by Professor Richard Owen (1810–1910), of the University of Indiana, and William Lowthian Green of Hawaii, an extraordinarily erudite Scot who described himself on the title page as "Minister of Foreign Affairs to the King of the Sandwich Islands." Owen's rather fanciful book, *Key to the Geology of the Globe, an Essay*, attracted attention at the time but had no lasting effect. Green's 1857 paper was published in the *New Philosophical Journal of Edinburgh*. His later (1875) book, *Vestiges of the Molten Globe, as Exhibited in the Figure of the Earth, Volcanic Action and Physiography*, contains concepts a full century ahead of his time (such as the global sinistral torsion along the Tethys discussed in Chapter 21), but they were not taken up and were largely forgotten.

In 1858 Antonio Snider-Pellegrini, a devout American Christian, published in Paris his book, *La Création et ses mystères dévoilés* (Creation and its mysteries unveiled), in which he anticipated Wegener's Pangaea, a single landmass occupying one hemisphere with ocean occupying the other hemisphere. This lopsided arrangement had broken up during the biblical flood, with the Americas drifting westward, opening the

Atlantic with its complementary coastlines. The whole book is a grand elaboration of the six "days" of *Genesis*. The Moon was expelled during the convulsions of the first "day." The fifth "day" produced Pangaea and the complementary primordial ocean. Noah's flood and the disruption of Pangaea occurred during the sixth "day."

Henry Wettstein published in 1880 in Zürich his book, *Die Strömungen der festen, flüssigen und gasförmigen Stoffe und ihre Bedeutung für Geologie, Astronomie, Klimatologie und Meteorologie* (The flow of solids, fluids, and gases and its significance for geology, astronomy, climatology, and meterology), in which he had the continents drift westward, driven by tidal forces acting on the viscous interior, thus opening the Atlantic. A rather similar concept was advanced later by E. H. L. Schwarz in the *Geological Journal* in 1912. Wettstein supported his theory by pointing to paleoclimatic and paleontological anomalies. In Germany and other countries where the German language was widely read, concepts of extreme mobility were common, but they mostly involved decoupling of the whole crust ("polar wandering") rather than relative drift of continents. Loeffelholz von Colberg (in 1895) and Kreichgauer (in 1902) are among these. In America, by contrast, prevailing conservatism discounted such mobilism.

Osmund Fisher (1817–1914) published *Physics of the Earth's Crust* in 1881, in which he followed George Darwin's theory of the separation of the moon from the earth. The Pacific basin was the birthmark. The great depression was largely filled from below, but also in part by convection currents rising under the oceans (especially under the Mid-Atlantic Ridge) and sinking under the continents, thus opening the Atlantic and Indian Oceans and the chain of scallop-like small seas of the western Pacific. W. H. Pickering also followed George Darwin in his 1907 paper in the *Journal of Geology*, the ejection of the Moon having occurred not in the Mesozoic but very early in the history of the Earth. W. Franklin Coxworthy published a book in London in 1890(?), *The Electrical Condition, or How and Where Our Earth Was Created*, advancing the theory that today's continents are the disrupted parts of a former single continent.

Frank Bursley Taylor (1860–1938), of the U.S. Geological Survey, addressed the Geological Society of America in 1908 on "The Bearing of the Tertiary Mountain Belts on the Origin of the Earth's Plan" (published 1910). This was probably the first clear presentation of continental drift. He amplified it in the *American Journal of Science* in 1926 in "Greater Asia and Isostasy," and again in the 1926 symposium on continental drift organized in New York by the American Association of Petroleum Geologists, where he suggested that the moon may have been captured in the Cretaceous Period, an event that increased the tidal forces and the polar flattening, and caused a former great continent to break up and drag toward the equator. Taylor was impressed by the apparent southward creeping of Eurasia, piling up the Alps, Carpathians, Zagros, Himalayas, and the offshore arcs of East Asia. The Mid-Atlantic Ridge marked the scar from which the Americas and Africa-Europe diverged. Taylor emphasized only the Tertiary orogeny (epoch of mountain-building), which implied that the well-documented older orogenies needed a different explanation. He did not consider paleoclimatic or paleontological evidence, nor did he attempt to reassemble the continents to their pre-drift positions.

In the September 19, 1909, number of the magazine *Je m'instruis*, R. Mantovani discussed continental displacement in a two-page contribution "L'Antarctide," with some ideas remarkably similar to those of Wegener a couple of years later.

Howard B. Baker presented a series of papers on continental drift, commencing in 1911 with "The Origin of the Moon," then "The Origin of Continental Forms," in four parts in 1912–14, and finally in 1932 his book, *The Atlantic Rift and Its Meaning*. His first paper anticipated much of Wegener, such as the breakup of a single primary continent, which split from Alaska to the Antarctic to open the Arctic and Atlantic Oceans, but that is all, as he did not discuss Asia or the Indian Ocean, and his guesses about Australia, New Zealand, and Antarctica were wide of the mark. His treatment of severed orogenic and paleontological links was thorough and convincing. Baker attributed the breakup of the primary continent to a catastrophe about 6 million years ago, when the eccentricities of the orbits of Earth and Venus caused such a close approach that the intense tidal forces tore a chunk out of Earth to form the Moon, leaving the Pacific basin as the scar. Earth lost most of her water, but captured more water when in the turmoil another former planet was disrupted to form the asteroids.

In retrospect, a common thread that runs through all this is the congruence of the coastlines facing each other across the Atlantic. In contrast, there has been a wide spread of opinion on what brought it about. Such accord and discord continue still.

Wegener

This brings us to Alfred Lothar Wegener, professor of meteorology and geophysics in the University of Graz, who was born in Berlin in 1880 and perished in a Greenland blizzard in 1930, while attempting to establish geodetic confirmation of contemporary widening of the Atlantic. Wegener is universally acknowledged as the patriarch of continental drift. Although it is clear from the foregoing that many aspects of "the Wegener theory" had been published by one or other of his predecessors, it was his masterly and comprehensive synthesis that shocked the geological world, and he merits the Domine status accorded him. Wegener drew together all that was then known in every relevant science (geology, geophysics, geodesy, biology, paleontology, oceanography, meteorology, and astronomy), throughout geological time, for every part of the world (Fig. 9).

Wegener was trained in astronomy, in which he gained his doctorate, but he then shifted his emphasis to meteorology, and married the daughter of the well-known meteorologist W. P. Köppen. I suspect that had he been trained as a geologist, he would never have embraced the concept of continental displacement. Exotic jumps are mostly made by interlopers from an alien discipline, who have not been imprisoned by orthodox dogma.

Wegener's first book, *Die Entstehung der Kontinente* (The origin of the continents) was published in 1912. His celebrated book, *Die Entstehung der Kontinente und Ozeane* (The origin of the continents and oceans) was published in 1915, with successive revisions in 1920, 1922, and 1928. The 1922 edition, which was translated into English by J. G. A. Skerl (in 1924) and also into French, Spanish, Swedish, and Russian, was the basis of the worldwide controversy. The fourth edition, incorporating much new data and eliminating some of the weaknesses of the earlier editions, was not translated until more than three decades later (1966), when some Americans had belatedly begun to recognize the validity of Wegener's work.

Wegener's case filled a book, the quintessence of which was three points. (1) The mean level of lands stands 5 kilometers higher than the mean level of oceans; the rocks of the ocean floors are some 15 percent denser than the rocks of the continents. Flat-topped icebergs, floating one-seventh of their thickness above the sea, model this relationship of continents in flotation equilibrium, about one-seventh of their thickness above the seafloor. (2) Continents can be fitted together, jigsaw fashion, with remarkable success, into a supercontinent that he named Pangaea. (3) If the continents' shapes had been cut from a newspaper, then when reassembled, the print-lines should read across. Wegener demonstrated this paleontologically (closely related land plants and animals are now separated by wide oceans), stratigraphically (similar sequences of strata are likewise separated and progressive character trends continue across the join when reas-

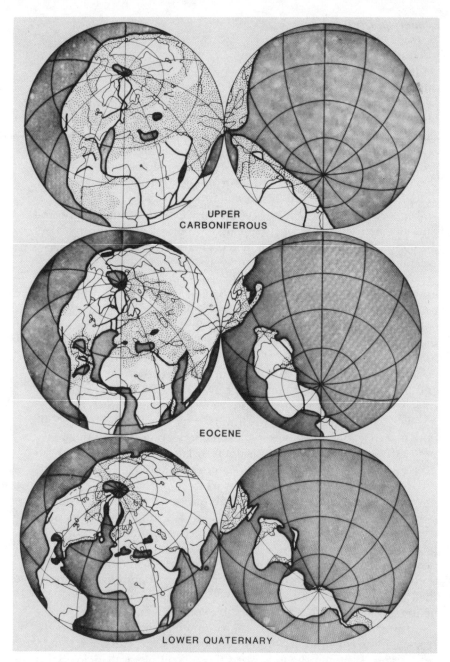

Fig. 9. Wegener's reconstruction of Pangaea at three different times. The light stipple indicates shallow seas. Modern rivers are added to aid identification. Stereographic projection; the grid is arbitrary.

sembled), paleoclimatologically (climatic zones, indicated by glaciation, coal measures, and desert salt deposits, for example, match across integrally but make absurd distribution patterns as currently located), tectonically (fold mountains are truncated illogically by the ocean but form single entities when reassembled), and so on.

It was not until the third edition (1922) of Wegener's book was translated into English, and burst like a bombshell on incredulous Americans and Britishers, that the status of the match across the South Atlantic was critically examined. My own personal inspiration, Sir Edgeworth David, who became deeply but cautiously interested, cut out the continents from a globe and pieced them together; "and would you believe it," he said, "the astonishing thing is that they do fit." David delivered a public address on June 12, 1928, in which he said that the whole theory looks fantastic at first sight, but a contemplation of the facts includes the hypothesis within the realm of probability.

As a student, on the suggestion of Professor Leo A. Cotton, I made cartographically valid stereographic projections, using both the 100- and 1000-fathom isobaths, not only of the South Atlantic fit (Fig. 10), but also several others including Australia–Antarctica, and Australia–India with the Bay of Bengal against Australia's northwest shelf. Cotton had suggested plotting on a normal equatorial stereographic net, and rotating the outline about an appropriate Euler pole, which is relatively easy on a stereographic projection, but I preferred to take the more arduous route of calculating a series of oblique stereographic projections, because I could work on a much larger scale, and Cotton's suggestion, although theoretically valid, accumulated progressive plotting errors. Common projections, including the stereographic, are illustrated in Fig. 32 and their use discussed in Chapter 12.

The continental-drift controversy raged with much heat during the 1920's. English-speaking geophysicists almost unanimously opposed it, probably largely because of the influence of Jeffreys, but in Germany more were willing to take it seriously, including Helmert, Albrecht, Förster, Milankovitch, and Gutenberg. The seismologist Beno Gutenberg (1889–1960) in 1927 used the distribution of earthquakes to revive George Darwin's theory of the ejection of the moon. The Pacific Ocean scar was left completely devoid of continental crust, while the rest of the earth's crust was dragged to the Pacific, and the new rift oceans thus formed had a thin layer of continental-type crust that had developed since the catastrophe. Gutenberg abandoned the lunar involvement in 1936 but retained his interest in continental drift, not-

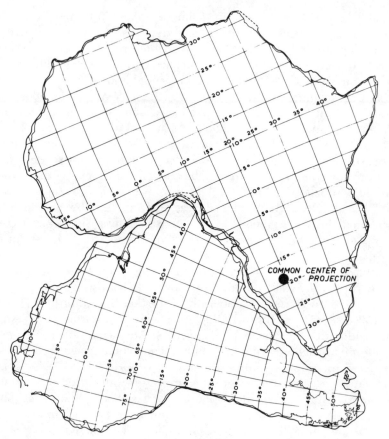

Fig. 10. Africa and South America as I fitted them at the 1000-fathom isobath on an oblique stereographic projection in 1933, before I realized that the earth is expanding. Although I placed the northwest angle of Brazil as close as I could into the Gulf of Guinea, I could not eliminate the residual gaps tapering in both directions resulting from the expansion of the earth.

withstanding the cynical pressure from his new colleagues in America, for the rest of his life. He contacted me for discussion of my 1956 symposium paper shortly before he died.

Geologists concerned with geological processes on a global scale tended to react favorably, such as Emile Argand, Rudolf Staub, E. B. Bailey, Reginald Daly, and Arthur Holmes, but geologists with more local interests, and laboratory-oriented geologists, generally rejected it. Many biologists, who had problems with the discontinuities in distribution of animals and plants (the leading British paleobotanist

A. C. Seward, for example), found more satisfactory solutions with continental drift. Opposition was strongest in North America, whereas many European geologists had been thinking along mobilist lines for decades. Geologists involved with the East Indies and those from Gondwanaland countries included many supporters: Reinhardt Maack and Beurlen in South America, Du Toit and Lester King in South Africa, Wadia and Krishnan in India, and Evans and later me in Australia. An important stream, confined to the German-reading schools, was stimulated by Wegener, but diverged from him by attributing the dispersion of Pangaea to gross expansion of the earth since the Paleozoic: this group included Lindemann (1927), Bogolepow (1928), Hilgenberg (1933), Halm (1935), Keindl (1940), and Egyed (1956). More about them later.

One far-sighted American who went against the flood was Professor Reginald A. Daly, of Harvard, whose 1928 book *Our Mobile Earth* influenced me as a student.

The AAPG Symposium

The American Association of Petroleum Geologists convened an international symposium on continental drift to coincide with their annual meeting in November 1926 in New York. The proceedings, published by the AAPG in March 1928, set the pattern for the next three decades. Wegener, who was then deeply involved with his geodetic work in Greenland, hoping to prove its westward drift relative to Europe, did not attend, although he sent two brief papers, one on the geodetic program and the other on the Carboniferous climate of North America. In Wegener's absence, the lead paper was given by W. A. J. M. van Waterschoot van der Gracht, then Vice-President of the Marland Oil Company.

The result of this symposium was that American geologists almost universally rejected continental drift, although contrariwise it was my study of this symposium volume, particularly van der Gracht's paper, that convinced me that the Wegener theory was probably valid. Much criticism was advanced against the theory, but much of it was trivial or superficial and where valid was directed against the physical mechanism. Most of the essential arguments for drift had remained intact.

Du Toit

In retrospect it is unfortunate that Alexander L. du Toit (1878–1948) did not attend the AAPG symposium. After many years of field work in the Geological Survey of South Africa, where he became the recognized authority on the regional geology, he received a grant from the Carnegie Corporation to travel extensively in South America to compare the geology across the Atlantic. From this resulted his 1927 benchmark paper, "A Geological Comparison of South America with South Africa." His own summary of it a decade later reads:

[In it] is set forth the host of correspondences between the sides of the South Atlantic in the shape of the Devonian system, Carboniferous glacials, Permo-Triassic strata, etc., and the significance is stressed of phasal variations away from the respective coasts. It is pointed out that the Mesozoic foldings of the Cape and Argentina meet at right angles the older structures that trend parallel to the two Atlantic shores, and, as Holmes has since phrased it, *the crossing begun in the one continent is completed in the other*. Such phenomena are explicable only on the assumption of drift. . . . The collective evidence points to the Falkland Islands as having formerly been situated between the Cape and Argentina, their stratigraphy and structure being almost identical with those of the Cape.

In 1938 du Toit published his magnum opus, *Our Wandering Continents*, quite the most thorough and comprehensive study since Wegener's. He strongly confirmed continental drift but departed from Wegener in one important way: whereas Wegener had restored the continents into a single supercontinent Pangaea, du Toit had two separate circumpolar continents, Laurasia in the northern hemisphere and Gondwana in the southern hemisphere; the ancestral Pacific Ocean was initially an equatorial ocean between them, taking its present form when Laurasia and Gondwana came together to form the Alpine-Himalayan orogenic belt.

The Gonds are an aboriginal Dravidian people of central India, of Negrito type. Some have said during recent decades that "Gondwanaland" is a tautology, because Gondwana itself means the land of the Gonds, but research by Dr. Fakhruddin Ahmad (a world authority on Gondwana matters) unearths a different story. After the Mahabharat, the great war between two jealous Aryan brothers in central Asia about 1500 B.C., the vanquished group went south into India, taking over the lands of the more primitive Dravidians, whom they contemptuously called Gowandawana ("bull's balls people," from *gow* bovine, *anda* testicle, and *wana* belonging to), and the name Gondwana has persisted to this day. It was first used in geology by H. B. Medlicott in a manuscript report to the Geological Survey of India and was adopted by Otto Feistmantel for the "Gondwana flora" in a Record of the Survey published in 1876. William Blandford, also of the Survey, coined the name "Gondwanaland" for the group of continents that he believed had been parts of a single greater landmass from the Carboniferous to the Jurassic. This concept was adopted by the famous Austrian synthesizer Edouard Suess (1831–1914) in his four-volume magnum opus *Das Antlitz der Erde* (The face of the earth) to denote the southern lands that shared the *Glossopteris* flora, late Paleozoic glaciation, and other singularities.

Du Toit's Laurasia-Gondwana pattern was adopted by Lester King and philosophically had much to commend it, because its symmetry suggested global convection currents within the primordial Earth rising equatorially where surface temperatures were highest, and dragging surface scum polewards to form the primary continents, where the colder convection flow turned down. (However, this preference vanishes on the expanding Earth model, where the initial lithosphere enclosed the whole Earth.) Subsequent work has unambiguously confirmed Wegener's Pangaea.

Du Toit's book appeared when general rejection of continental drift was at its maximum. However, he summarized the consensus among its few supporters as he assessed it:

(a) Two great parent masses throughout the Paleozoic—Laurasia and Gondwana;

(b) A fragmentation that began in the later Mesozoic and is still in progress;

(c) A dispersal of the fragments radially outwards and also equatorwards, with a tendency towards a westerly creep as well;

(d) Drift of crust relative to the polar axis, bringing about major climatic changes;

(e) Some distortion of the masses during their drift;

(f) Transfer of part of the Pacific waters to fill tension-basins making the other oceans;

(g) Recognition of drift as a process that has operated throughout geological time;

(h) A cause or causes that lie seemingly not outside but within the Earth itself.

Most of these still stand, but instead of two parent masses the single Pangaea has been restored, and in the third point the motion is not really equatorward but northerly with general dispersion.

The mechanism moving the continents was a major research problem aroused by Wegener's work. Mantle convection, then and now a lively topic, warrants a short digression here.

Flow and Convection in Solids

When molten lava, molten iron, and liquid water cool, they crystallize to the solid state in which the crystal grains are much the same in all directions. When an iron billet is rolled so that its length is increased perhaps fiftyfold to a long rod or a thin sheet, it becomes wrought iron as its crystals reform again and again with a pronounced grain. Likewise, when the ice in a glacier has flowed several kilometers, the glacier ice has a strong grain; it is indeed a metamorphic rock that has been recrystallized time and again as it flowed. The pressures within the earth are many times greater than in a glacier or rolling mill, and if the pressure is not equal in all directions the solid rock flows, until the pressures are equalized, by means of repeated recrystallization. The rock then has a pronounced grain and is called a gneiss or schist, with cleavage indicating the plane of greatest extension, and also lineation (alignment of needle-shaped crystals in the direction of maximum elongation). So slates, schists, and gneisses indicate that the rock has flowed in the solid state, usually tens of kilometers, or much more. Solid-state flow is of vast importance for the tectonic processes that shape the earth's surface.

Because the interior of the earth contains small amounts of radioactive elements, heat is continuously generated, and this flows out to the surface. As rocks vary in their heat conductivity and in their radioactive content, so temperatures rise higher in some places than others. This temperature rise feeds back on itself because the warmer rocks expand more, which makes them less dense and more buoyant and at the same time makes them yield more easily to pressure differences. (You heat a solid if you want to bend it without fracture.) So the hotter regions rise and spread sidewards, and the cooler regions sink and flow in underneath the rising column, giving rise to convective circulation. There is strong evidence that such convective circulation occurs continuously in the mantle of the earth.

Convection currents in the earth were proposed in 1881 by Osmund Fisher, in 1906 by Otto Ampferer, in 1928 by Rudolf Staub in Switzerland, and in 1935 by C. L. Pekeris, later of the Weizmann Institute in Israel, and particularly by Arthur Holmes in Durham and Edinburgh, by Felix A. Vening Meinesz in the Netherlands, and David T. Griggs in Harvard, as the main driving force in geotectonics generally, and latterly in respect to continental drift. Keith Runcorn of Newcastle adopted convection as the main cause of continental displacement; so also did Harry Hess of Princeton more recently. The

Continental Drift

flow is in the solid state, where viscosity or stiffness is many times greater than in a flowing glacier, so where the convection circulation rises and then turns aside, it would drag the overlying crust apart, forming tension rifts on the surface; where it flows horizontally under the crust, it might carry the crust along passively on its back, like a glacier carries a large boulder; where circulation turns downward it might cause compression in the surface rocks, but might also drag the subcrust down into the flow, thus undermining the crust.

All of the theoretical models of convection cells that have been published are grossly oversimplified. No allowance is made for variation in the viscosity, although there can be no doubt that the viscosity changes by a factor of perhaps 100,000 within the depth range of the postulated circulation, which implies that the rate of flow under a given driving force would vary by a similar factor. This would not prevent convection, but the pattern of the cells would be very different from that displayed in textbooks.

Another complication is that a large number of phase-change boundaries have to be crossed (where increasing pressure with depth causes minerals to recrystallize in a denser form, and vice versa). Some of these phase changes are known to be metastable, that is, minerals stay in the condensed form long after the pressure has been reduced (diamonds and garnets are well-known examples). As it is the change in density with temperature that drives the convection circulation, metastability should inhibit the circulation.

My own work very much curtails the role of convection circulation. In an expanding Earth, the descending limb of the convection flow would certainly be reduced, and would be entirely eliminated if the rate of upward flow balanced the rate of increase of volume and of surface area.

Decades in Contempt

From the 1930's to the early 1950's, Wegener's ideas were generally rejected as a fantasy—fascinating but false. "*Ein Märchen*, a pipe dream, a beautiful fairy story," chanted the American bandwagon. Papers that denied continental drift passed referees with little scrutiny; they were correct *a priori* because everybody knew that continental drift was wrong. Otherwise-great scientists such as Harold Jeffreys and George Gaylord Simpson got away with absurd solecisms, which were believed and cited as sound reasons for rejecting continental drift.

Simpson, for example, scorned the evidence for faunal ties between Africa and South America, citing the Triassic reptiles as having only a remote degree of kinship consistent with their present separation across a wide ocean:

	A	B	C
Families	100	89	43
Genera	82	64	8
Species	65	26	0

A: Percentage of recent Ohio mammals also occurring in Nebraska, 500 miles away.
B: Percentage of recent French mammals also occurring in northern China 5,000 miles away.
C: Percentage of known South American Triassic reptiles also found in the Triassic of South Africa, now 4,750 miles apart.

The naïveté of such an argument is stark. The faunal lists for the recent mammals are (for all practical purposes) complete. The probability is remote of finding a new species, still less a new genus or a new family. Among the Triassic reptiles, by contrast, a whole genus may be known only from a single bone. At the time Simpson wrote, not a single Triassic reptile was known from the whole Australasian octant of the globe from Malaysia to New Zealand. Since then a nearly complete skeleton has been found in Hobart, and others elsewhere, along with several amphibians. It would be absurd to suggest that the Triassic faunal lists are even 1 percent complete. If we took a random 1 percent from the Ohio and Nebraska mammals, the comparative percentages would drop drastically. Moreover, column C covers some 60 million years, so to make a valid comparison for column A, it would be necessary to take a random 1 percent from all the Ohio mammals known since the Eocene Epoch and compare them with a similar random list from Nebraska.

In 1929 appeared Sir Harold Jeffreys's prestigious book, *The Earth*—quite the most authoritative treatise ever on the physics of the earth, following the tradition of Osmund Fisher and Lord Kelvin. However, Jeffreys (1896–) was completely opposed to Wegener's hypothesis, and in regard to the alleged fit of South America into the angle of Africa, he wrote: "On a moment's examination of the globe, this is seen to be really a misfit by almost 15°. The coasts along the arms could not be brought within hundreds of kilometers of each other without distortion. The widths of the shallow margins of the oceans lend no support to the idea that the forms have been greatly altered by denudation and deposition."

Fig. 11. Hemispherical table with the same radius as the globe behind, and spherically molded tracing foil for accurate comparison of continental shapes. The foil sheet on the upper left of the globe fits so accurately that it does not need sticking down, and North America shows through. On the table, two foil sheets have been joined to cover the whole Arctic region.

From my many "moments" of accurate examination of this question that I had done, I knew this statement to be incorrect. I considered that the matter was rather trivial, that the true position would be generally realized, and that this criticism would fade away. But Jeffreys's prestige was so great that most workers accepted his pronouncement as final. Jeffreys repeated the statement in the second edition of his book in 1952, and to rub salt on the wound, Dr. George Martin Lees (my former chief in the Anglo-Persian Oil Company), in his 1953 presidential address to the Geological Society of London, listed this as one of his three crucial reasons for rejecting the Wegener hypothesis. So I sent Lees my stereographic projections of two decades earlier, together with the comparisons I had made on the spherical table (Fig. 11), proving that Jeffreys's statement was false. I added that "whether the continental drift hypothesis be true or false, this argument should never be used against it again." I asked Lees to arrange publication of this rebuttal, which he did.

When I went to England in the summer of 1960 as Tasmanian delegate to the third centenary of the Royal Society, Sir Edward Bullard invited me to lunch to discuss the Atlantic fit, which he then repeated with the aid of a computer. The Atlantic match has since been known as the "Bullard fit" and adopted generally.

Although any loose statement denigrating continental drift got easy passage to publication during that period, anyone unwise enough to speak for it was rejected by referees and editors with snide comments. Thus, in 1953 I sent to the American Geophysical Union a paper showing the earliest physically viable exposition of the passive transport of a continent toward an oceanic trench, beneath which the subcrust was progressively sapped by the down-going limb; it was rejected as "naïve and unsuitable for publication." Twenty years later, after plate tectonics had "arrived," I dusted off the manuscript and suggested that they might choose to publish it now as a historical document. Six months later the editor replied that the policy was not to publish a paper previously turned down, *however good it might be*!

CHAPTER 9

Sowing the Seeds of Revolution

WHILE CONTINENTAL DRIFT was in contempt, better equipment and techniques and increasing factual data and also conceptual advances were brought to bear on Wegener's model. This chapter and the next will examine them.

From 1930 until 1937 I had followed the Wegener model (underpinned by mantle convection), thence until 1956 the du Toit model of Laurasia and Gondwanaland, twin *Urkontinente,* and thereafter the expansion model wherein continents dispersed from a pan-global continental crust while each continent remained attached to its own subjacent mantle. Hence my perspective on the period between Wegener's death and his vindication differs from that of most of the textbooks as much as a black South African's view of contemporary society differs from an Afrikaaner's. This divergent retrospection needs to be expressed, if only to balance the historical record.

The revolution referred to in the title of this chapter has three phases. First is the mobilist revolution wherein the English-speaking scientific world accepted the relative motion of continents. This was immediately subsumed under the plate-tectonic revolution, much as the Mensheviks were succeeded by the Bolsheviks in 1917. To plate tectonicists the distinction may seem pedantic, but it is both the start and the heart of my two decades of dissent from their creed. The third phase of the revolution, which I believe to be imminent, will be the recognition of Earth expansion, with backtracking from subduction to the model which I have adopted since 1956, as defined above.

Paleomagnetism

In 1849 Achille Delesse discovered that some recent lavas were magnetized in the direction of the earth's local magnetic field, and Macedonio Melloni confirmed in 1853 that Vesuvian lavas were magnetized parallel to the ambient field. At the turn of the century, Giuseppe Folgerhaiter extended this to bricks and pottery, showing that during firing ceramics became magnetized parallel to the local field and that this magnetization remained stable (hence called remanent magnetization) even in pots buried in random positions for thousands of years. Hence the recent history of the geomagnetic field's inclination could be retraced from the study of ancient pots.

During the ferment of the late 1920's, Paul Mercanton reported to the French Academy that magnetic studies on the dispersed pieces of Pangaea could give objective tests of the Wegener theory. This idea was supported in 1940 by Beno Gutenberg, who had always had sympathy for the continental-drift concept, but, owing to World War II, it was not until the late 1940's and early 1950's that the English physicists Patrick M. S. Blackett (who with his astatic magnetometer had greatly improved measurements of very weak magnetizations) and Keith Runcorn and their students seriously investigated this proposition. They were stimulated by a report by John W. Graham, of the Carnegie Institution, that at least some sedimentary rocks also retained fossil magnetization directions through long geological ages.

At this time, a vertical 300-m borehole into the Jurassic dolerite on the eastern shore of Great Lake had just been completed by the Tasmanian Hydroelectric Commission (for which I was consultant), so I air-freighted to Blackett cores 30 m apart, with the prediction that they should prove to have been relatively near the pole of the time. In due course Blackett cabled that the magnetization of the cores was close to vertical, confirming my tectonic prediction. This work led, with the cooperation of Professors J. C. Jaeger and Keith Runcorn, to the invitation to Edward M. Irving, then a graduate student at Cambridge, to take up a research fellowship at the Australian National University. There he commenced a systematic study of the remanent magnetization of the Tasmanian Jurassic dolerite, whose Jurassic paleolatitude he found to be 80° S.

Meanwhile, one of my colleagues, Maxwell R. Banks, on study leave in Massachusetts, collected oriented samples for me from the Rhaetic dolerite on the west bank of the Hudson River, and I asked Dr. Reinhard Maack, of the Curitiba Museum, to collect from the basalts of the Parana Basin of Brazil. As D. Ian Gough and Anton Hales had been investigating the paleomagnetism of the Karroo dolerites at the Bernard Price Institute in Johannesburg since 1950, paleomagnetic confirmation of the validity of Gondwanaland was available to the 1956 Hobart symposium on continental drift (see below).

text continues on p. 113

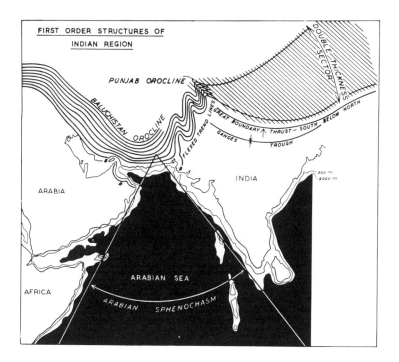
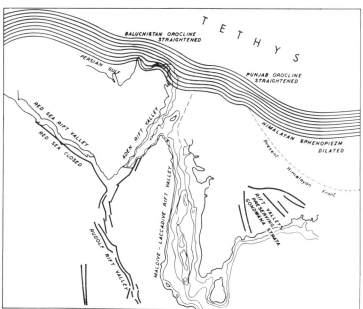

Fig. 12. The Baluchistan orocline as conceived by me in 1938 (before my recognition of Earth expansion).

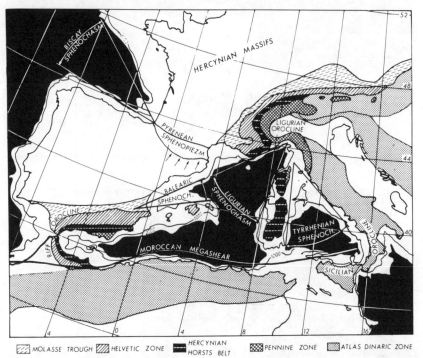

Fig. 13. Oroclines, sphenochasms, and megashears of the western Mediterranean, as conceived by me in 1938. The black areas are deeper than 2000 m. The line between that and the coast is the 1000-m isobath.

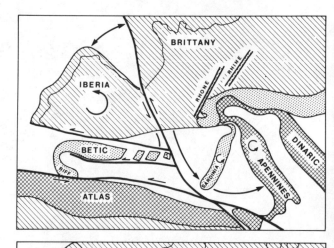

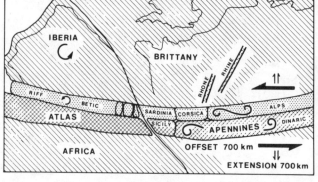

Fig. 14. Restoration of early Mesozoic western Europe by sinistral shift of 700 km and widening of the Mediterranean by 700 km.

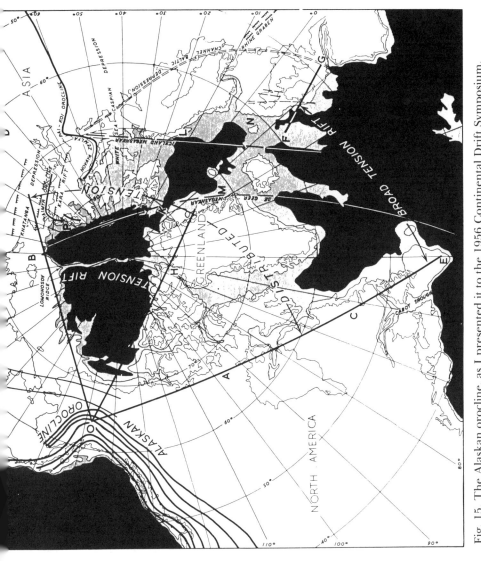

Fig. 15. The Alaskan orocline, as I presented it to the 1956 Continental Drift Symposium.

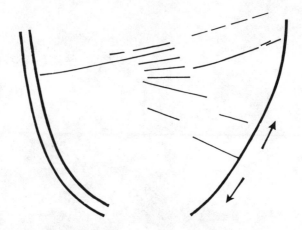

FRACTURE PATTERN OF NORTH ATLANTIC
SIMPLIFIED BY OMISSION OF TRANSCURRENT MOVEMENT

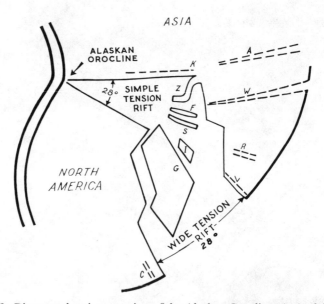

Fig. 16. Diagram showing opening of the Alaskan Orocline. A, Aral depression; W, White sea depression; K, Khatanga rift valley; Z, Novaya Zemlya–Pai Khoi coupled oroclines; F, Franz Josef Land; S, Spitsbergen; I, Iceland; R, Rhine graben; L, Lisbon scarp; G, Greenland; C, Cabot trough.

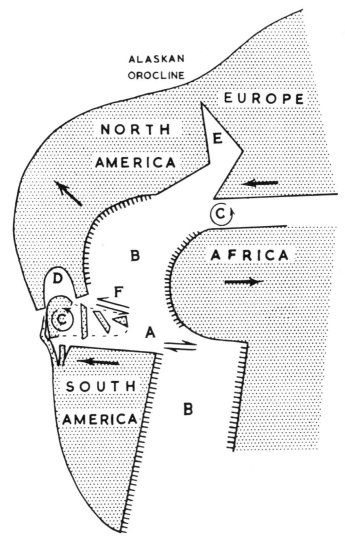

Fig. 17. This diagram that I prepared for the 1956 Continental Drift Symposium shows the relation of the Atlantic Ocean to the Alaskan orocline. The Americas swing like a gate about the Alaskan orocline away from Europe and Africa. At the same time, North America and Europe move west with respect to South America and Africa on the Tethyan torsion (see Chapter 21). A, dextral shear between Venezuela and the Gold Coast of Africa; B, Atlantic sphenochasm; C, Caribbean islands and Mediterranean continental fragments show sinistral rotation due to the Tethyan torsion (see Table 2); D, Central American blocks pulled out of America to form Gulf of Mexico; E, Arctic sphenochasm; F, sinistral shear of Tethyan torsion.

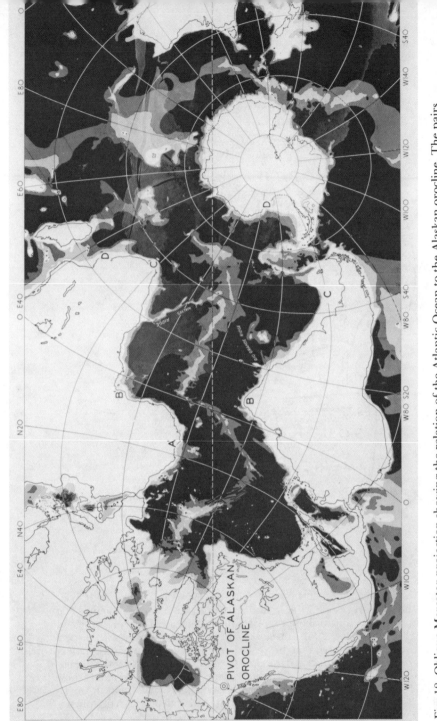

Fig. 18. Oblique Mercator projection showing the relation of the Atlantic Ocean to the Alaskan orocline. The pairs A–A', B–B', C–C', and D–D' indicate points formerly contiguous.

Oroclines

During the last 100 million years there have been active mountain-building belts right around the rim of Wegener's Pangaea, and also across its middle, from New Guinea through Indonesia, the Himalayas and the Mediterranean, northwest Africa, and the Caribbean. Characteristic of these belts were earthquakes, volcanoes, and great thicknesses of weak strata; the crust below was much hotter than normal and hence capable of yielding when stressed, and extensive deformation had occurred. These orogenic belts contrasted with the broad areas of the continents, where little deformation had occurred for hundreds of millions of years.

It seemed to me in the 1930's that if the continents had moved the long distances visualized by Wegener, then the orogenic belts between them, which were mobile and deforming at the relevant times, must record indelibly the relevant motions, and this should be obvious in the plan view of the surface. Even the most primitive physical maps of the earth's surface confirmed these inductions. I knew by 1938 that if I straightened the obvious bends, which I called oroclines (Greek ὄρος,

TABLE 2
Predicted and Measured Rotations on Oroclines

Rotated block	Rotation predicted from oroclines (degrees)	Paleomagnetic rotation (degrees)
North America to Europe	30	30
Africa to South America	45	45
Newfoundland	25	25
Spain	35	35
Italy	110	107
Corsica and Sardinia	90	50
Sicily to Africa	0	0
Arabia to Africa	4	7
New Guinea	35	40
Honshu (north and south)	40	58
Mendocino orocline	60	63
Puerto Rico to South America	45	53
Jamaica to South America	42	50
Hispaniola to South America	39	40
Colombia	Large	80
Appalachian arcs	20–40	29
Malay Peninsula	About 70	70
Seram (East Indies)	Large	98
Scotia arc	Large	90
India	70	70

mountain; κλίνω, to bend), restored the visible stretches, and closed the obvious openings, these processes alone reproduced Pangaea. Figures 12–18, on pp. 107–12, show some of the results. Most of what I published in my 1954 orocline paper was in the 1937 draft of my doctoral thesis, but I omitted it at the eleventh hour because I realized that these concepts were too radical for acceptance then, and would have cost me my degree. Even in 1954, publication was refused by the referees of the Geological Society of Australia. However, it was published in 1955 by the Royal Society of Tasmania and in 1963 was awarded the Gondwanaland Gold Medal as the most significant paper in the relevant triennium. Table 2 lists the oroclinal rotations published by me, and shows how they were subsequently confirmed when this became possible by paleomagnetism.

Ocean-Floor Spreading

The orocline reconstructions implied the growth of vast areas of new oceanic crust. For example, the Baluchistan orocline was complementary to the opening of the Indian Ocean (Fig. 12), and the Alaskan orocline (Figs. 15–17) was complementary to the opening of the Arctic and Atlantic Oceans. It was also obvious to me that the African rift valleys, the Red Sea, and the Atlantic Ocean formed a progressive series of a single spreading process. The mechanism for this was crucial to the whole process of the dispersion of the continents, which I emphasized in the discussion at the 1956 Hobart continental drift symposium. For an exact explanation I shall quote from the symposium report. (Fig. 12 in the report has been changed to Fig. 19 here.)

Dilatation fractures at shallow depth are vertical and normal to the tension (Figure 19a). However, a condition is soon reached in depth where the weight of overburden is such that pure tension failure cannot occur since the load exceeds the unconfined compressive shear strength and the first effect of stretching is to produce failure in pure shear. Because of this the surface vertical tension crack will eventually turn to an angle of about 58° (45° plus half the angle of friction) as shown in Figure 19b. One or both of the two possible fractures indicated may develop. Geological loads do not usually come into being suddenly, but develop over protracted periods. Since the rate of stress relaxation by creep flow increases with temperature, there will be a flow contribution to the strain, increasing with depth until a state is reached where all stress is relaxed by flow before it can reach fracture level. This is the lower limit of earthquakes for that rate of deformation. Hence the brittle shear fractures become more and more deflected by flow until they become dissipated in horizontal laminar flow (Figure 19c). Thus the downward succession in a

Sowing the Seeds of Revolution

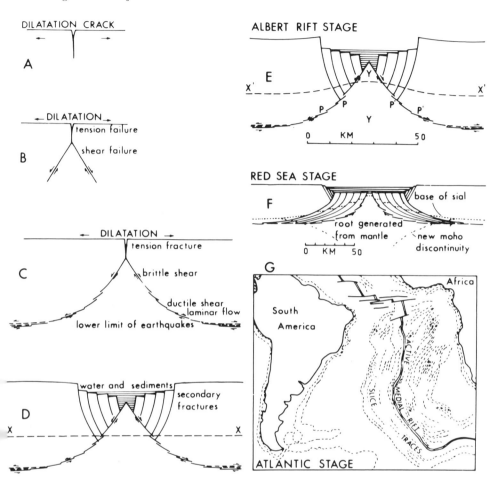

Fig. 19. Development of spreading ridges and paired growth slices with the youngest at the median rift (as presented to the 1956 symposium on continental drift).

stretching crust is surface vertical fractures, brittle shear fracture dipping at about 58°, ductile shear fracture at increasingly shallow angles, and finally horizontal laminar flow. In a crust with given physical properties and temperature gradient, the level at which ductile shear flattens out as laminar flow depends on the rate of stretching. For at a low rate of extension, creep will relax all the stress at comparatively shallow depth. At higher extension rates the stress in these same rocks will increase to fracture threshold before flow relaxation is sufficient to accommodate the extension.

Movement under the tension will produce Figure 19d, or a unilateral version of it. The secondary fractures are inevitable because otherwise the outward movement along the horizontal sole at the right and left would leave openings at the sloping fault surfaces.

However Figure 19d represents a very substantial departure from isostasy, owing to the deficiency of mass represented by the rift valley which, in a rift 30 km. wide and 4 km. deep, would amount to 300 billion metric tons per km of length. The earth's crust certainly could not sustain a differential load of this order. Since compensation in the cold upper crust is regional rather than strictly local, the whole area between X and X' (Figure 19d) commences to rise isostatically, concurrently with dilatation but lagging behind it. This produces a typical rift valley with raised but actively rising rim, and a negative gravity anomaly due to still incomplete adjustment (Figure 19e). The floor may be sinking or rising according as the subsidence caused by continuing dilatation is currently more or less rapid than the isostatic rise. . . .

It has already been pointed out that isostasy causes the line XX' of Figure 19d to bulge upwards to XX' of Figure 19e. This means that level for level the material at YY between and below the faults is warmer than the otherwise similar material outside and above the faults. Hence this central material yields more by flow under the same stress in the same time than material above the faults. Hence if stretching continues, the relief by flow extends inwards *below* the original faults, and faults such as PP and P'P' extend upwards until they meet, while the material above them is drawn away, allowing their junction to rise isostatically to reach the surface. Thus the rift widens by flow in depth with repeated slices developing below the older faults. This process can go on indefinitely to produce an ever widening ocean (Figure 19f). At all stages while this process continues, the marginal coasts may long since have ceased activity, but there will be an active median line of faulting and seismicity following a ridge composed of two tilted rims with an intervening narrow trough on the site of the latest extension. This trough has a mass deficiency and will be in process of rising isostatically. The raised rims will be due to the fact that regional isostatic adjustment is reached quite rapidly, but more local isostatic adjustment follows much more slowly and may never be fully attained. As local adjustment becomes more complete, the raised rim subsides. In any case it would be eliminated whenever progressive dilatation caused a further fracture to appear (such as PP in Figure 19e) since the former raised surface would then be in the area of slump settling (as in Figure 19d). Hence there will at any time be only one pair of raised rims even in a wide rift.

The rift mechanism produces identical morphology irrespective of whether it occurs in continental crust or across ocean floors. These characteristics fit precisely with accurate detail the median ridges of the Atlantic, Indian, and Southern Oceans.

Where the initial conditions are symmetrical, as when the rifting commences within a continent or within an existing ocean, the pattern develops symmetrically, with paired slices rising on opposite sides of the median rift. But if initial conditions are asymmetrical, as when the rifting commences at the junction of a continent and ocean, the slices may appear on one side only, because the contrast in physical conditions extends quite deeply into the subsurface.

I have taken pains to present this explanation mainly to show that plate tectonics provided neither the first nor the only mechanism for ocean-floor spreading that fits the data. My mechanism is as valid today as it was 30 years ago.

The Hobart Symposium

As mentioned above, I convened in 1956 an international symposium on continental drift at the University of Tasmania in Hobart. Up until this time, I had taken it for granted that the diameter of the earth had not changed significantly since the beginning (the question had not really arisen), but during the symposium discussions, anomalies that had bothered me came together and crystallized. I realized that reconstruction of Pangaea required a smaller Earth, and by the end of the symposium I had taken the major step to Earth expansion.

Among the symposium participants were Professor Lester King, of Durban, who had assumed the mantle of du Toit, Dr. J. W. Evans, who had long been interested in continental drift from the viewpoint of the distribution of insects, Dr. Reinhard Maack, of Curitiba, the leading authority on the regional geology of Brazil, Professor Kenneth E. Caster, of Cincinnati, who had spent many years studying the Paleozoic stratigraphy of South America and Africa, Edward Irving, who was already establishing leadership in paleomagnetism, Professor J. C. Jaeger, who had done so much to stimulate investigation of new fields in geophysics, Dr. Rudolf O. Brunnschweiler, who had become a mobilist under Staub of Zürich, Dr. James M. Dickins and Dr. G. A. Thomas, who had compared the Permian faunas of India and northwest Australia, Professor Alan H. Voisey, originally sympathetic to continental drift, but who had absorbed the American skepticism during his teaching assignments there, and Dr. Armin A. Öpik, the Estonian world authority on early Paleozoic trilobites, who was opposed to continental drift.

American geologists almost unanimously rejected continental drift; and because the strongest opposition had been centered at Yale, where the permanence of continents and oceans had been established by Dana and firmly followed by Charles Schuchert, I invited the chairman of the Yale geology department, Professor Chester Longwell, to come to Hobart as principal guest. The symposium made a deep impression on Longwell, particularly the success of the orocline method, the new evidence from paleomagnetism, and the symmetrical growth of ocean floors at the median ridges—so much so that at the end of the symposium he invited me to come to Yale as visiting professor for a year, "to stir the American pot." Longwell had a broader vision than most of his American contemporaries. Existing commitments prevented acceptance of his invitation until the academic year 1959–60.

Unexpected appreciation of the Hobart symposium also came from another quarter. Professor Eugene Wegmann, who had succeeded Emile Argand in Neuchâtel, sent to me the original colored drawings from Argand's benchmark paper at the 1923 Brussels International Geological Congress, "La Tectonique de l'Asie," with the note: "I hand you the mantle of Argand."

American Evangelism

In Yale I delivered complete courses in structural geology and global tectonics. But I also lectured in many other American universities, mostly under the American Geological Institute Visiting International Scientist Program: Brown, Columbia, Harvard, Wesleyan University, Lehigh, Princeton, Duke, North Carolina, Louisiana State, St. Louis, University of Cincinnati, and Ohio State, as well as Toronto, Western Ontario, McGill, Calgary, and British Columbia in Canada. As with Matthew's sower, some seeds did fall on fertile soil and took root, only to be choked off later when subduction weeds grew rank.

A quarter of a century later John Rodgers, reminiscing over the perennial controversy about the nature of orogenesis, wrote:

> My own contribution to the solution of this dilemma was negative; I left North America for a year to study the Alps. As a result, the Yale Geology Department could appoint a visiting professor for that year; we chose S. Warren Carey, and North American geology has never been the same since. He traveled all over the continent, he lectured in his inimitable now-you-see-it-now-you-don't style, he talked to anyone who would listen, and when he was through, no-one could laugh off continental drift any more.

Professor Walter H. Bucher, the patriarch of American tectonicists, who had been stung by my heresies, invited me to confront him in a debate at Columbia. The Schermerhorn Theater was packed as geologists and geophysicists gathered from far afield, and a most memorable night resulted. Geophysicists and geochemists marshaled be-

Sowing the Seeds of Revolution

hind the ghost of Kelvin to reject as really impossible the geological assault, and withdrew checked, but not mated.

Apart from Yale, my deepest involvement was with Princeton where I lectured several times in late 1959 and early 1960, including discussion of oroclines, the paleomagnetic evidence of large intercontinental movements, and ocean-floor growth by repeated insertion of paired slices at the mid-oceanic ridges as detailed in the Hobart Symposium (Fig. 19). Harry Hess, chairman of the Princeton geology school, and I cemented a warm friendship that deepened until his premature death.

The campaign culminated with a special session on continental drift sponsored by the Society of Economic Paleontologists and Mineralogists at the annual meeting of the American Association of Petroleum Geologists at Atlantic City on April 25, 1960. I was lead speaker, and with me on the panel were Keith Runcorn, Ken Caster, and William Gussow. The hall was packed, even the aisles and the walls. After the formal papers from the panel, the questions and discussion continued until long after midnight with few if any leaving, until the chairman had to terminate the meeting. The revolution to continental dispersion had begun!

CHAPTER 10

The Kuhnian Revolution

THOMAS S. KUHN, in his 1962 book *The Structure of Scientific Revolutions*, postulates that

in any field of science there are periods of relative tranquillity separated by revolutions during each of which there is a shift from almost exclusive support of one theory to almost exclusive support of another theory that is incompatible with the former. Each such shift is characterised by a switch in *Gestalt*—that is, there is a fundamental change in the way the *whole* field is comprehended. Each intervening period of relative tranquillity, each period of "normal science", is characterized by widespread adherence to whatever theory is current and by the engagement of almost all the scientists in the exploration and articulation of that theory and its ramifications.

During Kuhn's "periods of tranquillity" a few isolated heretics, including the forerunners of the next revolution, endure rejection and ridicule.

Really new trails are rarely blazed in the great academies. The confining walls of conformist dogma are too dominating. To think originally, you must go forth into the wilderness—Mendel in his Brno monastery, Hutton on his farm in Scotland, Darwin's years on the *Beagle*, Copernicus alone in Frauenburg. Newton conceived his gravitation law during a long exile to the country forced by the great plague of 1666; throughout his life he was an intellectual loner and did not bother to publish. Harry Hess told me he did his thinking at sea, and only worked it up at Princeton. William Lowthian Green, a century ago in Honolulu, preceded all of us, but is now forgotten. My own good fortune was to spend six years in primitive New Guinea. Without journals or peer pressure, I could think! I now welcome occasional visits to America and Europe, and bubble with the contemporary froth, but return to Tasmania to think. Nobody comes to Tasmania to go anywhere else, because it is not on the way to anywhere else.

The Kuhnian Revolution 121

Americans have often suggested that the flickering flame of continental drift was kept alive in a few outposts of the southern hemisphere because the Gondwana stratigraphy had more obvious bonds to link together the dispersed continents. Actually the reverse is true, because the Laurasian dispersion is less, and the oroclines, particularly, apply more obviously there. Suess, for instance, had been impressed by the striking similarity of the Paleozoic stratigraphy and tectonics between Europe and North America. Du Toit thought independently as he did because he was based in far-off Johannesburg, and made long field forages through South Africa and South America. Had he been chief geologist of the survey in the United States or Britain, instead of South Africa, he would never have escaped the dogma. The same applied to Reinhard Maack in Brazil, to M. S. Krishnan in India, and to me in Australia.

By the beginning of the 1960's the progressive gestation of several seeds had rendered the onset of the birth of the "new global tectonics" imminent and inevitable:

1. Accumulating evidence, from the work of du Toit, Maack, and others, supported Wegener's claim that bringing the continents back together was like matching the pieces of a torn newspaper—the lines read across.

2. Striking restorations followed from unwinding the oroclines I had demonstrated (Figs. 12–18).

3. The widely separated Mesozoic paleomagnetic poles of the continents only made sense when Pangaea was restored; this was largely the work of the British pioneers Runcorn, Blackett, Bullard, and Irving.

4. The crystallization of the 70,000-km global pattern of the mid-ocean ridges, with their central rift valleys, was led by Bruce C. Heezen of the Lamont Geophysical Observatory in New Jersey (Fig. 20).

5. The contrast between continental crust and suboceanic crust sharpened, already known from seismic and gravity measurements to be different. If the ocean basins were primitive, old sediments should still be there, but dredging, later confirmed by drilling to basement, found few sediments more than 100 million years old, and none as old as 200 million.

6. Least noticed, but most important of all, a generation of conservative geologists had passed on. As Max Planck wrote in his autobiography: "A new scientific truth does not triumph by convincing its opponents and making them see the light, but rather because its opponents eventually die, and a new generation grows up that is familiar with it." Wegener himself had to learn this truth the hard way. In a

Fig. 20. The circum-continental tension rift system, as developed by Bruce Heezen and others in the late 1950's. Numbers indicate the nine major lithosphere blocks.

1911 letter to his father-in-law (Professor Köppen) he wrote, "Ich glaube nicht dass die alten Vorstellungen noch zehn Jahre zu leben haben" (I don't think the old concepts have more than ten years to live). But a decade later he had to write to him, "Halten sie die Verschiebungstheorie schon auf der Schule gelernt, so würden sie sie mit derselben Unverstand in allen, auch den unrichtigen Einzelheiten ihr ganzes Leben hindurch vertreten, wie jetzt das Absinken von Kontinenten" (Had they already learnt the displacement theory at school, they would defend it throughout their lives with the same complete lack of judgment and false details as they now do with the sinking continents).

Cataclysmic events, whether they be outbreak of war, economic disasters, political revolutions, a fundamental switch in science dogma, or a heart attack, involve a gross cause and an immediate trigger. The trigger for the revolution from general rejection to general acceptance of the dispersion of the continents came from paleomagnetism.

Paleomagnetic Polarity Reversals

As paleomagnetic data multiplied, it became apparent that some rocks of different ages in various countries were magnetized in the opposite direction from what was expected—north and south magnetic poles were reversed—an effect that had been discovered by Pierre David and Bernard Brunhes at the beginning of this century. Much controversy ensued as to whether this meant that the earth's magnetic field had reversed many times in the past, or whether the rocks had reversed their magnetization while cooling. Indeed, it was

The Kuhnian Revolution

definitely established that some rocks did just that. But it turned out that such self-reversing rocks are rare, and that the earth's magnetic field has indeed flipped over repeatedly. Dr. Jan J. Hospers, a Netherlander at Cambridge, reported in 1950 that individual lavas in Iceland were consistent in being either normally or reversely magnetized, but that the polarity had flipped over at least twice during the last 60 million years.

As the precision of the potassium-argon method of measurement of the age of rocks was greatly refined, dated lavas were consistent the world over in recording reversals, so that specific epochs of "normal" (as it is now) and reversed polarity could be identified in chronological sequence (Fig. 21). Then the same chronology of magnetic epochs was recorded in bore-cores through the ocean-floor sediments.

Oceanographic mapping of the northeast Pacific by Victor Vacquier of Scripps Institution of Oceanography, and by Arthur D. Raff and Ronald G. Mason at Scripps, found that the magnetization of the underlying rocks formed a surprising and unexpected stripe pattern running roughly north-south. Each stripe, a few tens of kilometers wide, had its characteristic signature in profile, so that a sequence across several of them could be recognized where it had been offset many tens or even hundreds of kilometers along large fracture zones previously identified by their scarp lines and distinctive topography. Such stripe patterns of magnetization were soon found to be characteristic of spreading ridges (an example is shown in Fig. 63 in Chapter 18).

Frederick J. Vine of Cambridge and his superviser, Drummond H. Matthews, and Lawrence W. Morley of the Geological Survey of Canada independently guessed that the stripes represented reversals of the magnetization of the rocks, alternately enhancing or weakening the earth's general magnetic field. As basalt, newly injected into the axial rift zone of a spreading ocean ridge, cooled through the Curie point, it became magnetized by the prevailing field. During times of normal polarity, the new basalts would be normally magnetized, but when in due course the polarity reversed, the following new basalts would be reversely magnetized until the polarity flipped again, perhaps a million years or so later. Because the rift progressively widened with new basalt consistently taking the path of least resistance, splitting the most recently preceding basalt, the separating strips on opposite sides of the rift should be in mirror image. Vine proceeded to show that this was indeed so, in perhaps the most elegant geological demonstration I have ever seen (Fig. 22). The sequence of events lead-

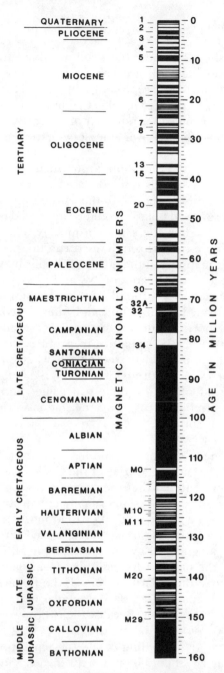

Fig. 21. The geomagnetic polarity time scale. During the periods marked in black, the north magnetic pole was in the present northern hemisphere. During the periods in white, the north magnetic pole was in the south.

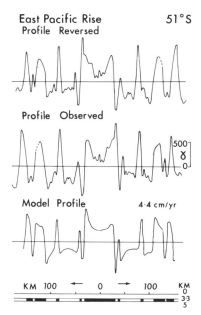

Fig. 22. Professor Fred Vine shocked orthodox geophysicists in 1966 when he compared an observed magnetic profile across a spreading ridge with the same profile reversed. The curves depict magnetic intensity (in gammas) relative to an arbitrary datum.

ing to this dramatic discovery makes a fascinating story, well told by William Glen in his 1982 book, *The Road to Jaramillo*.

The consequences of this research were profound. Not only was the spreading of the ocean floor established beyond reasonable doubt, but the age of each part of the ocean floor could be determined by matching its magnetic signature with the paleomagnetic reversal time-scale, just as ancient wood is compared to a tree-ring chronology. From the results, the rate of widening of the ocean could be measured not only now, but throughout its history! The spreading rate turned out to be a few centimeters per year, which meant that the whole of the floors of the present oceans could have grown in the last 100 million years, in keeping with the age of the oldest known oceanic sediments.

When I first saw Vacquier's map of the magnetic stripes, they matched so well the growth slices I had predicted a decade earlier (Fig. 19f) that it was immediately obvious that the slices should be differently magnetized, although my reason for the difference was not the same as Vine's. In Vine's model, basalt is injected more or less continuously, and the magnetic change occurs each time the polarity of the earth's field changes. In my model, as shown in Fig. 19, each slice begins its upward journey more than 100 km down, where the temperature is above the Curie point, and as it rises, always in the solid state, it cools below the Curie point and is magnetized according to the existing magnetic field; thus each slice is quite differently magnetized, but not necessarily tied precisely to the polarity reversal sequence, as they are in Vine's model.

I did not publish my solution then; I have rarely published anything until I have pondered it for a few years. I have contempt for the struggle for publication priority, for the publish-or-perish rat-race that rules American science, and for the pathological secrecy of ideas lest they be stolen. What does it matter whether Vine has priority over Morley, or vice versa, or even me? Likewise I scorn the ambitious drive to get into major international journals. My orocline paper was published by the Royal Society of Tasmania, my Earth asymmetry paper in the moribund Australian Journal of Science, my rheid paper in the first number of the nascent journal of the Geological Society of Australia, my paper on scales of tectonic phenomena in the first number of the newborn Geological Society of India, my tectonic approach to continental drift in the Hobart symposium volume, my isostrat paper in the Hobart dolerite symposium volume, and my universal null paper by the Royal Society of Tasmania. (These are listed in the Preface.) Each of these enunciated new geological principles that have subsequently been substantiated.

Plate Tectonics, a Shotgun Wedding

The progressive growth of the ocean floors at the mid-ocean rifts, now firmly established, was accepted with astonishment by the establishment. Unfortunately, this process was immediately married to another quite independent concept: concurrent complementary swallowing (subduction) of oceanic crust, mostly in the deep oceanic trenches that border the western Pacific (Figs. 23 and 24). The "new global tectonics" was born.

It is easy to see why this happened. English-speaking geologists almost universally still clung to the axiom that folding and overthrusting were caused by shortening of the crust, even though support for the thermal contraction theory was rapidly declining. (As late as 1963, George Lees's theme for his presidential address to the Geological Society of London was "the evolution of the shrinking Earth.") They also took for granted the axioms that the radius and the mass of the earth

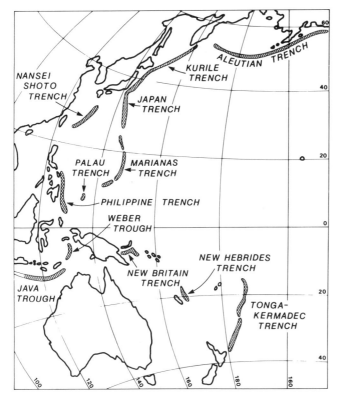

Fig. 23. Deep-sea trenches of the western Pacific Ocean.

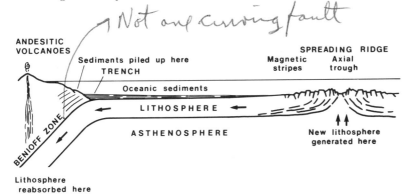

Fig. 24. The "conveyor-belt" model of ocean-floor spreading and simultaneous subduction into the trenches. Lithosphere material makes up the "plates" of plate tectonics.

had not changed significantly apart from the consequence of cooling just mentioned; hence, because the area of the oceans was rapidly receiving new growth at the spreading ridges, this must be compensated somewhere else. The obvious (and only) candidates were the trenches, which are associated with volcanoes, disturbance of gravity equilibrium, deep geosynclines rapidly filling with sediments, and strong folding and overthrusting, all of these at higher levels of intensity than anywhere else in the world.

Until the early 1950's, before I had realized that gross expansion of the earth was inescapable, I did exactly the same, and for the same reasons. I did have misgivings about accepting so much shortening in the trenches, which I suspected were really extensional structures. Any of my older students will confirm that what I taught them before the mid-1950's differed little from what the Kuhnian revolution had brought. They found little new in the "new" global tectonics.

In hindsight, it is unfortunate that the spectacular confirmation of ocean-floor spreading was done in Britain and America (made possible by the remarkable advances in American marine technology). In the USSR, the major tectonic school, led by Vladimir V. Beloussov, taught that orogenesis was a gravity-driven upwelling (diapiric) process that did not imply crustal shortening and could even involve some crustal extension. These concepts had deep roots and current interest in several European schools (for example, those led by van Bemmelen in the Netherlands and Ramberg in Sweden). Also, several people in various parts of Europe had recognized that the dispersion of the continents described by Wegener could equally be explained by Earth expansion. Thus the revolution to "new" global tectonics made one important advance in sweeping away the resistance to continental dispersion, but fettered as it was and is by the English-language dogma, it adopted the subduction myth instead of gravity-driven tectonism and hence missed the important complementary advance, the recognition of Earth expansion. I shall have much more to say about subduction in Chapter 13.

Ocean-floor spreading has now been investigated in all the oceans, and although much more surveying is still needed, especially in the Indian Ocean, the overall picture is not likely to be greatly modified. The pattern of magnetic stripes consistently confirms the Vine-Matthews model, though not always strictly symmetrically. Indeed, it would be most surprising if symmetry were universal because that implies symmetry of constraints; rifting commencing within a continent, like the African rift valleys, or within an ocean floor should approach

The Kuhnian Revolution

symmetry, but where rifting commences along a boundary of continent and ocean, asymmetry should be expected, and this will later be shown to be so.

Oceanic sediments have been found to be very thin and young near the ridges, but their thickness increases progressively away from the ridges, and the lowermost sediments become progressively older. J. Tuzo Wilson of Toronto observed that volcanoes are younger nearer the spreading ridges and become progressively older farther away, showing more advanced erosion, until they end up as flat-topped seamounts (guyots), many of which were discovered by Hess in his oceanographic cruises.

The broad downs of the ocean floors deepen progressively away from the spreading ridges from 2000 m or so to 5000 m. This is because near the ridges the newly risen rocks are still relatively hot, the temperature gradient is steep, and high temperatures are reached at relatively shallow depths, so that basalt (or its more coarsely crystalline equivalent, gabbro) is stable to quite considerable depth. But as the temperature drops the gabbro changes to the denser form eclogite, which has the same composition but in which the feldspar and augite are replaced by the denser minerals garnet and jadeite. The weight of the crust does not change, but it stands lower.

Transform Faults

Tuzo Wilson introduced the important concept of the transform fault to explain the at-first-sight anomalous direction of shift of marker strips where a spreading ridge has been offset.

The gross trend of a spreading ridge may have been determined by the direction of the rift that separated the continental blocks from which the spreading originated, but this direction may not remain the direction of extension required by current gross movements. So spreading ridges are found to jump sideways with a linear fracture zone connecting the terminal ends, thus maintaining the local spreading in the local extensional direction but also retaining the gross trend of the ridge.

The top diagram in Fig. 25 shows an active spreading ridge which has been spreading normally; C and C represent the most recently inserted strips of new oceanic crust, while the B and A pairs are progressively older ones. For reasons such as that just outlined, the spreading rift is about to jump sideways many tens of kilometers from Y to X, with a fracture zone connecting X and Y. Let us assume that X,

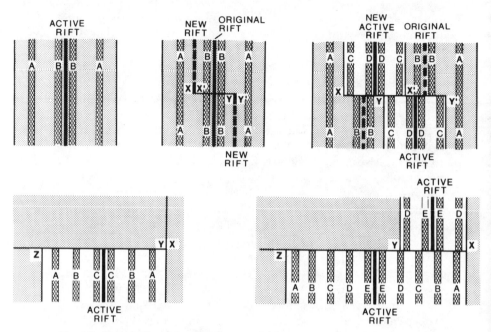

Fig. 25. Tuzo Wilson's transform fault concept (see text for explanation).

X', Y, and Y' are now marked on the crust for identification of displacements. In the middle diagram, spreading has continued with the D paired strips inserted along each leg of the offset spreading zone, and this has continued a further stage in the lower diagram with the insertion of the E pairs. Observe the following results of this system.

First, the overall spreading of the ocean floor has not been affected by the offset of the spreading rift zone. The outer A paired strips move outward as though nothing had changed. Second, the spreading rift is offset sinistrally by the initial leap, but the amount of offset remains constant without any increase during the continued spreading. Third, transcurrent motion is active only between the spreading rifts RR', and remains active there so long as spreading continues. But the sense of offset in this zone is dextral, the opposite to the sinistral sense of offset of the ridge; thus the right D strip of the upper leg of the active rift abuts the lower C strip in the middle diagram, but abuts the E strip in the lower diagram, its former C neighbor having been displaced far to the left. The upper block A + D together with the lower block A + B + C + D + E shift to the left as a single block relative to the rest of the diagram, that is dextrally. Fourth, the fracture zone segment between X and R and Y' and R' commence with dextral offset, but thereafter retain that offset but suffer no further movement. All this is confirmed by study of earthquakes, which only occur in the RR' segment, where they have sinistral displacement, and in the spreading axes.

The apparently anomalous offsets led Wilson to define the transform fault concept, which was valid in its original form; it has been extended from transferring motion between spreading ridges to offsets between trenches and between a ridge and a trench. However, the transform concept has been grossly overdone to the point where every large transcurrent fault (that is, a fault where the displacement has been horizontal along a fault trace, also called a megashear where the offset is large) tends to be called a transform fault, even where the transferred motion is not known.

Thus, the American Geological Institute glossary gives two definitions, the first being essentially Wilson's original restricted definition, and the second, quoting John Dennis and Tanya Atwater, as any plate boundary that shows pure strike-slip displacement. In a sense this, or an even wider definition, is valid because every transcurrent fault or megashear that does not complete a circle right around the globe or around a small circle must terminate in a compensating structure normal to it at each end, of which there are three possible categories: an extensional zone at one end and a foreshortened zone at the other end, both on the same side of the transcurrent fault; an extensional zone at each end on opposite sides of the transcurrent fault (Fig. 25 is of this type); and a foreshortened zone at each end on opposite sides of the transcurrent fault.

The real difference between a transcurrent fault and a transform fault is that in the former we have observed the offset between the rocks on opposite sides of the fault, but are not immediately concerned with or even aware of its terminating structures, whereas in a transform fault we are primarily aware of the spreading ridge or other structure, which we observe to be offset by a fault zone that transfers the motion of that structure laterally to a point whence it continues its original trend and function. Hence I suggest that we should be quite strict with our use of the term transform, and only apply it in this latter sense originally defined by Wilson.

More complex situations may arise. Let me name three. First is where a genuine transcurrent fault system that completes a global circle (such will be discussed in Chapter 21) crosses a spreading ridge, whose motion is transformed by it, so that both kinds of offset occur additively or subtractively along a major fracture zone. I will suggest later that some of the major fracture zones of the northeastern Pacific (Clipperton, Mendocino, etc.) are of this kind. Second, most plate tectonicists assume that spreading ridges necessarily grow symmetrically or nearly so, adding paired ridges on each side, whereas I will argue later that in the north Pacific, new ridges have mostly been added only on the western side of the spreading ridge. Third, in the lower pair of diagrams in Fig. 25, the initial spreading terminates on the fault XYZ in which the section to the left of Z is simply a transcurrent fault, the YZ section is partly transform and partly transcurrent, and the block on which X rests (which could be a continent) is not fractured at all. With continued extension, spreading may also commence along the edge of the X block, and the spreading ridge pairs DD and EE are formed as the new rift zone migrates to the left. The XY section of the fault is transform, the YZ part has both transform and transcurrent elements, and the section to the left of Z is simply transcurrent.

The idea of extensional openings complementing and terminating transcurrent faults, and vice versa, is not new. It was fully developed in the rhombochasm and sphenochasm concepts published by me in 1955 and again in the 1956 Hobart symposium volume, and also in respect to my explication of the sideways jump of the transcurrent shift of the Rocky Mountain Trench to the San Andreas group. In fact the non-stippled areas in the lower diagram in Fig. 25 are precisely rhombochasms in the terms of my definitions and descriptions.

The Revolution to End Revolutions

Revolutions in Earth dogma thus far include: from flat Earth to spherical, which was decided about 300 B.C.; from central Earth to central Sun, decided about 1550; fossils are remains of former life, decided about 1800; basalt sedimentary to igneous, decided about 1815; the reality of the Ice Age, decided about 1830; from 6000 to 4 billion years for Earth's age, decided about 1905; the revolution to acceptance of mobility of continents, decided about 1966; and the "plate tectonics" revolution to finally end all revolutions, or so its adherents think!

In retrospect, it is striking to notice only how recently many quite basic questions about the earth have been agreed upon, which suggests that dawning enlightenment still has some way to go. More often than not, geological and astronomical processes have been correctly

The Kuhnian Revolution

observed and described, but the physics and chemistry involved have not been known, and so wrong compromise and sophistry have resulted. Too often, geologists have been cowed by the "exact scientists," where they should have realized that valid data are uniquely available to them that could and should spawn new discoveries in physics.

Again and again, the powerful personality and the prestige of great contemporary leaders have frozen progress and rejected the most important new concepts. Thus Brahe was indeed a very great astronomer, yet he led the scorn of Copernicus. Newton, perhaps the premier scientist ever, removed Hooke's portrait from the Royal Society walls and ignored his farsighted contributions to geology. Werner was truly the greatest geology professor of his time, but his influence on the many great students who flocked to listen retarded fundamental geology for decades after his death. Von Buch, probably the most experienced field geologist of his generation, froze acceptance in Germany of continental glaciation until he passed on. Antoine Lavoisier, the leading French mineralogist and chemist of his time, rejected the evidence of peasants who brought him meteorites they had seen fall in a stream of light: "Stones couldn't fall out of the sky; there aren't any up there!" Baron Cuvier, the greatest comparative anatomist ever, could not conceive current processes achieving the vast events he observed in the strata. Lord Kelvin, the leading physicist of the nineteenth century, blocked for decades acceptance of the prodigious length of geological time. Bailey Willis led geologists in a great leap forward in structural geology but heaped scorn and ridicule on continental drift and was much to blame for the blindness of a whole generation of geologists. George Gaylord Simpson stands tall among contemporary biologists, yet his arguments to denigrate continental drift were sophistry.

The "new global tectonics" revolution of the 1960's was an enormous leap for conservative American geology. But it went only half way. The complete revolution to Earth expansion was too much for a mountain lion or quarter horse in a single leap, although natural for a kangaroo! The second half of the tectonics revolution is now just ahead. So let us turn to the current pariah—the expansion of the earth, and the concept that the universe is a state of zero.

PART THREE
THE EXPANDING EARTH

CHAPTER 11

Development of the Expanding-Earth Concept

THE FIRST HALF-CENTURY of germination of the theory of Earth expansion was retarded, not only by disinterest of the establishment but by the barriers of language. Up to the Russian revolution in 1917 there had been significant interflow between Russian and German science, but until recently the Russian and German papers in this field were ignored or unknown in the English-speaking world. The first translations of most of them were made by me decades later.

So far as I am aware, the first author to suggest from a cosmological viewpoint that the earth had expanded was I. O. Yarkovski, whose book, *Universal Gravitation as a Consequence of Formation of Substance Within Celestial Bodies* (in Russian), was published in Moscow in 1899 and in St. Petersburg in 1912. He conceived the transition from weightless matter (ether) to substantial matter to give birth to planets and stars. Through the decades since, this concept has been developed geologically by a small group in Russia in several papers and books, particularly by I. B. Kirillov, V. B. Neiman, and A. I. Letavin of Moscow and V. F. Blinov of Kiev. This group's work culminated in November 1981 in a conference in Moscow on Earth expansion and pulsation, organized by Prof. E. E. Milanovsky, of Moscow State University, and the Moscow Society of Naturalists. Over 700 specialists from Moscow and other regions of the USSR participated, and 20 papers were presented, divided between those who favored expansion and those who supported pulsation, that is, an alternation of expansion and contraction.

Ironically, unbeknown to the Russians, a well-argued case for Earth expansion had been published much earlier in London. Captain Alfred Wilks Drayson, of the British Royal Artillery (later a major-general), in his 1859 book, *The Earth We Inhabit, Its Past, Present, and Future*, concluded that the earth is expanding at a rapid rate. He cited a large number of surveys in which the distances between fixed points were greater in later surveys than in previous ones, and others in which the length of a degree had increased. Drayson believed the biblical account absolutely, including the implied date of creation some 6000 years ago. He postulated that the earth's diameter had doubled in that time along with the parallel increase in the number of rotations (days) in a revolution (year). Hence Adam's life of 930 "years" included only 34,000 days, or 93 of our present years, and in the same way the apparent longevity of other patriarchs was well within familiar life spans.

Captain Drayson produced truly impressive sets of measurements to back his conclusion. But one set was based on a quite naïve misunderstanding of precession and the latitudes of the tropics, also involved in his explanation of the above rotation-revolution ratio. Another set arose from the variation of the length of a degree, which had bothered the Cassinis and Bouguer as explained in Chapter 2. The breaks in the Atlantic telegraph cables, which he interpreted as indication of the impulsive widening of the ocean floor, are now explained by submarine slides triggered by earthquakes. But there still remains his list of accurate baseline surveys, which consistently show earlier distances shorter than later ones. Unless his conviction of expansion led to biased selection of examples, these remain surprising. He struck a sympathetic chord with me when he wrote: "The highest learning of the day has sometimes been wrong, for it was the most learned who ridiculed the miracles of our Saviour; it was the scientific who scorned the idea of satellites to Jupiter, of a Continent of America, of the sphericity of the earth, of steam, and many other matters within the compass of our memory."

About the same time as Captain Drayson, another remarkable man, William Lowthian Green, working alone in Hawaii, remote from the conceptual constraints of the orthodox establishment, conceived not only Earth expansion but also interhemisphere torsion whereby the southern continents were displaced eastward and separated from the northern continents by zones of profound crustal disturbance. The latter has only recently been accepted generally (see Chapter 21), and the former is only now on the threshold of a new revolution.

In 1909, Mantovani suggested Earth expansion to explain the similarity of the opposing Atlantic coasts, as mentioned in Chapter 8. In 1920, Hiram W. Hixon, an American, writing in *Popular Astronomy*, found the contraction theory unable to explain many geologic phenomena, such as the African rift system, the Great Basin of Nevada, the Colorado Plateau, and the preponderance of normal faulting. Like Lowthian Green, Hixon's understanding of the earth was far ahead of his time. He held that both epeirogenesis and orogenesis (formation of plateaus and fold mountains, respectively) were gravity-driven diapiric phenomena, caused by outgassing from an expanding Earth. He recognized that solid rocks would flow below the melting temperature, and that Earth expansion would cause slowing of the rotation speed and lengthening of the day.

Michael Bogolepow, of Moscow, published three papers in Russian in 1922, 1925, and 1928, followed in 1930 by a paper in German, "Die Dehnung der Lithosphäre" (The stretching of the lithosphere) in the *Zeitschrift der geologischen Gesellschaft*. Bogolepow proposed secular zonal motions in the mantle, a clockwise eddy-like underdrag in the southern hemisphere and a counterclockwise one in the northern hemisphere, caused by radioactive heating.

The storm whipped up by Wegener's book on continental drift led to the alternative explanation of his data by Earth expansion, but this developed in the German language, without English translations or interest. In 1927, Dr. B. Lindemann, of Göttingen, inspired by Wegener and probably unaware of the earlier papers in Russian, published *Kettengebirge, kontinentale Zerspaltung und Erdexpansion* (Mountain chains, continental splitting, and Earth expansion), in which he stated that the dominating phenomenon of the earth's surface is rifting and extension, with mountain chains marking the outflow of an expanding interior heated by radioactivity. His analysis shows perception and insight that warranted much more recognition than it received—the penalty for his heterodoxy.

As often happens to independent thinkers, Otto C. Hilgenberg, of Charlottenburg Berlin, had to publish himself his 1933 book, *Vom wachsenden Erdball* (On the growing earth). He dedicated the book to Wegener, but did not mention Bogolepow's earlier papers (but they were in Russian, which he did not read), and Bogolepow's paper in German appeared just after he had completed his book. Hilgenberg assembled the continents on a basketball-sized papier-mâché globe, the original of which I was privileged to handle when I visited him in 1964. All the oceans had been eliminated and the continental crust neatly enclosed the whole earth on a globe about two-thirds the diameter of his reference globe.

Hilgenberg was the first of a long line of such globemakers—Ludwig Brösske of Düsseldorf (1962), Cyril Barnett of London (1962), Kenneth M. Creer, then of Newcastle (1965), Ralph Groves of California (1976), and Klaus Vogel, of Werdau, East Germany (1977), of whom more later. They all produced globes of a little more than half the reference diameter, with the whole surface occupied by the assembled continents. They differed mainly on how they assembled the Pacific. Like the Russians, Hilgenberg believed that the mass of the earth as well as the volume increased with time, which he explained in terms of the ether; he suggested that energy of the ether flux was continually absorbed in ether sinks and was transformed into matter.

Dr. J. K. E. Halm, in his 1935 presidential address to the Astronomical Society of South Africa ("An Astronomical Aspect of the Evolution of the Earth"), denied the axiom that a cooling Earth would contract. On the contrary he argued from a theoretical analysis of the effective size of atoms that the earth has expanded by about 1000 km in radius. Irrespective of the validity of his model, he proceeded to show that expansion gives a much more plausible explanation of the data presented by Wegener. Like Lindemann's, Halm's model assumed constant mass, and he did not discuss Hilgenberg's proposal that the existing sialic (continental) crust originally enclosed the whole earth, but he was probably not aware of the German or Russian works. Halm adopted Wegener's Pangaea and explained the opening of the oceans in terms of his theory along with other broad geologic phenomena. He pointed to the implied progressive emergence of continents through oceans spreading over increased surface area without consideration of the greater water depth in the opening ocean basins, a point taken up below. Halm explained the Red Sea as an embryonic oceanic rift, regarded the Mediterranean as pulled apart, and interpreted the Gulf of Honduras as a yawning gape, "the jaws being hinged in the neck occupied at present by Mexico."

Josef Keindl was working in Vienna during the 1930's, but his book, *Dehnt sich die Erde aus?* (Is the earth expanding?), was not published until 1940. He was familiar with the German literature but not the Russian or English. Like Hilgenberg, Keindl opted for an original sialic crust completely covering the whole earth, subsequently disrupted to give rise to the expanding ocean basins. The source of the disruption and of all orogenesis had to be sought deep within the earth. Keindl argued mainly from geomorphology. The whole universe and everything within it is in a state of expansion, he concluded: normal stars, for instance, differ from white dwarfs in that the luminous gaseous envelope had been stripped from the superdense cores of the latter. Keindl, unlike Hilgenberg and the Russian expansionists, assumed constant mass and postulated a small superdense metastable core in the earth.

The Gravitational Constant

The late English cosmologist Paul A. M. Dirac pointed out in 1937 that any combination of physical constants that is a pure number is either about 1, or 1 followed by 20 zeroes, or 1 followed by 40 zeroes, or 1 followed by 80 zeroes. (If you divide a force by a force, or a distance by a distance, or a time by a time, the answer is a pure number.) The intervals between these groups is so very large (100 billion billion billion) that Dirac believed that they had to be meaningful. One such fraction included the gravitational constant G divided by the age of the universe. Philosophically, therefore, G should vary with the age of the universe, and should in fact get smaller with time, to keep the fraction the same. This meant that everything should weigh less as time went on, and this in turn implied that the earth should have expanded with time, because of relaxation of elastic compression and the change of minerals to less dense forms throughout the earth. Dirac, in retirement in Florida, returned to his "large numbers" philosophy in 1974.

Meanwhile Dirac's idea was taken up by others. Professor Robert H. Dicke, of the Princeton physics school, and his colleague Professor C. Brans developed this concept in a series of papers between 1957 and 1966, and also deduced diminishing G from Mach's principle (that the inertial mass of a body increases with the mass of the universe). On the suggestion of Hess, Dicke visited Yale in 1959 and discussed with me the geologic evidence for expansion, but the amount of expansion from decline in G was less than I believed had occurred, and its distribution in time was different.

Professor Pascaul Jordan, a Hamburg physicist, also followed Dirac in his 1964 book *Die Expansion der Erde*, an English translation of which was published in 1966. He tried to emulate Wegener by drawing together several disciplines, but with less success. Jordan corresponded with me about the possible causes of expansion.

In 1971 Fred Hoyle and J. V. Narlikar deduced a decline in G because the number of anomalous spectral red-shifts observed in globular clusters of stars had increased far beyond the probability that they could be explained by their chance occurrence at very different distances in the same line of sight from Earth.

In the mid-1970's, Dr. T. C. van Flandern of the U.S. Naval Observatory found that the moon's acceleration, determined by checking against atomic clocks the times of lunar occultation of stars, significantly exceeded possible tidal effects, and implied either that the sun's mass was increasing at an improbable rate or that G was diminishing.

These independent approaches to the decline of G converge on a rate of one ten-billionth (10^{-10}) per year, which is about the same as the upper limit which would have been detected in the orbit of Mercury.

Evidence from Mining Geology

In 1954 two American mining geologists, the brothers R. T. and W. J. Walker of Colorado, also had to publish their own book, *The Origin and History of the Earth*, because their conclusions were unorthodox. They wrote:

[The authors,] with a joint span of experience covering over fifty years of surface and underground observations, finding themselves confronted by more and more geologic evidence, which could not possibly be reconciled with [the contraction] hypothesis, were slowly and reluctantly forced to the opposite conclusion that the Earth was increasing in volume, and that the cause of this phenomenon must be some expanding mass at the center of the Earth. This idea once adopted, the phenomena of vulcanism and orogeny—heretofore inadequately explained—all fell into place like the parts of a jigsaw puzzle.

The Walkers reached this conclusion, it seems, wholly independently of the earlier writers on Earth expansion. Wegener's book, gross tectonics, and global morphology played no part in their conception, which came from smaller scale conventional mineralogy, petrology, and geologic structures.

Volume of Seawater

Professor Lazlo Egyed, who later hosted me at the Eötvos University in Budapest, concluded in 1956 that although the total volume of ocean water had increased during geologic time by more than 4 percent, paleogeographic maps of land and sea for the individual epochs since the Precambrian, compiled quite independently by Professor Henri Termier and his wife Geneviève of Paris in their 1952 *Histoire géologique de la Biosphère* and by N. M. Strakhov of Moscow in his *Outlines of Structural Geology* (in Russian), showed a progressive decline in the proportion of submergence of the continents individually and collectively. This implied that the surface area of the earth had increased and that the relative proportion of ocean basins to continental platforms also steadily increased.

Egyed calculated an average increase of 0.5 mm per year in the earth's radius which he assumed to be uniform, although Rhodes Fairbridge suggested later that the indicated expansion may well have accelerated since the Mesozoic, as indeed much other evidence supports. Egyed's estimate of the increase in volume of seawater is certainly much too low, because he did not know of the vast volume of water released with the submarine lavas in the ocean-floor spreading process. This additional water would strengthen his argument. On the other hand, the fact that the intervals covered by the successive paleogeographic maps are progressively longer for older periods biases the data the opposite way.

More recently Professor Uwe Walzer, of Jena, East Germany, has confirmed Egyed's conclusion by showing that when 23 paleogeographic maps of Eurasia published in Moscow by V. M. Sinitsyn in 1962 are measured by planimeter on an equal-area projection, the same result is obtained as Egyed got from the world maps of the Termiers and Strakhov.

My Conversion to Expansion

At the Hobart continental drift symposium in March 1956, I first realized that Pangaea could not be reassembled on the present-sized Earth without unacceptable anomalies, particularly the unavoidable gaping gore between Australia and east Asia (discussed in Chapter 12). I thereafter devoted myself to the study of expansion and its implications. I hunted out the German and Russian literature (there were no copies in Australia) and translated them. Quite early, the problem of the surface gravity on an Earth of the same mass and little more than half the size concerned me, but it was some years before I realized that the whole universe shared the same problem, and ever so slowly the null universe (Chapter 23) dawned.

Heezen and Wilson

Dr. Bruce Heezen, of Lamont Geophysical Observatory, became deeply impressed by the worldwide mid-oceanic rift system (Fig. 20), which he had done a great deal to explore. He stated at the Nice-Villefranche international colloquium of the Centre nationale de Recherche scientifique in May 1958:

Continental displacements can be effected in two very different ways. The way generally considered is continental drift. In drift, the continental blocks float laterally across the upper part of the mantle. Continental displacements can also be effected by the expansion of the interior of the earth. In this latter case, the original solid and undifferentiated crust breaks and individual fragments are eventually separated. In the case of continental drift one should find compression right along one side of a continent and extension right along the opposite side of the continent. In the case of displacement through expansion of the interior of the earth one should find extension in all the oceanic zones.

It seems almost impossible to reconcile the evidence of submarine topography with continental displacement by lateral drift. These same features can quite easily be explained by the hypothesis of continental displacement by internal expansion. [My translation.]

At that time Heezen favored the expansion model, but a few years later he abandoned expansion and joined the new bandwagon of plate tectonics.

Professor J. Tuzo Wilson, of Toronto, campaigned vigorously during the late 1940's for crustal shortening because of a cooling of the earth. After he had addressed the Royal Society of Tasmania on this theme, I spent several days trying to convince him of the validity of continental drift, but he firmly held his ground. However, I must have dented him somewhat, because when I went to Yale in 1959, he wrote inviting me to visit Toronto, and enclosed the draft of his 1960 paper for *Nature* in which he espoused Earth expansion. He had been stimulated by Dicke's support for Dirac's proposal that G had diminished with time, which would involve some expansion of the crust. Although he rejected the concept that the present continental crust completely enclosed the primordial Earth, which involved doubling of the radius, he considered that the mid-oceanic rifts, and the fact that the flow of heat through the ocean floors was about the same as that through the continents, and several other features, were compatible with more limited expansion.

A few years later Wilson, too, joined the plate-tectonics boom, and proposed what has become known as the "Wilson cycle," the paradigm for which was the history of the Atlantic Ocean as he reconstructed it: By 600 million years ago, an ancestral Atlantic was about as wide as the present Atlantic. During the next 200 million years, it had virtually vanished through the subduction of its floor to form the Caledonian and Appalachian mountain systems. About 200 million years ago, the precursors of the modern Atlantic appeared in the form of rift valleys through the Bay of Fundy, Maine, Massachusetts, Pennsylvania, and the Carolinas, which rapidly filled with sand, gravel, and basalt. By 100 million years ago, a new Atlantic rift ocean was opening rapidly, but not quite on the site of its ancestor, some of which had been left in Mauretania on the African side. Now the new Atlantic is as wide as its predecessor, and the Wilson cycle is complete.

A Pulsating Earth

Dr. A. J. Shneiderov, in 1943, 1944, and 1961, developed a theory of a pulsating Earth wherein cataclysmic expansions produced the oceans, and slower contractions produced orogenesis. The pulsation idea itself had a long history in Europe before that. According to Shneiderov, each contraction was less than the preceding expansion, yielding overall irreversible expansion. Shneiderov claimed that the earth had a nucleus of dense hot plasma, excited by a flux of cosmic subatomic particles ("radions"), the intensity of which was modulated by alignments of Earth with the sun, moon, and planets.

Several Russians still favor a pulsating model. Peter N. Kropotkin and Yu. A. Trapeznikov have proposed short-term fluctuations of G, annually, irregularly, and in longer pulsatory cycles of durations up to the age of the earth. The observed reversals of the magnetic field (Fig. 21) are attributed to such fluctuations. They invoke gravity variation as the prime mover in geotectonics, because gravity offers their only source for the implied rate of tectonic working, which they estimate at 10^{15} megawatts. Professors E. E. Milanovski and Ye. E. Khain, both of Moscow State University, and Dr. A. I. Letavin have also developed pulsating models. Dr. J. Steiner, of the University of Alberta, attributed a wide variety of large-scale geologic phenomena directly to the additive effect of the Dirac-Jordan secular decrease in G and a pulsation caused by the rotation of the galaxy with a period of some 200 million years.

Revival in the 1960's

The end of the 1950's saw an increase in interest in expansion among the unorthodox in many countries. Brazilian petroleum geologist P. Groeber, in his paper "La Dilatación de la Tierra" (The dilation of the earth), concluded that the surface area of the earth had increased by 27 percent during the last 250 million years. In 1962, Ludwig Brösske, of Düsseldorf, wrote "Wächst die Erde mit Naturkatastrophen? Die 'Expansions-Theorie'" (Does the earth grow through natural catastrophes? The "expansion-theory"). In the same year Professor Cyril Barnett, of London, reconstructed the continents on a globe two-thirds of the original size and remarked that "it is difficult to believe that chance alone can explain the fitting together of the continental margins." Later he recalled the resemblance of the southern continents to the petals of a flower:

A comparable pattern may readily be obtained by coating a rubber football bladder [that is, a sphere] with a continuous "crust" of damp paper and then inflating it. Linear fractures are produced enclosing three or more petal-shaped forms "aiming" towards the point of initial rupture. As the rupturing paper crust opens up like an opening flower bud, each primary fissure extends and divides peripherally into secondary fissures to form smaller but still tapering patterns with occasional complete separation of large paper "islands".

This bud-and-petal analogy is useful because it expresses the earth's hemihedral asymmetry, the antipodal relation of continents and oceans, the greater separation of the southern continents, and the apparent northward migration of continents with respect to the relatively southward-moving parallels of latitude as the southern hemisphere (the opening calyx) opened more widely than the northern (Fig. 26).

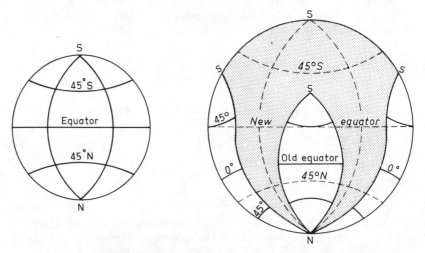

Fig. 26. The "opening bud" analogy, drawn with the south pole at the top, showing more expansion and more new ocean in the southern hemisphere. Note that the parallels of latitude appear to move southward across the lands, giving the impression that the lands move northward, but really they still rest on the same mantle, as new oceanic material is inserted between them.

Professor Rhodes Fairbridge, of Columbia University, gave a comprehensive review in 1964 of the literature on, and the evidence for, an expanding Earth. He pointed out that all the ocean basins are youthful, and that theoretical consideration of the gravitational constant, mantle-core evolution, and geodetic consequences of mass displacements, polar shifts, and paleogeographic development, all converge to support youthful expansion of the globe. In a further review (1965), Fairbridge could find no evidence that could justify the ocean trenches as compressional phenomena, nor did orogenic belts call for primary crustal compression. On the contrary, he interpreted the deep-sea trenches as contemporary prototypes of eugeosynclines (see Chapter 18), as tension gashes in a crust extending continuously at an increasing rate.

In 1965 Professor Kenneth Creer, then of Newcastle, also reassembled all the continents on a smaller globe, and like his predecessors thought the fit of the continents too good to be due to coincidence, and required explaining. Creer, deeply immersed in the current paleomagnetic dogma refuting late expansion, thought that the expansion must have occurred very early in Earth history, and attributed it to a cosmological cause.

In the same year, Professor Arthur Holmes, then of Edinburgh, reviewed the development of the expanding Earth concept and favored decrease in the gravitational constant, coupled with phase changes (change of minerals to less dense forms at lower pressure) throughout the whole interior of the earth as the prime cause, with convective circulation in the mantle as the probable mechanism.

Also in 1965, Dr. H. C. Joksch stated that a statistical analysis of the heights of all parts of the earth's surface showed the highest frequencies at 0.5 and 0.2 km above sea level and 4.5 km below sea level. He suggested that lighter material (sial—from *si*licon and *al*uminum that make up light minerals) has been separating out from the interior (like cream from milk), and that the areas now forming the +0.5-km level originally enclosed the whole primitive earth. This was broken into the separate initial "continents" with what is now the +0.2-km areas in the gaps. Underplating of sial from the interior continued until the main expansion, which again ruptured the crust and left the present continents separated by the −4.5-km areas, the present oceans.

Sam Elton, a Los Angeles geophysicist scorned because of his unorthodox writings, wrote in 1966:

Instead of invoking the continental drift theory, we are going to suggest that the continents recede from one another due to the uniform expansion of the earth . . . the continents do not drift, but are rather carried apart by the general expansion of our planet. It is quite possible to follow this picture back in time—drawing all the continents closer together and eliminating the oceans until the earth possesses only about one quarter of the surface area; that is, an area equal to the present continents.

Dr. Raymond Dearnley, of the British Institute of Geological Sciences, in three papers in the mid-1960's, deduced an expanding Earth model from a reconstruction of early (Precambrian) orogenic belts, which he assumed to be a surface expression of convection cells in the mantle. He proposed that the earth's radius was 4400 km 2750 million years ago, and 6000 km 650 million years ago, compared with 6378 km today. This implies that the rate of expansion has been slowly declining, whereas I will argue that the rate has been increasing exponentially.

In 1967, Professor Bruce Waterhouse, then of Toronto, preferred a model of an earth exploding in size through the last 200 million years, with the continental crust never significantly detached from the underlying mantle, but carried apart by the insertion at spreading areas of new material derived from the mantle, coupled with considerable transcurrent shift. His views have much in common with those of Heezen and Fairbridge and have been strongly supported by more recent work, discussed in Chapter 22.

Dr. R. Meservey of the Massachusetts Institute of Technology showed in 1969 that the movements of the continents around the Pacific Ocean postulated by all the models of continental drift or plate tectonics were topologically impossible unless the earth was expanding, a conclusion I had already reached at the Hobart symposium. Meservey pointed out that although the perimeter of the Pacific is less than a hemisphere, the continents surrounding it have all moved apart *in the direction of the perimeter*, and therefore the area within it must have greatly increased, if the earth's radius has not changed. But according to plate-tectonics models, this area has greatly decreased. Meservey emphasized that to transform any of the configurations proposed for the early Mesozoic to the present configuration, consistently with the seafloor growth indicated by paleomagnetism, is impossible unless the earth has greatly expanded.

Important developments in the theory of Earth expansion have occurred in the last decade. These include: the *Atlas of Continental Displacement, 200 Million Years to the Present* produced by Dr. Hugh Owen of the British Museum; the report of the international symposium on Earth expansion, which I convened in the University of Sydney in

February 1981, which contains 51 contributions from 17 countries; the special conference on Earth expansion and pulsation, organized by Professor Milanovski in Moscow State University in November 1981; and the commencement of the NASA program of measurement of intercontinental distances with an accuracy of a couple of centimeters. Within a decade this program should establish the validity of expansion.

CHAPTER 12

The Earth *Is* Expanding

SO WHAT IS THE EVIDENCE for expansion? When given its due, the evidence is found to be considerable, and it lies in several domains of empirical fact. Fossil rock is a good place to begin.

The Arctic Paradox

Fossil assemblages give a good indication of where the climatic zones lay on the continents during past ages. Such paleogeographic information may be independently confirmed by paleomagnetism. When molten lava solidifies, the temperature drops through the "Curie temperature," at which minerals become magnetized in the direction of the ambient magnetic field of the earth. The Curie temperature varies for different minerals from a couple of hundred degrees to over 700°C. If the solidifying lava happened to be at the magnetic equator, the direction of its magnetization would be horizontal, but if it were at the magnetic pole, it would be magnetized vertically. Hence, provided care is taken to "wash" out subsequent changes in its magnetization, and to correct for any tilting the rocks have suffered since cooling, such lavas record the latitude in which they solidified, and the direction in which the pole then lay. Where the magnetization of the rocks was spread over a few thousand years, a group of such measurements compensates statistically for the movement of the magnetic pole about the rotation pole, and indicates the ancient direction of true north as well as the latitude of the samples.

Tropical fossils (with a diversity of brachiopods, corals, and fusulinids), together with quite independent paleomagnetic data, establish that during the Permian (roughly 245–280 million years ago), the equator crossed North America through Texas and New York. The present equator crosses Brazil. Hence North America is now some 35

The Earth Is Expanding

degrees nearer the north pole than it was during the Permian. Similarly, European fossils and paleomagnetic data indicate that the Permian equator lay a few degrees south of France. The present equator is in central Africa. Hence Europe is now some 40 degrees nearer the north pole than it was during the Permian. Likewise, central Siberia is now about 20 degrees nearer the north pole than it was during the Permian. So, since the Permian, the continents have converged on the Arctic, which consequently should have contracted by some 5000 km. Did it? Just the opposite—throughout that time, the Arctic has been an *extending* region, opening the Arctic Ocean. This is impossible, unless the earth is expanding.

Fossils and paleomagnetic data from the Triassic (200–245 million years ago), Jurassic (145–200 million years ago), and Cretaceous (66–144 million years ago) all independently produce the same paradox: the continents have converged on the Arctic since each of those periods, but in progressively diminishing degree for each successive period. Each data set proves that gross expansion has occurred progressively.

The Paradox of Paleopole Overshoot

Paleomagnetic inclinations averaged over thousands of years reproduce statistically the contemporary rotation pole. The position of this pole is usually stated as a latitude and longitude within a small oval, commonly a few degrees in diameter, that expresses the 95-percent statistical probability that the ancient pole lay within the oval.

In the early 1970's, paleomagneticians reported that if all the positions of the magnetic pole indicated by rocks magnetized during the last 2 million years throughout the world were combined statistically, they reproduced a rotation axis—a paleopole—within a small oval enclosing the present rotation axis. But the paleopole thus determined from any *one* region was a little beyond the mean paleopole determined from all regions (see Fig. 27). This is just what the expansion theory predicts, because angular distances on the earth's surface—which are what paleomagneticians measure—get progressively longer in kilometers as the radius increases. For example, one degree on the earth's surface is a bit more than 110 km, but one degree on the moon is only 30 km long. The paleomagnetic measurements indicate the number of degrees to the pole, and as the pole positions were plotted assuming constant radius and degrees of present length, the poles plotted from each region must overshoot a little if the earth has expanded since the time of the magnetizations.

Fig. 27. Dr. Roderic L. Wilson, of Liverpool, found that the Quaternary poles indicated by the palaeomagnetism of any one region lay a little beyond the present rotation pole. Open circles are the centroids of sampling sites, and filled circles are the corresponding calculated poles.

This result is shown more clearly when the regional rays on Fig. 27 are plotted on a single meridian; the statistical paleopole for the last 2 million years is within the lowest of the three ovals in Fig. 28. The middle oval shows the statistical pole indicated by all the measurements from the rocks between 2 and 7 million years old, and the upper oval is the paleopole of 7 to 25 million years ago. These ovals, by definition, should fall at the center of the diagram, but they overshoot because the present radius of the earth was assumed for each plot. This means that the length of a degree is progressively less as we move back in time, at least for the latest 25 million years.

The Earth Is Expanding

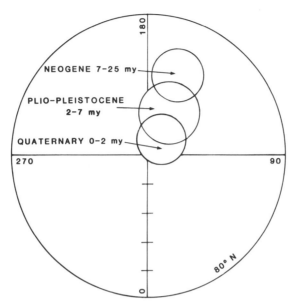

Fig. 28. Roderic Wilson and Michael McElhinny found that the overshoot of the poles increased with age. Each circle indicates the statistical probability of the pole position for each age (95 percent probability).

Paleomagneticians, wedded as they are to their constant-radius creed, seek to explain these anomalies by invoking *ad hoc* departures from their central-dipole model of terrestrial magnetism, but these overshoots merely bring the Arctic paradox, just discussed for the Permian and Mesozoic, right up to the present day. From the figures available, the overshoot would be some 25 degrees for the Permian, 20 degrees for the Triassic, 16 degrees for the Jurassic, 12 degrees for the Cretaceous, 6 degrees for the Miocene, 3½ degrees for the Pliocene, and 1½ degrees for the Pleistocene. These figures indicate that the expansion of the earth has been progressive for the last 200 million years, but they do not give a reliable estimate of rate of expansion. That would require many more sampling sites for each interval, uniformly distributed with respect to latitude.

The Pacific Paradox

Figure 29 shows three versions of Pangaea as it was supposed to be at the beginning of the Mesozoic, 200 million years ago: Wegener's in 1915, mine in 1945, and one by two plate-tectonics adherents,

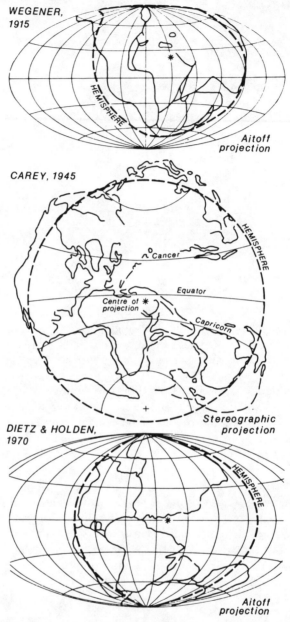

Fig. 29. Pangaea as reconstructed by Wegener in 1915, by Carey in 1945, and by Dietz and Holden in 1970, assuming that the earth's radius had not changed. In each reconstruction the heavy broken line indicates a hemisphere, whose center is marked by an asterisk.

Robert S. Dietz and J. C. Holden, in 1970. Each is based on the assumption that the earth's radius has remained constant. On each, the heavy broken line is a hemisphere. The rather distorted shape of the hemisphere in Wegener, and to a lesser extent in Dietz and Holden, are artifacts of the projection, a topic I take up a bit later. The Aitoff projection used by Wegener and by Dietz and Holden represents the whole earth, as though you slit the globe down the back and flattened it out, so that the outermost edges on both sides represent the sides of the slit. The stereographic projection I used for the middle figure shows only one hemisphere. Each version has a gaping sector about 50 degrees wide between Australia and China. This is unavoidable if you try to reconstruct Pangaea while keeping the earth's radius constant. In the Dietz and Holden version, this gape looks wider, but they noted that New Guinea, New Zealand, and southeast Asia were "omitted for cartographic convenience." In Wegener's version, this gape looks narrower because he stretched south Asia to undo what he called the "Lemurian compression," which he believed had foreshortened this region.

The general pattern of Pangaea has changed little from Wegener's early conception of it to the form currently accepted by plate tectonics. In each version, Pangaea (including its inherent gaping sector) just spills over a hemisphere. In Wegener's hemisphere in Fig. 29, the ocean lune on the left side of the hemisphere is about equal in area to the part of Pangaea that spills off the right side. The rest of the globe was occupied by the ancestral Pacific Ocean, which must also have been roughly a hemisphere in area, assuming constant earth radius. Since then, the area of "Pangaea" has nearly doubled through the insertion of the Arctic, Atlantic, Indian, and Southern Oceans between its parts, plus the opening of the Tasman Sea and all the little seas of east Asia. This means that the Pacific should have been greatly reduced in area, not quite to zero, but to an area equal to a hemisphere from which the enlargement of Pangaea (the sum of the areas of the other oceans) had all been subtracted! This is clearly not so.

The paradox is not eased by the assumption of subduction within the area of the Pacific or around its margins. Nor can the paradox be relieved by including the gaping sector mentioned above and claiming that it has been closed by subduction, because on the present globe the distance between continental Australia and continental China is still 47 degrees, about what it was on Pangaea. The gross enlargement of Pangaea was not compensated by shrinkage of the Pacific, and can only be explained by expansion of the earth.

The Pacific-Perimeter Paradox

The Pacific Ocean is roughly circular, but is somewhat skewed because the northern continents have been displaced westward relative to the southern continents by an equatorial torsion, which will be discussed in detail in Chapter 21. When this offset is removed, the Pacific becomes even more circular (Fig. 30).

The Pacific rim is formed by the continents of Asia, North America, South America, Antarctica, and Australia. Comparing the gaps between the continents around this rim with the distances that separated them in Pangaea, it is obvious that the distance between the pairs has increased. The relative position of North and South America in Pangaea has been determined by their fit against Africa, published by me 30 years ago, and subsequently confirmed by several investigators using computers. All agree that central Mexico then abutted northwestern Venezuela; but in the dispersion of Pangaea this distance has increased by 2500 km, which lengthened the Pacific rim by this amount. In Pangaea, Antarctica wrapped around Africa and nudged Madagascar, and the movement to its present position lengthened the Pacific rim there by 3200 km. In Pangaea, Antarctica fitted into the Great Australian Bight; dispersion of Pangaea to the present configuration lengthened the Pacific rim here by a further 3500 km. It will be shown in the following pages that the gaping sector in Pangaea between Australia and southeast Asia is an artifact, elimination of which means that the distance between Australia and China has been increased by about 3800 km during the Pangaean breakup. There is also some extension between Siberia and Alaska, but this is difficult to quantify in relation to the Pacific rim.

These figures show that the Pacific rim has lengthened by 13,000 km or more. This figure is too high, because the skewing by the equatorial torsion mentioned above has added to the lengthening in both East and West Indies, but even allowing generously for that, the Pacific rim has increased by at least 10,000 km during the dispersion of Pangaea.

Fig. 30 (*facing*). The Pacific-perimeter paradox. The upper figure shows the present separation of formerly contiguous points on Pangaea. In the lower diagram the skew caused by the Tethyan torsion (see Chapter 21) has been removed. The dextral circum-Pacific torsion (also discussed in Chapter 21) extended the Pacific perimeter between China and Australia by 3800 km. The perimeter of the Pacific has *increased* by a little more than a third, when according to plate-tectonic theory the area of the Pacific has been *reduced* by the sum of the areas of the Arctic, Atlantic, Indian, and Southern Oceans.

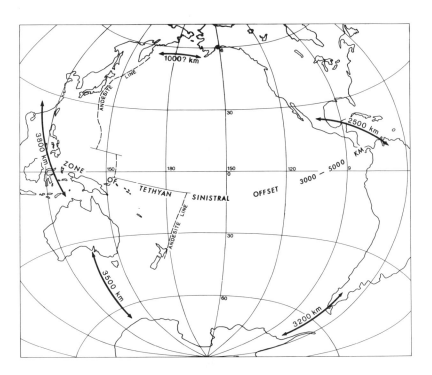

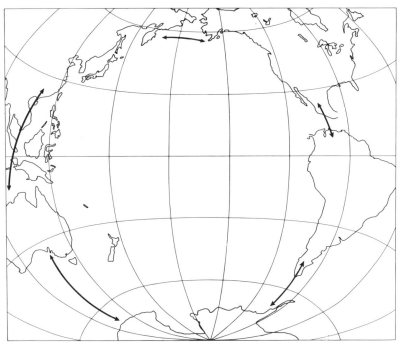

Here is the paradox. After correcting for the equatorial torsion, the Pacific was nearly circular at the beginning and the end of the breakup of Pangaea. Yet its perimeter has been increased by about one-third, while the area within that perimeter should have been decreased by more than a third. This absurdity arises from the assumption that the earth's radius remained constant. These facts can only be reconciled by a large increase in the radius of the earth during the dispersion of Pangaea.

The Gape Artifact

The gaping gore between Australia and southeast Asia has been mentioned in previous paragraphs. When I was trying to make an accurate reconstruction of Pangaea more than 30 years ago, I had a table built with a hemispherical top the same size as my 30-inch globe, and also a mold so that I could mold large sheets of plastic tracing foil to fit the globe (Fig. 11). These aids enabled me to transfer accurately one part of the globe to fit against other parts, such as the closing of the Atlantic Ocean by fitting the Americas against Africa. As I proceeded to reassemble Pangaea, an unwanted open gap appeared between the India–Antarctica–Australia group and Asia, tapering from zero in the Mediterranean to 50 degrees between New Guinea and Indonesia (Fig. 29, middle).

I knew that this gape was wrong, because from the earliest fossils 600 million years ago until the breakup of Pangaea, the east Asian faunas have linked through to their close relatives in Australasia. This is confirmed by the Early Cambrian archeocyathids; the Middle Cambrian trilobites of the *Redlichia* fauna; the Ordovician calymenid trilobites, cephalopods, gastropods, conodonts, and stromatoporoids, and the *Selenopetis* fauna (the North China and Australian faunas had closer affinities than either had with European faunas); the Llandovery corals; the Devonian brachiopods; the Carboniferous foraminifera; the Permian blastoids; and the *Glossopteris* and Cathaysian floras, which mingled in New Guinea, Sumatra, Thailand, China, and Turkey. Finally, just before the breakup of Pangaea, the Late Triassic monotid clams firmly bind the Tasman province to China and East Siberia. As new studies continue, additional links keep cropping up; for example, C. R. Johnston found it necessary to ferry Timor back and forth across this fictitious gape to satisfy the alternating proximity demands of Asia and Australia; N. W. Archbold and his colleagues had to do the same to New Guinea to match the Permian brachiopod faunas with those of Indonesia. B. A. Stait and Clive F. Burrett found that the Ordovician cephalopods of the Shan Mountains of Thailand, Burma, and Malaysia required close proximity with northwestern Australia, and P'an Kiang found that the Early Silurian fish indicated a similar continuity.

Confident that the gape was false, I started a reconstruction of Pangaea from Australia–Indonesia without any gap, but as I proceeded to assemble the other continents, a new gape appeared, widening to 50 degrees, between the Americas, which was also false. Whatever I tried, I always ended with a gaping gore from about the middle of the assembly to a 50-degree gap at the periphery, opposite where I had started. Finally, after months of frustration and anguish, I realized that my troubles arose because I was trying to reassemble Pangaea on a spherical table the same size as my globe, whereas I should have been using a table of smaller radius, because the earth had expanded significantly since the time of Pangaea. I was trying to button a waistcoat over an enlarged belly! Every seamstress knows to insert a tapering gore into a skirt to increase the flare. I had been working on continental drift for a quarter of a century, taking it for granted that the earth's radius was constant.

The India Enigma

During the Paleozoic era, from 600 to 200 million years ago, the reconstructions adopted by all plate-tectonic models assume that the obtuse angle of the Bay of Bengal from Sri Lanka via Madras to Calcutta wrapped around the Enderby Land angle of Antarctica, and the west coast of India lay against the east coast of Africa. However, India has close faunal and paleogeographic ties, not only with Antarctica, Madagascar, and Africa, but also with China, Tibet, and Kazakhstan (from which it was separated in their assembly by thousands of kilometers of ocean), with Afghanistan and Iran (from which it was also alleged to have been far distant), and also with Australia.

Much solid evidence indicates that the obtuse angle about Madras, mentioned above, should really wrap around the angle of northwestern Australia, not Enderby Land. R. G. Markl and R. L. Larson independently reported that this was indicated by the magnetic lineations on the floor of the northeastern Indian Ocean. Several authors have correlated the ancient basement rocks of India and northwestern Australia in their petrology, geochemistry, iron ores, and diamonds, and in their tectonic grain. Dr. Curt Teichert, who was not a supporter of continental drift at the time he wrote, was surprised to find the close similarity of the Cretaceous faunas of Trichinopoly in India and the Manilya River of Western Australia, which are close together when the Madras angle is brought against northwestern Australia. Thirty years ago, two Australian paleontologists, George A. Thomas and James M. Dickins, were studying the early Permian faunas of the Lyons Conglomerate (as it was then called) of northwestern Australia. The Permian strata of the Indian peninsula were deposited almost exclusively in fresh water, but one thin marine bed had been found in a branch railway cutting at Umaria, which would be very close to the Lyons Conglomerate if India lay next to northwestern Australia.

So I went to Umaria, collected a suite of fossils, and brought them back to Thomas and Dickins for comparison with their fauna. My collection proved to be identical with a marine horizon of the Lyons sequence, not only in species, but in oddities. Both contained plates of the unusual crinoid *Calceolispongia* (which had earlier been thought to be shark's teeth!), and trivial details of ornamentation on a gastropod were the same. Professor Fakhruddin Ahmad, of Aligarh, has reconstructed the Permian paleogeography of India and Australia (Fig. 31). The evidence for this relative position in Pangaea is compelling, but plate-tectonic adherents still cling to the Antarctica position for the good reason (for them) that a coherent assembly of Pangaea cannot be made with India and Australia so placed. I tried repeatedly to do so on my hemispherical table. But the latent error was not in the India–Australia correlation, but in adopting the same radius for the earth of Pangaea as the present radius. Dr. Kenneth Perry's computer-controlled regression to 76 percent of present earth radius (discussed in Chapter 20) likewise puts the Madras coast of India against Australia's northwest shelf.

This same problem recurs with India's other relations. The plate-tectonic reconstruction has India separated from Afghanistan and Iran by thousands of kilometers. Yet Arthur A. Meyerhoff and Curt Teichert reported that several ancient geological features connect India and Iran:

> Detailed mapping from the central Indian shield to southeastern Afghanistan and Iran show that several distinctive rock formations and faunal zones extend from central Iran and Afghanistan on to the Indian shield (Madhya Pradesh). These stratigraphic units and paleontological markers, which are continuous, include the Proterozoic–Cambrian Hormuz Salt Series, two of the *Productus*-bearing beds, and the Aptian–Albian *Orbitolina* zone and associated warm-water shoal carbonates.

These facts cannot be satisfied by the plate-tectonic model.

Likewise, the constant-radius constraint has India and China sepa-

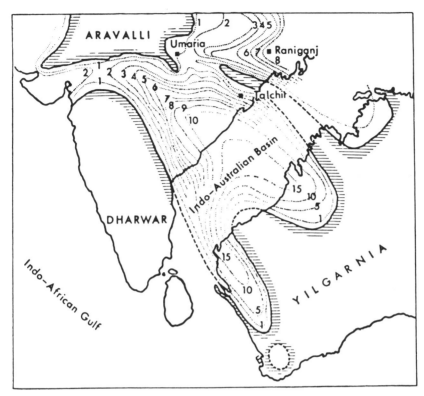

Fig. 31. Professor Ahmad's reconstruction of Indian–Australian paleogeography during the Permian Period. Numbers on the contours indicate in thousands of feet the thickness of strata that accumulated. The basin was always shallow and nonmarine, but its floor steadily subsided by as much as 3 km in the central area, allowing the temporary invasion by the sea on five occasions.

rated by thousands of kilometers of open ocean, yet *Lystrosaurus*, a reptile with size, shape, and habitat rather like a hippopotamus, wandered back and forth from India to China, along with other fellow-travelers, insects, and plants. Chinese geologists have now reported that "Gondwana" faunas and rock associations characteristic of India and her southern neighbors extend across the Indus–Tsangpo line (claimed by plate-tectonicists to mark the suture where thousands of kilometers of former ocean was subducted) far across Tibet and China. Howard A. Meyerhoff and Arthur A. Meyerhoff, father and son, insist that "India has been part of Asia since Proterozoic or earlier time. This is a geologic fact, which nothing can change."

India's problems are tied to the entire area involved in that part of the Pangaean reconstruction. If you satisfy India's bond to northwestern Australia, an unacceptable gap appears between India and Africa, and between India and Afghanistan and Tibet. If you satisfy the latter, the gaps dismember the whole Gondwanaland association, which all accept as valid. Lester King tried to solve the problem by detaching Iran and Afghanistan from Laurasia and so filling the gap caused by holding the India–Australia link, but clearly the close association of India, Afghanistan, and Iran with Asia should not have been broken. A. Ray Crawford tried to solve the enigma by holding the India–Arabia–Africa links, and detaching Tibet and South China from Laurasia to fill the gap left between India and Australia. Of course, Tibet and South China certainly do have close bonds with Australia, as I emphasized above in relation to the gaping gore artifact, but Crawford's solution breaks their Cathaysian links, which are also real.

The simple conclusion is that, so long as the reconstruction is attempted on an Earth of present radius, it is impossible to satisfy valid proximity demands of one insistent Indian neighbor without wide blank spaces between India and other former neighbors, whose proximity demands are equally compelling. When the reconstruction is made on an appropriately smaller Earth, the whole enigma vanishes. When Michael W. McElhinny and his coauthors concluded from their paleomagnetic measurements that before the dispersion of Pangaea Malaysia lay a few degrees north of the equator, they announced that therefore Malaysia could not have been part of Gondwanaland, which was so much further south. What they should have said was that Malaysia could not have been part of Gondwanaland if the present earth's radius be assumed for the past.

Missing Archean Crust

Dr. Andrew Y. Glikson, of the Australian Bureau of Mineral Resources, has studied intensively the petrology and geochemistry of the most ancient rocks—the foundation platforms older than 2 billion years on which later systems of strata were deposited. These make up about 80 percent of the present-day continental foundations and add up to about a quarter of the earth's crust. Glikson asks: What was the nature of the crust that occupied the other three-quarters of the earth's surface 2 billion years ago? He considered four conceivable answers.

First, the unknown three-quarters was essentially the same as the known quarter. This was rejected because that three-quarters would have to have been swallowed back into the mantle, which is geochemically unacceptable, and physically improbable because of the lower density and implied buoyancy of such crust.

Second, the known quarter was originally spread over the whole surface at a quarter of its present thickness, and has since been progressively thickened and reduced in area by successive compressional foldings. Glikson also rejected this proposition on a number of grounds. Given the pressure-temperature combinations at which particular minerals crystallize or transform into other minerals, Glikson found, the ancient continental crust could not have been significantly thinner than it is today. Ubiquitous extensional phenomena belie the suggested compressional foreshortening; furthermore, the original geometries of dikes, intrusions, and the very extensive blankets of overlying sedimentary rocks are commonly retained. Finally, paleomagnetic measurements on these old foundation rocks contain no suggestion that the crust converged through the large angles implied by this proposition; on the contrary, they suggest that the angular size of the continental foundations remained generally stable.

Third, the currently known quarter existed then, as now, as distinct plates that moved relative to each other, through repeated spreading of oceanic crust in some places, compensated by subduction back into the mantle of oceanic crust in other places. Glikson pointed out that this proposition raised severe geochemical contradictions, and was also in conflict with much paleomagnetic evidence.

Fourth of Glikson's hypotheses was that all the present continental foundations were grouped 2 billion years ago into a single supercontinent, and the rest of the earth's surface was oceanic. This model conflicts least with paleomagnetic and paleogeographic data from the later Proterozoic rocks and with plate-tectonics theory. Glikson rejected it because if the implied spreading and subduction processes operated in the oceanic hemisphere, then by a conservative estimate about 500 million cubic kilometers of rock would have accreted against the continent, much as subduction complexes and island arcs do today (according to plate tectonics). However, no such vast areas of island-arc rocks of Proterozoic age are known, nor do volume considerations allow for such large masses beneath later sedimentary troughs (geosynclines).

Glikson concluded that the probable solution of the enigma of the missing ancient crust is simply that it is not missing at all—it never existed. Instead, the known ancient foundations then completely enclosed a much smaller Earth, and have since been separated and dispersed by the growth of oceanic crust between them.

Missing Ophiolites and Flysch

A similar conclusion was reached by Dr. Keith Crook of the Australian National University, from a quite different kind of data, the apparent variation in abundances of ophiolites and flysch. Ophiolites are a group of igneous rocks derived from the earth's mantle that come to the surface during the early stages of actively subsiding troughs (geosynclines), which subsequently develop into fold-mountain systems (orogens). Flysch is a particular association of sediments that is rapidly deposited during the active stages of the development of geosynclines and orogens. Crook has written:

Ophiolites and flysch, two mobile belt associations indicative of the oceanic realm of the earth's crust, are well represented in the rock record of the last billion years. Their abundances, expressed as area of outcrop per million years, decrease exponentially with increasing age during this interval, in a manner consistent with predictions based upon the probability of their preservation. Although ophiolites are virtually unreported from the one billion to 2.5 billion years rock record, possible Archean analogues, the greenstone belts, are represented throughout the pre-2.5-billion-year record. Similarly, flysch is rarely reported from the one- to two-billion-year part of the record, but is not uncommon in terrains older than 2.0 billion years. These abundance patterns depart so grossly from predictions based upon preservation potential as to demand an explanation. Among several conceivable explanations, two are of particular interest. The first proposes that the one- to two-billion-year oceanic record is obscured by a combination of intense metamorphism and, more generally, burial beneath Late Precambrian, Palaeozoic and perhaps Mesozoic mobile belts. This implies fundamental differences between post-Palaeozoic ocean margins and post-Archaean pre-Mesozoic ocean margins. The second explanation accounts for the missing one- to two-billion-year oceanic record by postulating the virtual absence of an oceanic realm during this interval of Earth history. This implies global continental crust, and gross expansion of the Earth during the past one billion years.

Cartographic Precision

Two centuries before Christ, Eratosthenes knew that drawing part of the spherical earth as a flat map must distort it, so he devised projections to show correctly the ancient world from Spain to India. A century later, Ptolemy developed several ways of precisely depicting

The Earth Is Expanding

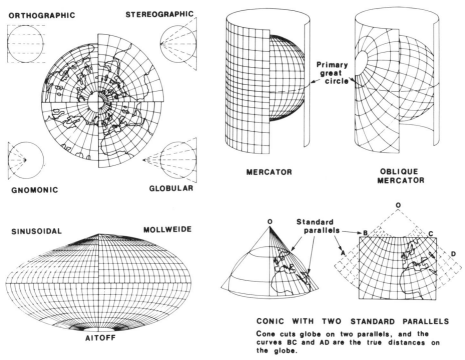

Fig. 32. Map projections.

global data on flat surfaces. Indeed, we can choose a way of constructing a map to preserve those aspects that interest us (Fig. 32). An orthographic projection is the view we would have of the earth from the moon. It only shows one hemisphere; angles measured from the center are correct, but no others are; an area near the rim of the map represents a very much larger area than the same area near the center; shapes near the rim are squeezed narrower as radial distances are represented by shorter lines, shrinking to zero at the rim. Many systems of projection are available on which equal areas on the map represent equal areas on the globe, but these projections grossly distort shape (such as the sinusoidal and Mollweide in Fig. 32).

The Mercator and stereographic projections and some others are orthomorphic, which means that they preserve shapes because angles and the scale start off correctly from any point, but the scale increases away from the base line in the Mercator (usually the equator), and away from the midpoint of the stereographic. Any circle on the globe, however large or small, still remains a circle on a stereographic map, which is not so on any other projection. The line between any two points on a normal Mercator map cuts the meridians at an angle that is the correct bearing to take one to the other point (a property useful to sailors); this loxodrome or rhumb line is not the shortest possible course—that route lies on a great circle. On the gnomonic projection, great circles *are* straight lines, which makes it useful to aerial navigators.

The gnomonic, stereographic, orthographic, and some others are azimuthal projections, because angles from the center are not distorted. An azimuthal projection also can be equal-area, but not then orthomorphic. The whole surface of the earth can be represented on cylindrical projections (including the Mercator) and on a variety of oval projections, which may preserve equal areas, but not angles, distances, or shapes. There are also conic and polyconic projections, some of which are widely used for many general purposes because they were easy to draw before we had computers. Though they distort shapes, they do not distort much, areas are not equal but not grossly out, and angles and distances are not true but near enough, so they make a happy compromise so long as you do not want to make precise measurements or to cover the whole globe.

Through naïve or careless use of such projections, much of the cartography in connection with continental drift and plate tectonics has been inaccurate or incorrect. Sometimes this does not matter, and a sketch or diagram may convey a concept not so easy to explain in words. But some quite wrong conclusions have followed. Besides, precise cartography can throw up unexpected but significant discrepancies, as some following examples show.

More than half a century ago I used oblique stereographic projections for all my work on continental drift. This precision enabled me to prove conclusively that Sir Harold Jeffreys was wrong when he ridiculed the fit of Africa against South America "on a moment's examination of the globe" (see Chapter 8).

For several years, Dr. Hugh Owen of the British Natural History Museum has been making a precise geometrical analysis of the ocean-floor spreading patterns during the last 200 million years. He has plotted his own oblique projections and eliminated any artifactual source of error. In addition to important papers, Owen has produced a comprehensive atlas showing the implied continental distribution through that time span. Owen has found that when the earth's radius is assumed to have been constant,

The Earth Is Expanding 167

areas of anomalous oceanic crust appear, together with misfits of continents one to another in the reassembly of Pangaea. The progressive appearance of these anomalous areas of continental crust, anomalous in that they are required to be present by the spherical geometry of the reconstructions, but of whose former existence no record remains in the spreading patterns of passive-margined oceans, suggests a progressive increase in surface curvature back in time.

Dr. Kenneth Perry of Boulder, Wyoming, has set up a computer program to generate the Mid-Atlantic Ridge by progressive ocean-floor spreading. He commenced with the "Bullard fit" and the Dietz–Holden reconstruction of Pangaea (see Fig. 29), and proceeded to other recent reconstructions. He found that for any of the models of Pangaea, irrespective of its geographical position, irrespective of dispersal paths of individual plates, and irrespective of the time frame, it is not possible to generate ocean-floor structures that have the observed geometrical configuration of the Mid-Atlantic Ridge by plate motions *on an Earth of constant radius*. Assumption of a progressively increasing radius is necessary to generate the observed Mid-Atlantic Ridge from Pangaea. (A picture of Perry's reconstruction appears as Fig. 76 in Chapter 20.) The work of Owen and of Perry illustrates the discrepancies in the plate theory that show up with precise cartography.

Recently Dr. Fakhruddin Ahmad, of the Indian Academy, measured the arc between the Permian north pole in eastern Siberia and the Permian south pole southeast of Durban, South Africa, on a reconstruction of Pangaea that closed the artifact gap discussed earlier and found that this indicated a Permian radius of about 55 percent of the present radius, approximately the same as Vogel and Perry found by quite different routes.

NASA Geodetic Measurements

Writing in 1972, soon after Apollo 15 had placed cube-corner reflectors on the moon, I urged the immediate commencement of intercontinental measurements by laser ranging interferometry of critical distances on the earth, which would within a decade prove the expansion of the earth.

The American National Aeronautics and Space Administration has now been measuring intercontinental distances for a few years by three independent methods: lunar laser ranging, a similar procedure

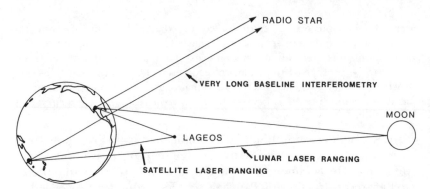

Fig. 33. NASA satellite geodetic ranging system.

using artificial satellites, and very long baseline interferometry (VLBI for short), with errors of only a few centimeters (Fig. 33). This is precise enough (with repetition over a few years) to measure relative movements of the size predicted by the plate-tectonic and earth-expansion theories, and to judge between them. A fourth method, using the multi-satellite global navigation system, gives promise of attaining similar accuracy with less effort. In laser ranging, pulses are reflected back to receiving telescopes from cube-corner reflectors already placed on satellites and on the moon. The distance from reflector to telescope is measured by the time delay, knowing the velocity of light. In the VLBI method, signals from extragalactic radio stars are recorded simultaneously by two or more stations, and the difference in arrival times of identifiable wave-fronts measures the component of the distance between the stations in the direction of the radio star.

Results have only now (1986) begun to emerge, but they should crystallize more and more sharply with each year. A critical measurement is the distance from Easter Island to the All-American Observatory in northern Peru, but this came late into the program, so conclusive results there are still a few years away. According to plate tectonics, this distance should be shortening by 10 cm per year, whereas according to Earth expansion it should be lengthening.

The preliminary results from NASA indicate that the chord distance from Europe to North America is increasing by 1.5 ± 0.5 cm per year, North America to Hawaii is increasing by 4 ± 1 cm per year, Hawaii to South America by 5 ± 3, and South America to Australia by 6 ± 3, and the distance from Hawaii to Australia is decreasing by 7 ± 1 cm per year. The other displacements so far released are not large

The Earth Is Expanding

enough to be significant. These results support Earth expansion, but not the plate-tectonic theory, which is denied by the radius increase implicit in the data. Widening of the Atlantic Ocean is predicted equally by both.

Dr. W. D. Parkinson has pointed out to me that it is important to remember that the distances and their increments reported by NASA are chords, not surface distances. Figure 34 shows the relations of surface arc D, chord C, earth radius R, and the angle subtended at the earth's center ϕ, for any two NASA stations A and B. In the triangle AMO, $C = 2R \sin \phi/2$. Differentiating against time gives:

$$\frac{dC}{dt} = 2 \frac{dR}{dt} \sin \frac{\phi}{2} + \frac{dD}{dt} \cos \frac{\phi}{2}$$

in which the derivatives of C, R, D, and ϕ are the rates of change of these quantities. If we have three NASA stations on a great circle, and the rates of change of chord for each pair are inserted, we have three *simultaneous* equations, which can be solved to give the rate of change of Earth's radius. If the plate-tectonic theory is correct, this should come out at zero. But it doesn't.

Parkinson has solved these equations for the three stations in Arizona, Hawaii, and Canberra. These arcs intersect at Hawaii at an angle of 159°. The difference from 180° does not affect the result seriously. Parkinson reports that the NASA data indicate that during the

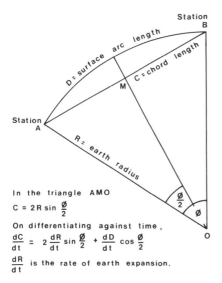

Fig. 34. Geometry of NASA intercontinental measurements.

period of their measurements, the radius of the earth has been increasing at 2.8 ± 0.8 cm per year.

Dr. Parkinson's calculation is pure geometry, quite independent of the presence or absence of any spreading ridges or subduction zones, or of any theory of expansion or otherwise. It simply gives the rate of radius change implicit in the changes in the chord lengths. If we used three independent pairs of stations, the equations would not be simultaneous, and the result could be interpreted in terms of changes within the arcs without uniquely determining the rate of change of Earth radius.

These calculations were made using the basic numbers reported by NASA. If the calculation is biased by using in each case the limiting number least favorable to radius change, a constant radius is still just possible within the error limits. This escape should disappear with more observations. Compare this with Fig. 28, where the discrepancy in the length of a degree could just be escaped in the data for the last 7 million years, but not when the time span is increased.

Although it cannot be assumed that the rate of expansion is uniform, especially for such short periods as a couple of decades, nevertheless this very preliminary measurement of radius increase of 2.8 ± 0.8 cm per year is of the right order of magnitude. The mid-value is a 17.6 cm per year increase in circumference, which would imply between 12,600 and 22,600 km increase since the middle Cretaceous. This agrees well with the rate of change in the length of a degree shown by paleopole overshoot during the last 25 million years, and it agrees with the rate of convergence of America and Europe on the Arctic since Permian time.

This rate of increase nicely accounts for all new ocean floor since the middle Cretaceous *without any subduction*. Adding the equatorial opening of the Pacific (approximately 120 degrees) to the opening of the Atlantic (approximately 45 degrees) gives an increase in the Tethyan equator of 18,300 km, a sum that lies in the middle of the expected range. The Southern Ocean between Australia and Antarctica has widened by about 3300 km since the Cretaceous, which agrees well with this expansion rate, because Australia–Antarctica occupies about one-sixth of a great circle and would expect an increase of between 2300 and 3800 km if expansion were distributed uniformly around the great circle. Likewise India–Antarctica occupies about four-tenths of a great circle, so extension of between 5000 and 9000 km would be expected. The actual separation is a little more than 7000 km.

Care is needed in interpreting the length of the Australia–South America chord because each continent has moved north, increasing the length of the parallels between them, irrespective of any radius change.

Still another trap confuses the interpretation of NASA measurements in relation to radius change. Whenever new crust 100 km wide is inserted at a spreading ridge, the angle subtended at the center of the earth by each degree around the great circle is reduced by 9 seconds to accommodate the added segment. As all NASA chord measurements ultimately involve the angle subtended by that chord at the center, any continental block or stabilized oceanic crust will *appear* to shorten if constant radius is assumed.

Consider the great circle through Tokyo, Hawaii, and the All-American Observatory in Peru. Nearly 8000 km of new crust has been inserted in the Peru–Hawaii segment during the last 100 million years, and the East Pacific rise is still spreading, whereas no new crust has been inserted between Honolulu and Tokyo in that time. According to the above rate of expansion, the great circle has increased by $17,600 \pm 5000$ km whereas the Honolulu–Tokyo segment of this great circle has remained constant in length, which means that the angle subtended by this arc at the earth's center has nearly halved. Hence if the analysis of the NASA measurements of this arc assumed constant radius, the Hawaii–Tokyo distance would *appear* to be shortening by about 6 cm per year although in fact the distance had not altered.

Similarly, the Hawaii–Australia chord would *appear* to be shortening by a little more than this, while the true chord length had not changed. Another factor also enters here. Hawaii and North America lie to the north of the Tethyan torsion discussed in Chapter 21, while Australia and South America lie to the south of it. The western boundary of the Pacific has been offset 5500 km by this shear (see Fig. 80 in Chapter 21), which means that the eastward movement of Hawaii relative to Australia is of this order, so, allowing for the angular divergence, the shortening of the chord between Hawaii and Australia from this factor could exceed 3 cm per year. The rate of extension of the Hawaii–Peru chord would also be increased by a like amount.

Like the Hawaii–Japan and the Hawaii–Australia chords, the widths of stable continents such as North America and Australia would falsely appear to be shortening. Indeed, preliminary results reported by D. C. Christodoulidis and his NASA colleagues in the 1985 *Journal of Geophysical Research* suggest that this is indeed so, contrary to the tenets of plate tectonics. According to the mean of nine measurements on four chords, stable North America (that is, east of the Rocky Mountain Front) *appears* to be shrinking at 1.2 cm per year, and from the mean of four measurements on one chord, stable Australia *appears* to be shrinking at 2.4 cm per year. Although continued measurements are needed to establish that these apparent shortenings exceed error limits, the bias so far certainly favors expansion rather than constant radius.

The Earth *Has* Expanded

In the introduction to his 1986 Presidential Address to the geology section of the British Association for the Advancement of Science, Professor Derek V. Ager wrote: "As a palaeontologist naturally I prefer the evidence provided by fossils, especially the Mesozoic brachiopods which I have studied for some 35 years." At the end of the address he concluded, "I find it difficult to accept different explanations for the same phenomena which occurred in the various great oceans of the world. On balance, I prefer to think that all the oceans have been expanding since early Mesozoic times and that therefore the hypothesis of an expanding earth is inescapable."

The Arctic paradox, the paleopole overshoot, the failure of the Pacific to compensate for the growth of the Arctic, Atlantic, Indian, and Southern Oceans, the gross enlargement of the Pacific perimeter while its area should have been grossly shrinking, the anomalous "gape" artifact, the enigma of India's former neighbors, the thermal evidence that continents have not moved over their underlying mantle, Glikson's "missing" Archean crust, Crook's "missing" Proterozoic ophiolites and flysch, the "missing" young ocean crust indicated by Owen's precise cartography, Vogel's and Perry's generation of the present continental distribution by radial outward motion, and Perry's reconstruction of the generation of the mid-Atlantic ridge—so many wholly independent kinds of investigation—all indicate Earth expansion.

But, no matter how compelling the proof, so heretic a concept inevitably suffers the scorn and ridicule of the establishment, as has been the fate of every major advance from traditional dogma. So, when in January 1979 I was invited to address the Geological Society of London on the expansion concept (by a group who were confident they could demolish it), one of the most senior British geologists pointedly walked out as soon as my address was announced. Robert Muir Wood reported the meeting for *New Scientist*:

The Earth Is Expanding

With the idea that when things get sleepy you bring on the magicians and clowns, the Geological Society recently convened to discuss "The expanding Earth theory," or EET. Proposing the notion were a bombastic Tasmanian professor of geology (Warren Carey) and a less flamboyant English geophysicist (Dr Hugh Owen). During the course of a theatrical afternoon they elaborated their case that the Earth had swollen by some 20 per cent in the past 200 million years.

Wood continued with a series of statements *ex cathedra* and attacks *ad hominem* and concluded: "Such ideas as the expanding Earth, with its biological metaphoric supply from growth and pregnancy, provide entertainment for a cold winter afternoon."

The reply I submitted to *New Scientist* ignored the insults and rebutted Wood's arguments, but *New Scientist* did not publish it.

In April 1984, at the end of my honorary invitational lecture at Yale, on Earth expansion, another professor rose and shouted "Bullshit." Rejection of my argument did not disturb me, but the obscenity, which illustrates the excesses of emotion that are often directed at heresy in academia, shocked me. Pythagoras suffered such excesses with his round Earth, Copernicus with his heliocentricity, Hutton for the great age of the earth, Darwin over evolution, and Wegener over continental dispersion. Earth expansion, valid though it can be shown to be, must likewise endure scorn and derision.

CHAPTER 13

The Subduction Myth

WHEN AMERICAN mapping and dating of the paired spreading ridges in the Atlantic Ocean finally established that the ocean is widening with—for a geological process—astonishing speed, this new fact had to be reconciled with the general assumption that the earth radius is essentially constant, so that an area equal to the oceanic increment had to be excised somewhere else. The solution adopted and called "subduction" was identical with what I had submitted to the *American Geophysical Union* in 1953 (but which had been rejected as naïve), namely that the excess lithosphere was consumed in the oceanic trenches. The alternative solution that the earth surface was increasing at the rate of the new crust increment was not even considered. This subduction process was combined with another axiomatic dogma, almost universally believed in the English-speaking world, namely that folding, thrusting, volcanism, seismicity, and orogenesis generally are due to horizontal shortening of the earth's crust. Again, the alternative, that all these phenomena express a bursting outward from an expanding interior, was not considered. Thus the myth, that subduction is a firmly established fact, became universally believed and all processes were interpreted on this premise. This chapter sets out to show that the subduction assumption leads to irreconcilable anomalies.

The Africa Enigma

Like all continents, Africa is surrounded by its ocean-floor-spreading rift zone, shaped like an inflated caricature of Africa with more than twice its area (Fig. 20). Newly formed oceanic crust increases in geologic age from very recent at the rift, through the Tertiary and Cretaceous Periods, all of it having been added during the last 100 million

The Subduction Myth

years. Without Earth expansion, an equivalent area of crust would have to have been consumed and removed. The Atlantic accretion implies subduction somewhere east, and Africa's Indian Ocean accretion implies subduction somewhere west. Somewhere within Africa, plate theory demands a subduction sink that has swallowed an area of crust greater than the area of the whole of Africa. Where is it? Such just does not exist! On the contrary, between the Atlantic and Indian Ocean spreading ridges there is only extensional ocean floor, and on land the great rift valley system of latitudinal extension is itself a nascent spreading ridge!

To sustain the excuse for the absence of subduction in Africa, Antarctica has to accept subduction in compensation for the spreading

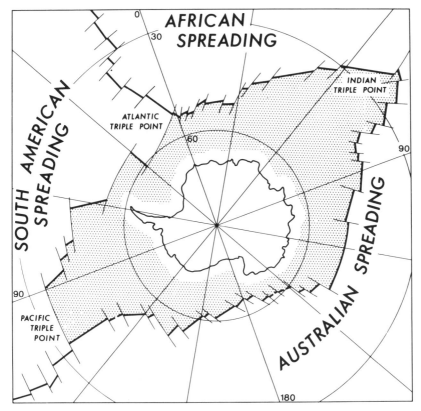

Fig. 35. Spreading ridges surround Antarctica, but there is no subduction zone within Antarctica to swallow the new lithosphere growth (stippled).

between the Indian Ocean triple point and the South Atlantic triple point (Fig. 35). But Antarctica, too, is surrounded by its own peripheral growth, a growth that doubles its area, quite apart from any additional contribution from Africa. According to the subduction theory, all of this has to be swallowed within Antarctica, where no sign of such subduction exists. The only trench in the vicinity is the small South Sandwich trench, which is at right angles to the Antarctic growth, and in any case its function according to plate theory is to subduct local South Atlantic crust. The sophism of evading Africa's lack of an interior subduction zone by transferring the problem to her neighbors fails totally with respect to Antarctica, which also lacks an interior subduction zone, and indeed would have to transfer to Africa one sector of her own problem.

The Peru–Chile Trench Anomaly

To escape this dilemma, plate theorists adopt a worse contradiction by declaring that Africa is a special case, and that Africa's latitudinal subduction problem is to be transferred to the Pacific trenches, as well as southward to Antarctica. The Peru–Chile trench then has to swallow more than 1600 km of Africa's Atlantic spreading, plus more than 1400 km of South America's Atlantic spreading, plus 3700 km of eastward spreading from the south Pacific, meaning that some 7000 km of lithosphere was underthrust below the Andes, mostly during the 50 million years since the Paleocene. (The current subduction rate estimated by plate-tectonicists, 9 cm per year, would suggest 4500 km of subduction in that time.)

This stupendous underthrust has to crop out in a trench where the sedimentation during that time shows no disturbance whatever (Fig. 36, left), and has to be driven below the Andes, a region that has risen in a state of *tension* throughout that time, as Dr. H. R. Katz and others have reported. Professor William F. Tanner, of Florida State University, writing in the *Bulletin of the American Association of Petroleum Geologists*, summarized the field evidence in 1974:

Carlos Ruiz, for years director of the national geologic survey of Chile, stated on more than one occasion that the Andes Mountains of Chile exhibit specifically an east-west tensional style. J. C. Vicente (1970), of the University of Santiago, Chile, has detailed a similar position. Ramirez (1971) pointed out that the main structure in west-central Colombia is a graben (oriented roughly north-northwest and south-southeast). Cobbing (1972) emphasized that the structure of Peru must be contrasted with the theoretical model of marginal thrusting, and concluded that the basic deformation was rifting. Carter and Aguirre (1965) described the horst-and-graben structure and the Cenozoic history of extension of Chile. Zeil (1965) stated that there had been no folding in the Chilean coastal range since the beginning of the Triassic; his structural cross-sections show normal faulting and graben development. My own field work, primarily in Colombia and Chile, leads me to the same conclusion: the fundamental tectonic style is extensional (east-west).

Moreover, where are the accumulated scrapings of oceanic sediments stripped off this 7000 km of oceanic lithosphere as it was driven down below the continental crust? They are not there; indeed, some parts of the trench are quite empty (Fig. 36). In other parts, the turbidite fill in the trench is derived from the land, and is quite different from the deep-sea oozes that should have piled up there.

Dr. David W. Scholl and Dr. Tracy L. Vallier of the United States Geological Survey pioneered the work that established the absence of accumulations of oceanic sediments anywhere around the allegedly subducting Pacific rim. They tried to escape the dilemma their own work raised by suggesting that the oceanic sediments were dragged down under the continents along with the subducting slab. But surely these soft oozes would be squeezed out rather than drawn in below the full weight of the 100-km-thick continental crust! Is it really credible to postulate the outcrop of 7000 km of underthrusting there? or even 1000 kilometers? or 100? or even 10? Would not the outcrop of a *one*-kilometer underthrust be obvious in these soft sediments? According to the plate tectonicists' own estimate of the current subduction rate, 1 km of subduction would have occurred since the Pleistocene! The subduction model for these trenches is clearly wrong.

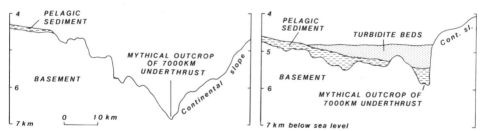

Fig. 36. According to the subduction theory, vast volumes of soft sediment from the ocean floor should have piled up in the ocean trench as 7000 km of ocean floor was drawn down below South America. But some parts of the Peru–Chile trench are empty (as at 28° S, left), while other parts are filled with undisturbed Tertiary sediment (as at 37° S, right), where according to the subduction hypothesis the 7000 km underthrust should crop out.

The Kermadec Trench Anomaly

Eastward from Africa, only ocean-floor spreading is postulated between Africa and Australia and between Australia and the Kermadec trench, which stretches from New Zealand to Tonga (see Fig. 23). So the eastern part of the African enigma has to be accepted by this trench. Unless the earth expands, the Kermadec trench would have to swallow 6000 km of oceanic lithosphere to compensate for the widening of the Indian Ocean between Africa and Australia, plus 2000 km of Australia's allotment in the Tasman Sea, plus 5000 km coming from the East Pacific Rise on the other side of the trench. Thus the Kermadec trench would have to subduct 13,000 km of lithosphere (one-third of the circumference of the earth!) during the last 150 million years.

Here again, where are the 13,000 km of oceanic sediment scrapings? With no associated continent to hide them under, even that implausible evasion is denied. Most of the floor of the Kermadec trench is bare rock, with no sediment whatever. Indeed, the nature and volume of trench fills generally are correlated with the supply of sediment from nearby land, not with the alleged amount of subduction. The Chile trench is empty adjacent to desert lands. The Kermadec trench is empty because there are no lands to feed it.

Where are the colossal andesitic volcanoes resulting from the processing of so much alleged subducted lithosphere? Is this really credible? Lest it be claimed that the above 13,000 km of subduction down the Kermadec trench is an exaggeration, let me point out that the *Plate Tectonic Map of the Pacific Region*, issued in 1982 with the support of the leading plate tectonicists, shows the subduction rate into the Kermadec trench as 10 cm per year, one of the highest anywhere, and only marginally less than the Japan trench (10.5). The south Pacific and Indian Ocean spreading goes right back to the Early Cretaceous, which would mean 14,000 km of subduction, if the 10-cm rate were uniform.

Himalaya and Tethys

Through nearly 500 million years from the Cambrian to the Eocene, the area now occupied by the Himalayas was a shallow sea, and for the latter half of that time this seaway extended continuously from Spain to New Guinea and New Zealand (and during the Permian and Triassic before the opening of the Atlantic, it continued on to Central

The Subduction Myth

America). Edouard Suess called this seaway Tethys. When the plate tectonicists attempted the reconstruction of Pangaea, their assumption of constant earth radius forced them to accept an ocean more than 6000 km wide between India and Asia, so they remodeled Tethys to fit. The Himalayas are claimed to result from the collision of India into Asia as India moved north and the intervening oceanic crust was subducted, the Indus–Tsangpo line being the suture that marks the collision boundary.

This concept is denied by the facts. Dr. Augusto Gansser, one of the leading authorities, if not the leading authority, on Himalayan geology, wrote in 1979: "Plate-tectonic models have had India drifting thousands of kilometers, while all the field relations suggest India and Eurasia could never have been far apart." Dr. Jovan Stöklin, another authority on the Himalayas, has written: "The geology of the Himalaya fails to indicate a Tethys ocean of Paleozoic–earliest Mesozoic age, and in this respect supports the theory of Earth expansion." Dr. Fakhruddin Ahmad, of the Indian Academy, showed at length that only shallow and intermittent seas covered the Himalayan region during the relevant times, then went on: "As no oceanic Tethys existed, and India and Angaraland [Suess' name for the central Siberian block] did not collide, the Himalayas could not have been born of collision nor of subduction, but resulted from vertical uplift." The genesis of the Himalayas and other orogenic mountain belts by vertical tectonics is discussed in Chapter 18.

How is it that plate tectonicists continue to cling so religiously to the Himalaya "compression" as their very avatar of subduction orogenesis in the face of all the contradictory field evidence? Alexander Pope gave the answer more than two centuries ago in his *Essay on Criticism*:

> The ruling passion, be it what it will,
> The ruling passion conquers reason still.

Quite apart from the fact that India did not converge with Asia but was always part of Asia, without an intervening deep ocean, the timing of the uplift and folding of the Himalaya conflicts sharply with the subduction concept. According to the subduction theory, the site where a subducting slab is forced down into the mantle is marked by intense compression, folding, overthrusting, seismicity, volcanism, and the rise of mountain belts, all concurrently with the subduction. The rate of subduction is matched by the rate of growth of new oceanic crust. The age of new ocean floor is indicated by its paleomagnetic signature, so from this the rate of growth of the Indian Ocean

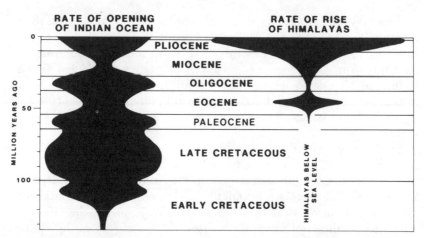

Fig. 37. Comparison of rate of opening of Indian Ocean and rise of Himalayas.

floor in the wake of India's alleged movement (and hence the rate of subduction in front of India) has been plotted in Fig. 37. From this it is clear that the folding and uplift of the Himalaya does not correlate with the rate of growth of the Indian Ocean. The rate of spreading of the Indian Ocean was highest between 50 and 100 million years ago, but during this time the Himalayan region was a shallow sea receiving quiet sedimentation, with no sign whatever of folding, thrusting, igneous activity, and general turmoil, which should have accompanied rapid subduction there.

What then is the Indus suture? In Gansser's words: "This suture represents one of the most outstanding structural elements on Earth, beautifully exposed, but not readily accessible. It has stimulated many theories on its origin—theories which often contradict known field facts." It is indeed part of the greatest structure on Earth—the Tethyan torsion, which displaces the northern hemisphere with respect to the southern hemisphere. This is discussed in Chapter 21.

The Myth of the Iapetus Ocean

Current orthodoxy claims that more than 600 million years ago, ocean-floor spreading had produced an ocean called Iapetus a couple of thousand kilometers wide, roughly but not exactly on the site of the present North Atlantic Ocean, which separated North America and Africa–Europe, much as the North Atlantic does today. This ocean was filled with sandstone, shale, and limestone during the early Paleozoic (600 to 400 million years ago), but during the Middle Devonian (some 370 million years ago), earlier ocean-floor spreading was replaced by subduction, which dragged the ocean floor down into the mantle, bringing northern Africa and Europe back against North America, and squeezing up the sediments to form the Caledonian fold mountains of Scandinavia, Scotland, and Ireland, and the Appalachian Mountains of eastern North America, which completely obliterated the Iapetus Ocean. It was not until the Rhaetic Epoch, about 200 million years ago, that a long chain of rift valleys, which rapidly filled with coarse sediments and basaltic lava, initiated the opening of the modern Atlantic Ocean. The axis of this new rifting had rotated some 30 degrees counterclockwise from the early Paleozoic trend, so most of the early fold mountains were on the eastern side of the new ocean in Britain and Scandinavia and most of them on the western side in North America, although residuals of the Appalachians were left in northwestern Africa and parts of the Caledonides on the east coast of Greenland.

No one disputes that the Caledonian and Appalachian fold mountains in Scandinavia, the British Isles, Greenland, northwestern Africa, and eastern North America were a single mountain system that was folded during the Devonian (also at some other times), and fragmented and separated during the Mesozoic opening of the North Atlantic. But what I do deny is that the Iapetus Ocean ever existed.

The primary reason for postulating Iapetus was the orthodox assumption that fold mountains result from subduction of earlier ocean floor. Hence in this view a significant seaway had to be postulated there before the subduction. Moreover, this appeared to be strongly confirmed by the fossil faunas and paleogeography. Whereas clear evidence of Ordovician continental glaciation was found throughout the northwestern bulge of Africa, limestones with subtropical faunas were traced along the western slopes of the Appalachians, apparently much too close to the frigid ice sheets unless there was wide separation between Africa and North America at that time. The Ordovician trilobite faunas on the west side of the axial zone of the Appalachians and Caledonides were also incompatible with those adjacent to them on the east side. Trilobite genera occupying similar ecological environments were so different genetically that it seemed impossible that they could have lived so close together for so many millions of years. They needed to have been separated by a barrier such as a wide ocean a thousand kilometers or more wide.

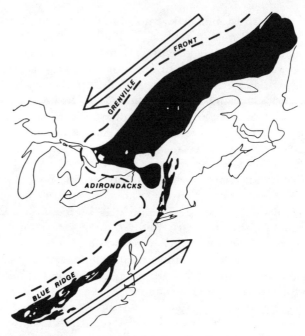

Fig. 38. Sinistral drag suggested by the distribution pattern of the Late Proterozoic Grenville rocks in eastern North America.

Certainly these trilobite faunas could not have been so near each other as they are now found. But the separation was 3000 km *along* the axial line of the Appalachians and Caledonides, not across it. Restoring this separation brings the trilobites of Pennsylvania back to face those of Scandinavia, with which they *are* compatible.

I first recognized this gross shear along the axial zone of the Appalachians from the S-shaped distortion of the "Grenville front," the boundary of the Late Proterozoic fold-mountains that preceded the Appalachians (Fig. 38). Later, firm confirmation came from paleomagnetic measurements.

During the early Paleozoic time interval when the Iapetus Ocean is alleged to have existed, lavas and other suitable rocks were magnetized by the contemporary magnetic field, so that these rocks recorded their latitude and the direction to the pole. Such rocks, irrespective of whether they were in North America, Africa, or Europe, would indicate the same position for the pole. But if during the Devonian Period the hypothetical Iapetus Ocean was closed by the subduction of its floor, and the Africa–Europe plates collided with the North America

The Subduction Myth

plate, the pole indicated by Africa and Europe would come to differ from the pole indicated by the America plate by the amount of the closure and in the direction of the closure.

Dr. W. A. Morris pioneered the investigation of this question, and found that indeed the poles had separated by some 30 degrees, but the separation was in the direction of the Appalachian–Caledonian axis, and not transverse to it as it would have to be if the alleged Iapetus Ocean had been subducted. Throughout the Early Paleozoic Era up to and including the Early Devonian (about 370 million years ago), southern Britain lay adjacent to North Carolina (Fig. 39). The Ordovician trilobite fauna of Pennsylvania was adjacent to the Baltic faunas. During the Middle Devonian (a time when the Appalachian and Caledonian belts were strongly folded), Africa and Europe were sheared 30 degrees counterclockwise, bringing northwestern Africa opposite North Carolina. Morris's conclusions have since been confirmed by other paleomagneticians with some variations in detail; for example, some think that the shearing displacement continued for 50 million years into the Carboniferous. This is quite acceptable, because both the Caledonian shearing torsion and the later Tethyan torsion movements probably lasted something like 100 million years. Caution is necessary with paleomagnetic data because some rocks have been remagnetized several times, making it essential that the geologic date of each separate magnetization be identified.

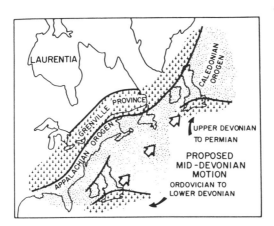

Fig. 39. Before 370 million years ago, the British Isles were adjacent to North Carolina. Middle Devonian torsion moved them more than 3000 km to a position opposite Greenland, bringing East Africa adjacent to the Carolinas, as the reconstruction of Pangaea finds it. (From W. A. Morris.)

The alleged anomaly between the northwest African glaciation and the subtropical limestones of the western slopes of the Appalachians during the Ordovician is also eased by the displacement *along* the Appalachian axis, but on an expanding Earth, it was not anomalous anyway. If the pole be assumed to have been in the center of the records of glaciation and the earth's radius assumed to have been 0.7 of the present radius, the most northerly glacials in northwest Africa would have been in latitude 43° (the same as the most northerly Quaternary glacials at sea level in Tasmania), and the subtropical Ordovician limestones would have been in the latitude of the present Great Barrier Reef of Queensland. The evidence for Iapetus, and the mythical ocean itself, vanish.

The Zodiac Fan Anomaly

Wherever a large river reaches the sea, its load of silt usually continues far out to sea, transported as intermittent turbidity currents. These flow across the ocean floor much as rivers do on land, scouring deep channels that may continue for hundreds of kilometers, until the flow loses speed through flattening gradient, and there the mud and silt accumulate as a growing fan, shaped like the back of a shovel. Such fans are essentially part of the delta of the river, but are called fans when the surface is far below sea level.

Three oceanographers of the United States Geological Survey, Andrew J. Stevenson, David Scholl, and Tracy Vallier, recently (1983) drew attention to the anomalous position of the Zodiac fan on the Aleutian deep-sea plain in the northeastern Pacific Ocean, south of the Alaska Peninsula (Fig. 40). The Zodiac fan, about a million square kilometers in area, contains nearly 3 million cubic kilometers of fine-grained mud, which was deposited there during the early Tertiary until at least 24 million years ago; so it is a fossil fan. The maximum thickness is more than 600 m, thinning out toward the edges. The Zodiac fan is not really a single fan but four successive superimposed fans, each with its own well-developed distributary channel system. The Taurus fan rests on the Aquarius fan, which in turn rests on the Sagittarius fan, which dates back to about 40 million years ago. The fourth fan, called Seamap, also rests on the Sagittarius, but its relation to the other two is not yet clear, although the authors suggested that it was the last of all to be deposited.

The question is, where was the river that fed the Zodiac fan? The authors estimated that a drainage area of at least half a million square

The Subduction Myth

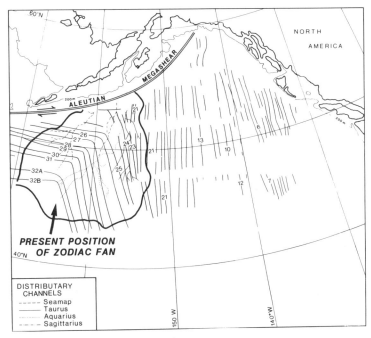

Fig. 40. The Zodiac fan in the Gulf of Alaska. Numbered continuous lines are magnetic anomaly stripes whose ages are shown in Fig. 19.

kilometers is implied, or half the total area of Alaska. Pollen and spores from sediments in the fan imply a nontropical climate similar to or cooler than that of Alaska today. The authors write: "Tertiary plate-motion models requiring large amounts of relative convergence along the Aleutian Trench are judged unworkable, as such reconstructions require the Zodiac fan to have formed 1,000 to 3,000 km from the nearest continental landmass and separated from it by topographic barriers, requiring that the drainage be inflated in size manyfold to overcome anticipated sediment losses during such a lengthy transport."

This anomaly is due solely to the gratuitous assumption inherent in the plate-tectonic theory that gross subduction of Pacific Ocean floor has occurred along the Aleutian trench, and this imaginary crust has to be reinserted to reconstruct the position 40 million years ago. In contrast, the expansion theory holds that the inferred crust, now held to have been subducted, never existed, and that all the Pacific Ocean crust that ever existed is still there. To restore the Zodiac fan to its position at the beginning of the deposition of the Sagittarius member,

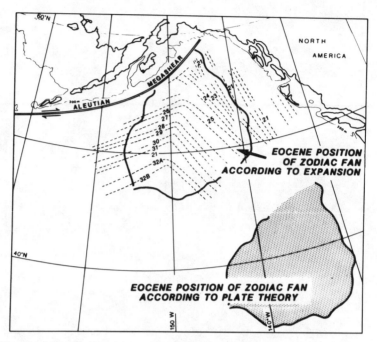

Fig. 41. Alternate Eocene positions of the Zodiac fan. The broken lines indicate the magnetic anomaly stripes after removal of anomalies younger than the fan (1 to 18).

all crust with magnetic polarity intervals 17 and younger has to be excised, because that crust has been inserted since by ocean-floor spreading. When this is done (Fig. 41), the enigma vanishes. To reconstruct the paleogeography exactly, other large adjustments have to be made, because during the 40 million years since the beginning of the Sagittarius fan there has been major dextral movement along the North American coast and inland at least as far as the Rocky Mountain Trench. However, none of these movements separate the Zodiac fan from its wholly adequate drainage area. Which is credible, subduction or expansion?

Why No Residual Ocean Floor?

Plate tectonics assumes that great oceans, covering more than half the earth's surface, have always existed. More than 200 million square kilometers of such ocean floors is assumed to have vanished since the Paleozoic, and as much again several times during previous eras.

Surely we should expect that at least *some* remnants of these old oceans have escaped subduction *somewhere*? But no ocean floor is known anywhere older than 150 million years. The reality is that all the ocean floor that ever existed is still there!

Subduction *Is* a Myth

All of these anomalies—the absence of subduction within Africa or Antarctica where it should be, the latitudinal extensional regime in the southern Andes when they should be under intense compression from the underthrusting Pacific slab, the nonsense of postulating the outcrop of a 7000-km underthrust in the Peru-Chile trench, the absence of the inevitable gross accumulation of oceanic ooze scrapings in any of the trenches where they should be, the absence of a great chain of andesitic volcanoes along the Kermadec ridge where they should be if crust equal to one-third of the earth's circumference had been subducted there, the absence of Himalayan orogenesis and vulcanism during the epochs when the subduction theory requires it, the myth of the Iapetus Ocean, the total absence of any pre-Jurassic oceanic crust, and the stranding of the Zodiac fan in the open ocean without a sediment source—all arise by trying to compensate the observed ocean-floor spreading by subduction instead of by Earth expansion. Subduction is a myth!

Procrustes was the fabulous robber of Attica who fitted his victims to his bed by lopping them shorter or stretching them. This is precisely what plate tectonicists do to the trenches, and many geologists have pointed out the arbitrariness of the plate-tectonic view of trenches. Thus, Rhodes Fairbridge, of Columbia University, wrote:

> There is no reason to view the island arc-trench [as] a compressional phenomenon at all. Could it not be a tensional feature? Many of the trenches are known to be faults with a strikeslip element, and the morphology of the least sedimented troughs suggest a graben-like character. The present oceanic trenches may thus be readily viewed as the modern equivalents of eugeosynclines of the past which often coincided with a major fracture.

Dr. Trevor Hatherton, a New Zealand geophysicist, has written:

> The special problem of the trench is the conflict between the theoretically compressional nature of plate boundaries and the apparently tensional nature of the trench itself. Indeed why the collision of two "rigid" plates should produce a trench is so far unexplained. Such earthquakes as are observed beneath trenches have a fault-plane solution best satisfied by normal [tension] faulting.

Professor William Tanner, of Tallahassee, Florida, objected that the numerous writers on the trenches have started with the "fact" that trenches were compressional structures on overthrust surfaces and have interpreted their data within this framework, even though their own data were more amenable to a tensional environment: "The sea-floor spreading *hypothesis* may, for some geologists, require compression in the vicinity of trenches, but the *data* require horizontal tension." Later he summed up his conclusions:

1. Compression is of little significance in creating and maintaining the major structures that underlie trenches, island arcs, and adjacent basins; instead, these features, in a strip up to 1700 km wide, are caused by primary regional tension.

2. There is no "down-going slab", whether driven by pushing from the rear, or pulling from a sinking front edge.

3. The only important motions to be accepted for island-arc and trench areas are horizontal tension and strike-slip.

4. Many authors conclude that the "down-going" concept is correct, in order to maintain the hypothesis, but this conclusion generally is contrary to their own data and should not be accepted.

Russian Academician P. M. Sychev reached similar conclusions:

Unfortunately the objective examination of the data obtained and their interpretation are nowadays more and more often substituted for by a preconceived theory. . . . The available data on island arcs and trench systems are in poor agreement with the concept of "new global tectonics".

Dr. Hugh Wilson, an American consulting petroleum geologist, in his review of Cretaceous sedimentation and orogeny in nuclear Central America, concluded:

The plate-tectonics advocates have produced a concept based on well documented expansion criteria and complemented by a hypothetical subduction process.

Dr. David Scholl and Dr. Michael S. Marlow, United States Geological Survey oceanographers, admitted:

Quite clearly the understanding we have of the tectonic setting of trenches comes more from concepts of global plate motion than from ideas generated by data gathered in the studies of trenches themselves.

Dr. B. W. Brown asked:

What is the essential difference between a statement beginning, "Because the continents drifted 2.5 cm per year . . .", and one that might have begun, "Because the flood of Noah drifted the icebergs . . ."? I fail to see any difference;

The Subduction Myth

both are deductive arguments, arguing from the specific to the general. Both arguments begin with conceptual principles. One could challenge this statement by remarking that there are "scientific" data in support of the first principle but not the second. But one could equally argue that within the fundamental *framework of the respective principles involved* there is as much support for the one as the other; fundamentalists certainly find that to be the case.

Dr. Wolfgang Krebs, of the Braunschweig Technical University, West Germany, wrote:

> Foredeeps and deep-sea trenches, crystalline belts and volcanic arcs, and intermediate furrows and interarc basins, respectively, are equivalent structures . . . which represent the top of diapir-like upwelling material from the asthenosphere. These subcrustal asthenoliths are characterized by crustal thinning, extension, inversion structures, high mean heat flow, deep earthquakes, positive gravity anomalies, extrusion of mantle-derived tholeiite basalts, and intrusion of ultra-mafic massifs. The forces caused by the rising asthenoliths are primarily vertical; horizontal stresses of secondary origin are gravity controlled. . . . Global vertical tectonics explains the evolution of island arc-trench systems and mountain belts in a much simpler way than the hypothesis of plate tectonics.

Krebs's conclusions are here given in the technical language of geologists; readers who are not familiar with such terminology will find the concept developed at length in Chapter 18.

CHAPTER 14

Criticisms of Earth Expansion

OBJECTIONS TO THE CONCEPT of Earth expansion have been raised on the following grounds. (1) If the earth were originally half its present diameter, gravity acceleration at the surface would be so high that it would have shown up unmistakably in ancient geologic processes. (2) The total volume of seawater would submerge all lands to a depth of 2 km or more. (3) Paleomagnetic data show that the earth's radius has not changed since the end of the Paleozoic. (4) Growth lines on fossil corals indicate that the number of days in the year 400 million years ago and at intervening times are consistent with the tidal retardation of the earth. (5) Whereas change in the moment of inertia of the earth implied by expansion is inconsistent. Plate tectonics adequately integrates a vast array of disparate data. What physical mechanism could explain such gross expansion? Finally, other planets should show similar expansion, and analogous phenomena should have been seen on the sun and stars.

Let us examine each of these objections in turn.

Surface Gravity

The criticism that the gravity acceleration at the surface would be increased fourfold on an Earth of half the diameter, and that everything would be four times as heavy, applies to the models proposed by Lindemann, Halm, Keindl, Egyed, and all others who assume that the mass of the earth has remained constant, but does not apply to the models adopted by Hilgenberg, the Russian pioneers (Yarkovskii, Kirillov, Neiman, and Blinov), and by me.

why not?

The Volume of Seawater

Egyed uses the total volume of seawater as his principal argument for Earth expansion, using paleogeographic maps of the Termiers and Strakhov (see Chapter 11), but he assumes a constant slow rate of expansion consistent with his data. Because I conclude that the rate of expansion has increased exponentially, and that the most rapid expansion has been in the most recent time, the volume of seawater would be a difficulty if I adopted the naïve assumption that the earth got its ration of water early in geologic history, and that it has been constant thereafter. On the contrary I believe that seawater, in step with the core and mantle of the earth, has increased exponentially with time. The freeboard of continents relative to oceans has not changed significantly.

The keynote of William Rubey's masterly presidential address to the Geological Society of America (1979) was that the whole of the waters of the oceans had been exhaled from the interior of the earth, not as a primordial process, but slowly, progressively, and continuously, throughout geologic time. As the generation of the ocean floors depends fundamentally on the same process as the outgassing of juvenile water, it would be expected that the volume of seawater and the capacity of the ocean basins both increased in a related way. But they would not necessarily be precisely in phase. Several variables are involved, some with feedbacks and time delays. There should be times when the capacity of the ocean basins increased more rapidly than the total volume of seawater, and vice versa. The former would result in general emergence, and regression of the sea from the lands, the latter a transgression of the seas over the lowlands. This could happen on the gross scale of the order of whole geologic periods, down to scales as short as a few years.

During the last decade, improvements in the quality of seismic-reflection profiling have enabled the recognition of short- and long-period oscillations of sea level, and the correlation of them between sedimentary basins, even between continents. Indeed, not far ahead we may expect correlation of individual beds globally as the fluctuation of sea level becomes more precisely known. Figure 42 shows the variation of sea level on the epoch scale as reported by P. R. Vail and R. M. Mitchum in 1977. In the past, such "eustatic" cycles were generally attributed to waxing and waning of continental glaciers. But it is clear from this figure that the sea-level cycles continue quite regard-

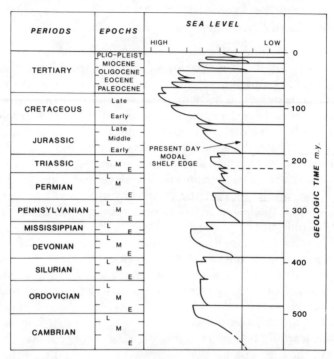

Fig. 42. Global supercycles of sea level, as reported by P. R. Vail and R. M. Mitchum.

less of contemporaneous glaciation (such as the Permo–Carboniferous and Quaternary) or global high temperatures (such as the Triassic). The cycles reflect the balance of total seawater volume against the total capacity of the ocean basins, as each increased with time.

Paleomagnetism

Paleomagneticians, particularly, have been obsessed with the dogma of constant Earth radius. First, they "proved" this by showing that the 20 degree angle subtended at the earth's center by two points in Europe and Siberia was the same as that indicated in the Permian at those points by the difference in paleomagnetic inclination (these points being then on the same paleomagnetic meridian). Unfortunately for them, when correction is made for the Ob sphenochasm, which opened during the Tertiary, the 20 degrees of paleolatitude represents the length of the Permian arc *before* the opening of the

Criticisms of Earth Expansion   193

sphenochasm, which in fact proves that the earth has greatly expanded since the Permian.

Next, attempts were made by several paleomagneticians to measure the ancient radius by what was called the minimum-scatter method. Consider all the paleomagnetic readings from a single geologic period on a continental block without sphenochasms or other gross distortions since that period. As each reading points to the pole of the time, any pair not on the same ancient meridian should give the distance to the pole by triangulation. But as each reading has an uncertainty of a few degrees, the distance to the pole would have a similar uncertainty, which would become very large if the angle between the two readings were small. This uncertainty could be narrowed down by taking a large number of pairs as well as by choosing widely separated points. An elaborate computer program was set up to take each pair with a range of assumed Earth radii (with a bias in favor of pairs with a wide angle, for the reason just mentioned), and to determine the scatter of pole positions derived from each assumed radius. That radius with the minimum scatter would be the probable ancient radius of the earth. The computer found in all cases studied that the present radius was the most probable, and that therefore the earth had not expanded (Fig. 43).

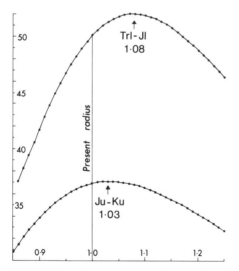

Fig. 43. Michael McElhinney's computer printout of statistical scatter (left axis) of Triassic–Jurassic and Jurassic–Cretaceous paleopoles derived from paleomagnetic data as the radius of the earth (bottom axis) is assumed to vary progressively between 0.8 to 1.3 times the present radius.

This operation should go down in computer history as a supreme example of GIGO—Garbage In, Garbage Out. In the discussion of map projections in Chapter 12, it was shown that it is impossible to draw a map of the spherical Earth surface on a plane surface without distortion of area, angle, or shape. This is only a special case of a general proposition that surfaces cannot be transferred between spherical surfaces of different radii without distortion of area, angle, or shape, the plane surface being part of a sphere of infinite radius. The three angles of a triangle on a plane surface add up to 180 degrees, but the three angles of a triangle on the surface of a sphere add up to something between 180 and 540 degrees, according to the solid angle subtended at the center of the sphere. If this is omitted from the computer program, the computer will find the least scatter with the present radius.

If a sector from a smaller sphere is flattened onto a larger sphere, it will have to either stretch or tear. The assumption fed to the computer that the continental block does not deform when the radius is changed is garbage. Some of the paleomagneticians concerned realized that the problem was serious, but hoped to minimize it by adopting the centroid of the reading points; but this does not eliminate the error, especially as the program was biased to favor the more widely separated points. That bias would give less scatter on the polar intersections but would increase the angle and distortion errors to a maximum.

Chapter 20 discusses the hierarchy of extension by which the whole lithosphere adjusts to the changing radius, with the largest polygonal adjustments determined in the first place by the thickness of the mantle down to the fluid core, then by the depth to the yielding asthenosphere, thence to smaller and smaller polygons, and ultimately to joints a few centimeters apart. So in addition to the errors of the geometrical artifacts, the minimum-scatter method ignores the billions of minute adjustments on regional ruptures and on smaller and smaller scales right down to ordinary joints. If each joint yielded only one-thousandth of a degree, the error could be 10 degrees in a kilometer! As joints are inherently systematic, forming in response to the pervasive stress field, they are of necessity additive. The net result of this continuous adjustment of the continental slab to the changing radius is such that the assumption of the present radius *should always* give the minimum scatter in the paleomagneticians' program.

The omission of provision for millions of tiny adjustments on the joints recalls the popular puzzle of the vanishing square (Fig. 44). Here are two rectangles, 12 squares by 5 squares. Triangle B is clearly

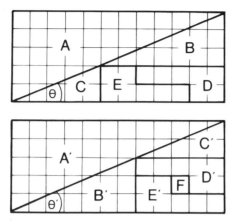

Fig. 44. The vanishing square puzzle.

the same as B', 7 squares by 3; triangle C is the same as triangle C', 5 by 2. Blocks D, D', E, and E' are clearly identical. Then where does block F come from? The answer is that triangle A is larger than triangle A' by the area of the F square; the tangent of angle θ is ⅖, which is not quite the same as the tangent of θ', which is 3/7. There are 12 extra tiny bits along the apparent hypotenuse of A that add up to square F. Similarly, by ignoring millions of tiny adjustments along millions of joints, paleomagneticians pretend to prove that the earth has not expanded.

Growth Lines of Fossil Corals

During the 1930's, Dr. Ting Ying Ma, a Chinese stratigrapher whose ideas were ahead of his time, found that the growth ridges of the coral *Favia speciosa*, which he believed to be seasonal, varied in prominence between equatorial and higher latitudes. He proceeded to study fossil corals of many ages in different countries as indicators of paleolatitude, and hence of continental drift. Concurrently in Annam (then part of French Indochina), Armand Krempf counted 112 of these annual growth ridges in *Favia speciosa*, and observed a waxing-and-waning cycle repeating every 18.6 years, which he believed was the record of the cycle of lunar nutation.

In 1937, John Wells, of Cornell University, noted that between annual ridges, which were about a centimeter apart, there were finer ridges, which he suggested recorded the lunar months, and within the annual ridges there were about 360 very fine growth lines, which he suggested were circadian (Latin *circa die*, about a day), caused by the effect of daylight on coral growth. Later, Wells reported that an Ordovician coral from Ohio had 412 of these circadian growth lines per year, a Silurian coral from Gotland had about 400, several Middle Devonian corals from New York ranged from 385 to 405, an Early Carboniferous coral from Wales 398, and two Late Carboniferous corals from Texas had 380 and 390 growth lines respectively. Several other workers followed Wells's lead and reported similar results.

Taking the known retardation of the earth's rotation by the tidal drag of the moon—2 seconds per century—and projecting it back 400 million years gives 399 days in the Devonian year, and other counts fit quite well. Therefore, it was argued, the earth has not expanded, because that would involve a very substantial increase in moment of inertia, and hence the retardation in the earth's rotation would far exceed that explained by the moon's tidal friction.

I am skeptical about the validity of the counts. The growth lines vary in spacing from a few microns to nearly zero, and it is often difficult to decide whether one should be counted or not. Under these circumstances it is notorious that total counts come out at what the investigator thinks they should be. The subjectiveness of such counts is highlighted by a report by R. G. Hipkin, an Edinburgh University geophysicist, that he counted 253 ridges and later 359 ridges in a repeat count of the same specimen. And recall the sophists who reworked their data to get a result in line with Kelvin's short age for the earth (Chapter 6). Besides, it is quite certain now that corals do not grow every day. Floods make the water murky with suspended matter and vary the supply of nutrients, and may suspend coral growth for several days. Strong winds over shallow water stir the bottom sediments and again suspend coral growth. Each of these factors tends toward too few estimated days in the year, perhaps substantially too few.

But even if we accept these counts as valid, this does not deny Earth expansion. If a body of constant mass expands uniformly, its moment of inertia increases. If a body is undergoing differentiation (core becoming denser and shallow zones becoming less dense) without expansion, the moment of inertia decreases. If a body of constant mass is both differentiating and expanding, its moment of inertia may stay the same or change either way. I argue as well that neither mass nor energy is constant (see Chapter 23).

The Blinkers of Dogma

Since the Kuhnian revolution to plate tectonics in the 1960's, the validity of this dogma is taken for granted by all "respectable" scientists. When a new fact appears, it is automatically interpreted in terms of this ruling dogma, even though it may be equally or better explained otherwise. Indeed, if it were explained in terms of Earth expansion, the report would certainly be sent back by journal referees for rewriting, if not rejected outright as naïve. The American journals were the worst offenders during the 1930's, 1940's, and 1950's in rejecting anything smelling of what is now called plate tectonics, and since the revolution they have been the worst offenders in rejecting as naïve anything based on any alternative model.

The magmatic-differentiation sequence from the trench to the volcanoes of the orogenic arc, the so-called "paired metamorphic belts," the ophiolite zones, and the petrological succession of elemental abundances, which have become pillars of the subduction theory, are more simply explained under Earth expansion. When higher seismic velocities were found under the Benioff zone, this was immediately seized upon as evidence of the cold down-going lithosphere slab, whereas such velocity increase is predicted by the expansion model (see Chapter 19). The same applies to the beryllium-10 anomaly, which has been cited by some geochemists as conclusive evidence of subduction. The beryllium of the lithosphere consists entirely of the beryllium-9 isotope, but beryllium-10 is produced in the upper atmosphere by collision of cosmic ray particles with nitrogen or oxygen. As the half-life of beryllium-10 is only one and a half million years, its existence is transient, but it reaches the surface in rain, and traces of it are readily detectable in surface waters and sediments. It is also found in volcanic rocks of some orogenic belts but not in the basalts of the oceanic spreading ridges. The geochemists argue that the orogenic volcanic rocks acquire this isotope from the subducted slab, which has dragged down some sediments containing beryllium-10, whereas the spreading-ridge basalts, derived directly from virgin mantle, have no such source of it. The evidence is strong that volcanic rocks in relevant orogens have been derived at least partly from sediments, but this does not prove subduction. Such incorporation of sedimentary rocks into nascent lavas is much more certain in expanding-earth orogenic processes, developed in Chapter 18, than in the subduction model. In the latter, the pelagic sediments are thin and accumulate slowly, the volume of sediments subducted would not be great, and the time required for the slab to reach melting depth is long in relation to the half-life of the beryllium isotope. In contrast, in the expansion orogen the whole of the eugeosynclinal sediment many kilometers thick is involved, and the time span between sediment deposition, magmatic assimilation, and volcanic extrusion is very much shorter. Far from establishing the validity of subduction, the beryllium-10 anomaly favors earth expansion.

An extreme example is the interpretation of the so-called "blueschists." These are metamorphic rocks containing the blue amphibole glaucophane, along with other minerals such as lawsonite, aragonite, and the ubiquitous quartz, which characteristically occur near the outer edge of the orogen, nearest to the trench. Laboratory experiments have produced this assembly of minerals under conditions of relatively low temperature and very high confining pressure. To achieve these conditions, the subduction theory drags trench sediments down to great depths, then brings them back for tens of kilometers *against the direction of flow* of the coldest part of the adjacent down-going slab! If I find an erratic boulder at the edge of a glacier, I know for certain that it has come from further up the glacier. There is no possibility whatever of it having been brought there from *down* the glacier. Nor is there any possibility of the blueschists ascending several tens of kilometers up against the flow of the down-going slab. To increase the absurdity, this relatively low-temperature part of the lithosphere is certainly a million times as viscous as glacier ice.

The plate tectonicists have reached this absurd position because of their subduction model, but also because of misinterpretation of the laboratory experiment on which the conditions for crystallization of the blueschists was based. The laboratory experiments involved *static* confining pressure and varying temperature. But the blueschists are *schists*. The name itself implies not static confining pressure, where the stress-difference is zero, but very large stress-difference, that is, a very large pressure in one direction, but a very much smaller confinement in a direction at right angles to the maximum pressure, so that the rocks flow toward that direction and crystallize as schists.

High confining pressure with no stress-difference produces granulite, tectonically isotropic marbles, and quartzites, with minerals typical of that environment, such as garnets, pyroxenes, calcite, and quartz. The same rock in an environment of large stress-difference recrystallizes as gneiss or schist, made up of *schistophile* minerals, which are different from those of the granulite. Schistophile minerals, such as the micas, the amphiboles, and aragonite, commonly are denser in one crystal direction than another. The laboratory experiment to produce the blueschist mineral assemblage had to use very high pressure with low temperature, but this is *not* their environment. They really form in conditions of relatively low temperature but very high stress-difference, and not necessarily with deep burial. These are exactly the conditions that apply in the diapiric orogen inward from the Benioff zone, according to the expansion theory. As the motion there is upward at all times, there is no absurdity of bringing these rocks up against the flow.

Other Planets

I agree that if a cosmological cause is invoked for expansion of the earth, it must be applicable generally. But it would be naïve to assume that all bodies would be at the same phase or stage. The earth's surface now is bimodal, containing continents and major ocean basins. But this condition only appeared during the Mesozoic, that is, only during the last fortieth of known geological history (see Fig. 101 in Chapter 22). In the early Proterozoic, the earth was probably very much like Mars now, with a great rift zone extending halfway around the equator. Still earlier, it probably looked like Mercury does now, heavily cratered, but with a polygonal tensional fracture system that cuts across the older craters.

Mercury's polygonal fracture system has been interpreted as a compressional system, on the grounds that the fractures seem to dip under the higher side. I am not sure that the alleged overhang is correct, because, given the physical properties of known rocks, their strength could not sustain the weight of any overhang visible from a satellite so many thousand kilometers away, and I have studied good prints of these photographs for deflection of the fault traces in the slopes into craters, without conviction of the direction of dip. However, if the dip is indeed toward the elevated side, this is exactly what should be expected on this scale from vertical tension fractures, as the rising block spreads under its weight. Still earlier, the earth may have looked like the moon does now. At some future time it may look like Neptune, and still later like Jupiter. In comparing the expanding Earth with other members of the solar system, it is naïve to assume that the planetary evolution process should be without physical thresholds.

Consider a simple analogy of the state of water on three terrestrial planets: on one all the water is in the form of ice, a second (rather like Earth) has all the water in liquid form, and on the third (more like

Venus) all the water has been vaporized into the atmosphere. The change from the first to the second and second to the third type would happen suddenly as the temperature reached critical thresholds. Similar thresholds may exist in the evolution of planets. In Chapter 23 I will argue that the expansion of the earth was real from the start but increased very slowly for the first 37 hundred million years of known geological time, and only 100 million years ago it was still only 60 percent of its present size. Then a threshold was reached and expansion since has been very rapid; if this continues, Earth would progress rapidly to a state more like Neptune and join the family of great planets. In that family, Jupiter seems to have reached another threshold toward stardom, and may rapidly advance to become an average star, where another stable state would be reached. From the dawn of contemplation of the earth and universe man has assumed gratuitously that all matter that *is* always *was* and ever *will be*: that Sun, Moon, and planets are permanent unchanging entities. But a few thinkers now question that axiom and, like acorn to oak, replace fixity by process. This concept is explored in Chapter 14.

All the planets have their axes tilted at various angles to the ecliptic. In my 1976 book, I argued that the obliquity of the earth's axis to the ecliptic is a most important geologic variable, caused by asymmetric expansion. Jupiter's Great Red Spot likewise has the appearance of a tumor or upwelling convection cell, which is certainly asymmetric with respect to longitude.

On Earth, for example, during the early Mesozoic, an enormous area of Pangaea developed a mantle blister, which led to partial melting in the upper mantle and produced the vast volumes of tholeiitic magma that ascended as dolerite and basalt over all of southern Africa, much of South America, Antarctica, India, and Tasmania. The isotherms must have been much shallower than elsewhere over all this area, and hence phase changes to less dense rocks must have reached quite deeply into the mantle. As gravity insists that each radial sector of the earth should have the same weight, this vast region must have stood significantly higher than average, which in turn gave it a higher contribution to the moment of inertia. The bulge would induce an axial wobble, and as the rocks of the upper mantle slowly yielded, the earth's axis would creep until the rotation was about the axis of maximum moment of inertia. The blister would also be attracted differentially by the sun and moon, giving an external contribution to the torque.

Again, the Tertiary spreading of the ocean floor was not symmetrical, and this in turn must have induced further wobbles, migration of the rotation axis, and variation of axial tilt, or obliquity. Precession depends on obliquity, and because the rotational bulge of the fluid core is less than the surface bulge, it would precess at a slower rate if it were free to do so, and the core, as we know from magnetic observations, does drift slowly westward because of this. The resulting internal friction slowly damps down the obliquity. So the obliquity of the earth's axis is repeatedly increased by perturbations caused by asymmetric expansion, and is continually damped by internal friction.

I submit that the obliquity of the rotation axes of the planets records their asymmetric expansion. The moon bulges on the side facing the earth, and the large mare depressions are all on this side, a fact hitherto unexplained. As the figure of the moon is determined isostatically by its own gravitation, this implies that the side facing the earth has less dense materials. It also raises the question whether the mare plains, in contrast with the impact craters, are expansion krikogens (see Chapter 18). Earth's gravity would ensure that as the moon's rotation rate was reduced by tidal friction to become synchronous with its revolution, the bulging side would permanently face the earth.

PART FOUR
VERTICAL OROGENESIS

CHAPTER 15

Gravity Rules the Earth

THE STAGE HAS BEEN REACHED to face a fundamental difference between plate tectonics and Earth expansion—the nature of the process which gives birth to fold-mountain belts, the process called orogenesis (Greek ὄρος, mountain; γένεσις, origin).

According to plate tectonics, orogenesis is caused by horizontal compression where plates collide and squeeze the material between them into a crumpled and overthrust complex. The motion in the lithosphere slab is downward (Fig. 24), as is the motion in the mantle below where convection sinking is believed to drive the process. But the orogenic zone itself rises many kilometers.

The expansion concept is the antithesis—crustal divergence, not convergence. The orogenic zone is where extension is concentrated, not only in the crust, but far down into the mantle, and even though the surface begins by subsiding because the crust stretches and thins, the motion in the roots of the orogen is upward at all stages, and continues to be upward for hundreds of millions of years afterward as the surface mountains are eroded away. Orogenic zones, like oceanic spreading ridges, are fundamentally widening zones; indeed, where the rate of extension in the mantle below an orogen increases enough, it becomes an oceanic spreading zone.

Textbooks in English universally explain orogenesis in terms of crustal shortening and compression, so the mechanism I will present, although far from new, will be unfamiliar to most readers. Therefore I must go back to the beginning and show how these two contrary concepts originated, then in the next four chapters develop from first principles vertical orogenesis on an expanding Earth. Gravity is far and away the ruling force in the earth, and all tectonic movements are driven by gravity to restore gravity equilibrium where it has been disturbed by thermal inequalities, differential loading by glaciers or lavas, erosion of highlands or sedimentation of depressions, or some other cause.

When in the 1960's the establishment was shocked into the realization that Wegener's concept of gross relative movements of the continents was correct after all, it was unfortunate for geology that the English-speaking world believed that orogenesis was fundamentally a compressional phenomenon. For from the union of the newly validated ocean-floor spreading with this illegitimate faith, the spurious subduction concept was born. The pitfall was that a large proportion of the folded and faulted strata mapped by geologists in the field do indeed imply substantial horizontal contraction throughout their study area, to the full depth of the sections they draw. But this genuine foreshortening on the district scale of field geology, I will show, is a secondary corollary of *vertical movement in a primary extensional environment* on the large scale of geotectonics. As a result, the recognition that the real cause of ocean-floor spreading is Earth expansion has been delayed by a generation. It is therefore necessary to go back to the beginning and examine critically just what folding, thrusting, and mountain-building really are.

Vertical or Horizontal?

In 1815, Sir James Hall read a paper to the Royal Society of Edinburgh "On the vertical position and convolutions of certain strata and their relation with granite." He pushed together the sides of a pile of paper, and also clay models, to produce folds or convolutions on a small scale like the large folds he saw in the rock strata, and deduced that horizontal pressures had laterally compressed and folded the strata. In 1852 the brilliant French geologist Elie de Beaumont (1798–1874) published a three-volume book on the origin of mountain systems, in which he attributed the folding of mountain chains to the contraction of the earth as it cooled. This hypothesis, soon known as the contraction theory, became the accepted dogma for the next century. From these beginnings, strongly supported by Professor James Dwight Dana (1813–95) of Yale and by Professor Bailey Willis of Stanford in the first quarter of this century, the conviction matured that folding generally, and particularly the great belts of folding, thrusting of piles of rock strata over other rocks, and uplift characteristic of mountain ranges like the Alps, Appalachians, and Himalayas, were the result of horizontal compression of the earth's crust.

Meanwhile the alternative idea, that vertical disturbances of gravity equilibrium were the primary cause of folding and orogenesis, took root in Europe contemporaneously with Hall's proposal. Although never universally adopted, this concept has been persistently advocated as a minority view ever since, mainly among German-reading geologists, along with a significant group of Italians, and since World War II the concept has been strongly developed in Russia. Pioneers of vertical tectonics were Gillet-Laumont (in 1799), Scrope (1825), Schardt (1823), Kuhn (1836), Naumann (1849), Herschel (1856), and Bombicci (1882). All of these thought that a primary uplift was caused by gravity acting vertically, and that the folding of strata was due to compression under downhill gliding pressure, without any shortening of the underlying basement crust.

Reyer (in books of 1888, 1892, and 1894) broadened the concept by suggesting that the gravitational upthrust caused folding directly, as well as secondary folding from the spreading pressure of the uplifted zone. He backed his conclusions with experimental models in which he recognized that to match correctly the behavior of very large bodies (such as a mountain belt) over a very long time with a small laboratory model in a few hours, systematic changes in the deformational properties of the model materials must be made. Reyer's experimental methods were much later adopted and widely developed by Vladimir Beloussov in Russia and Hans Ramberg in Sweden, with impressive results.

Gravity tectonics was invoked by Schardt, in 1898, to explain structures on the north slopes of the Swiss Alps, and by a succession of Italians investigating extensive down-slope displacements recognized in the Apennines; these included Bonarelli (1901), Anelli (1923 and 1935), and Signorini (1936). Dal Piaz reviewed this work in 1943 in a paper read before the Academy of Science of Turin.

Professor C. Eugene Wegmann, who had succeeded Emile Argand in Neuchâtel, applied to mountain ranges the mechanics of salt-dome diapirs (salt bodies which being less dense than other strata are driven upward by gravity in columns a kilometer or more in diameter, piercing the overlying strata). In Germany, the more fundamental approach started by Reyer was amplified by Steinmann, and particularly by Haarmann, whose "oscillations" theory inspired R. W. van Bemmelen, a Netherlander from Batavia, to develop his "undation" theory (that is, "wave" theory) in a series of papers throughout the 1930's and again after World War II. According to van Bemmelen, gravity was the ultimate cause of all rock deformation and operated on all scales, from relatively local uplifts to "mega-undations" of continental dimensions, causing in turn horizontal movements on a similar range of scales to restore gravity equilibrium.

In the 1950's and 1960's, two independent schools developed Reyer's experimental methods. The Russians, led by Vladimir Beloussov, modeled a wide range of fold and fault structures, taking care that the deformation properties of their test materials conformed to the constraints imposed by the small scale and the short experimental time, so that the pattern of motion of their models truly indicated the behavior of the real earth. Beloussov concluded, as others had before him, that gravity-driven vertical movement was the fundamental cause of folding and that the horizontal shortening was secondary, without any implication of crustal shortening. Contemporaneously, Hans Ramberg in Uppsala designed a centrifuge to shorten the duration of his experiments by increasing the "gravity" force 50 times in his models, and thus he successfully matched structures large and small. His results agreed fully with Beloussov's, and he concluded that the great fold-mountain belts such as the Alps and Caledonides were large diapirs driven upward by gravity.

The experimental and theoretical work of Reyer, van Bemmelen, Beloussov, and Ramberg emphasizes the dominance of gravity in orogenesis. Indeed, the larger the scale, the shorter the time required for stress differences to be relaxed by flow. As the earth rotates below the moon, the earth's attraction of the surface rocks is countered in part by the moon's attraction, so a tidal bulge in the solid earth follows the direction of the moon. This is an elastic deformation, but if a load is applied for times longer than a year, permanent plastic yielding begins to be noticeable, and over a millennium the yielding is complete, as though the solid earth had no strength at all. The shape of the earth is entirely determined by gravity, the radius being some 20 km greater at the equator (where gravity is partly countered by the centrifugal force) than at the poles (where the latter is nil).

Isostasy

When a plumb bob (or a spirit level, which acts the same way) is set up near a mountain range, the excess mass above the horizon attracts the plumb bob slightly toward it from where the bob should rest if the mountains were not there, and surveyors must correct for this in high-grade work. When Sir George Everest, surveyor-general of India, was making his primary triangulation of India, he was astonished to find that the Himalayas did not seem to be pulling their weight. The plumb bob was deflected toward the mountains, but only by a third of what it should be when Newton's law of gravitational attraction is applied to their mass above the horizon.

Sir George Airy (1801–92), the Astronomer Royal at Greenwich, in a brief paper to the Royal Society (1855), replied that the strongest known rocks could not bear the weight of the Himalayas without failing and subsiding under the load. Hence the Himalayas, and indeed all mountain ranges, must be underlain by less dense rocks to a depth of tens of kilometers, so that the mountains and their underlying foundations are in buoyant equilibrium, like icebergs, which float with only one-seventh or less of their thickness showing above the sea. The less dense foundations of the Himalayas should exert less attraction on the plumb bob than rocks generally below the horizon, which partly countered the excess attraction of the mass above the horizon, and Airy's calculation agreed with Everest's observations.

Airy's observations led to the enunciation of the principle of isostasy: namely that the whole of the earth's crust is in buoyant equilibrium and that the weight below any area of the surface is equal to the weight below the same area anywhere else, just as the weight of a ship or an iceberg is the same as the weight of the water displaced by it. Because rocks have more strength than water, and a "viscosity" so very much greater, the time needed for significant buoyant adjustment is much longer, and loads only small in area may be supported indefinitely by the strength of the rocks. Nevertheless, time available in geology is long and areas are large, so in most things geologic, isostasy substantially prevails. For example, the weight of the ice sheets during the last ice age depressed the earth's surface by hundreds of meters, which was recovered slowly after the ice melted; and the weight of the water in the former Lake Bonneville (which has now evaporated down to the relatively small Great Salt Lake of Utah) similarly depressed the crust, and the shorelines formed then, still strikingly visible around what were islands in the lake, warped upward out of level as the crust recovered from its burden. In addition, just as the accumulation of snow in Antarctica and in central Greenland causes glaciers to flow outward, so there is gravity force to drive the mountain rocks to spread outward. This is exactly what has happened and is still happening, as we shall see later.

The Significance of Scale

Everyone would agree that in stepping from geotectonics to astrophysics, we must beware of the effects on our mental models of the great change of scale. Yet the whole range through global geology to astrophysics involves only three orders of magnitude (that is, 1000 times), whereas *within* the field of rock deformation, structural geology, and geotectonics, the linear scale varies through 15 orders of magnitude—one million billion times (Fig. 45). The scale of time is as wide as the scale of size (Fig. 46). Two kinds of trap await us when jumping such orders of magnitude.

First, a small correction may become significant, even overwhelming, when length or time becomes large. For example, the additional term I found necessary to add to Newton's law of gravitation for cosmic distances (see Chapter 23) eventually dominates the behavior of galaxies. Where parameters involve different powers of size or time, change of scale affects one property more than another. Thus the most successful airplane design, if faithfully copied at twice the size, could not fly. The weight would have been increased eight times, but the wing area to support it only four times. A double-size airplane can of course be built, but not to the same proportions.

Since its foundation by Hutton two centuries ago, igneous petrology has been based on models in the wrong scale of size—thinking in terms of solutions, macromolecules, melting points, eutectics, instead of colloids. A cube of rock one centimeter wide has a *surface* area of 6 square centimeters, but a one-centimeter cube of clay has a *surface* area of 6000 square kilometers, indeed ten times this because clay particles are sheets, a scale differential of a hundred billion! The *surface* energy in geosynclinal sediments is so great that it dominates their chemical behavior. (See "The Granite Controversy," in Chapter 4.)

Second, a threshold to grossly different behavior may intervene. Such thresholds are the melting and boiling points and the critical temperature, at which surface tension becomes zero and liquid must become gas irrespective of pressure or of space to expand into. Such gas-phase water may be as dense as normal water but would have no capillarity to restrict its seepage through the tiniest openings. Similarly, the yield stress at which plastic deformation (*sensu stricto*) commences, the threshold of rupture, the sensitivity index of soils (where a load-bearing stratum transforms to a slurry), the critical mass for the onset of fission, the pressure-temperature thresholds where specific minerals recrystallize to denser phases or where one set of minerals changes to a different set of minerals with different physical properties, and the dimensionless numbers that determine the change from laminar flow to turbulence, or of heat conduction to convective circulation, or of magnetohydrodynamic phenomena, are thresholds of quite different behavior.

An instructive example of discontinuous change in behavior of liquids as scale increases is given by the transition in liquids from laminar flow, where all particles follow parallel streamlines, to turbulent flow, where eddies form. (Still another threshold is involved in the flow of gases.) Osborne Reynolds demonstrated in 1833 that the threshold for turbulence depends on vl/ν where v is the velocity of flow, l is the distance to the wall, and ν is the kinematic viscosity of the fluid, but the physical reason for this relation was not understood until I derived this relation from first principles three decades ago. (I presented the explanation in my 1976 book, on p. 102.)

The 15 orders of magnitude in the size range of geologic phenomena takes the scale of mass and volume through 10^{45}, a number too large for comfort, so it is not surprising that naïve concepts have resulted from the transfer of mental models from field geology to global phenomena. Hazards are less, but still real, if we confine our thinking to only three orders of magnitude at a time, as in the categories in Fig. 46.

The smallest field is rock mechanics, ranging from one-tenth nanometer (one angstrom unit) to 10 micrometers, where we are concerned with static and kinematic ideas of stress and strain, strength, the concepts of elasticity, viscosity, plasticity, and flow, the forces between slip surfaces in crystals, and the deformation of single crystals and of polycrystalline aggregates. We work in the office on theoretical analysis and induction, with the electron microscope as our eye.

The next field is petrofabrics, ranging from 10 micrometers to 1 centimeter, and the microscope stage is our arena. We are concerned with individual mineral grains within rocks rather than the relations between rocks, and hence we study rock fabrics and textures, schistosity, foliation, lineations, augen, mylonites and gouges, and joints. Interstitial fluids, seepage pressures and flow nets, the consolidation of sediments, the behavior of soils, and the phenomena of freezing of water-saturated materials all lie in this domain.

In outcrop geology, on a scale from 1 cm to 10 m, we are concerned with everything we can see in a single exposure. The hammer, compass, clinometer, tape, hand lens, and camera are our tools. The mylonites and gouges of petrofabrics are replaced by breccias, melanges, and chaos—rock fragments instead of mineral grains. We study categories, repetitions, and associations of strata, their dips and strikes and facings, their termination at ruptures and intrusive contacts, and their minor folds and drags; we do comparable studies of igneous and metamorphic rocks. We abstract the results of the petrofabric scale and integrate them here in terms of rock-deformation structures.

From 10 m to 10 km is the range of structural geology in the narrow

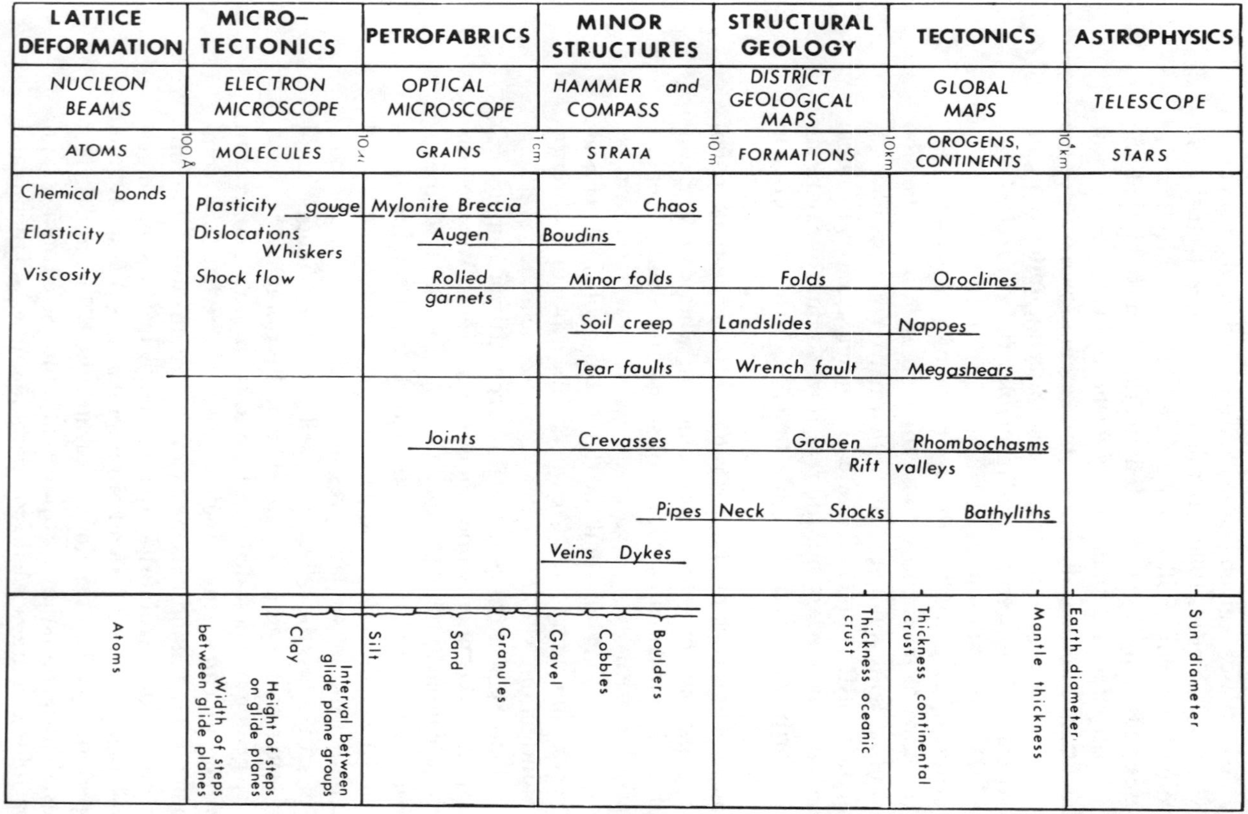

LATTICE DEFORMATION	MICRO-TECTONICS	PETROFABRICS	MINOR STRUCTURES	STRUCTURAL GEOLOGY	TECTONICS	ASTROPHYSICS
NUCLEON BEAMS	ELECTRON MICROSCOPE	OPTICAL MICROSCOPE	HAMMER and COMPASS	DISTRICT GEOLOGICAL MAPS	GLOBAL MAPS	TELESCOPE
ATOMS	MOLECULES	GRAINS	STRATA	FORMATIONS	OROGENS, CONTINENTS	STARS
	100 Å	10 µ	1 cm	10 m	10 km	10⁴ km
						Sun diameter
						Earth diameter
						Mantle thickness
				Oroclines		
				Nappes	Megashears	
					Rhombochasms	
				Folds	Rift valleys	Bathyliths
			Landslides	Wrench fault	Graben	Stocks
						Thickness continental crust
				Pipes	Neck	
			Soil creep			Thickness oceanic crust
			Minor folds			
			Tear faults	Crevasses		
				Veins Dykes		
						Boulders
		Augen	Boudins			Cobbles
		Rolled garnets				Gravel
	Plasticity gouge	Mylonite Breccia				Granules
		Chaos				Sand
		Joints				
Chemical bonds	Dislocations Whiskers					Silt
Elasticity	Shock flow					Interval between glide plane groups
Viscosity						Clay
						Height of steps on glide planes
						Width of steps between glide planes
						Atoms

Fig. 45. The size of bodies and structures involved in geological processes ranges from a hundred-millionth of a meter to a hundred million meters. Gravity is of quite negligible importance at the lower end of this range, but is the only force that matters at the upper end. Nuclear forces and interatomic forces (including elasticity, viscosity, plasticity, strength) rule the lower end of the range but are insignificant at the upper end. Thresholds of strength, of state (solid, liquid, gas), flow (laminar or turbulent), convection, and magnetohydrodynamics further partition the fields of behavior.

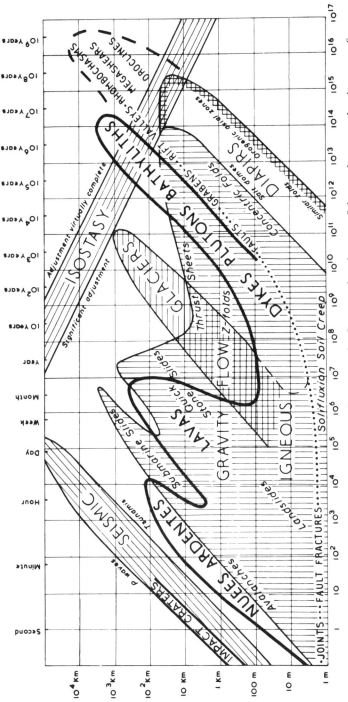

Fig. 46. Time fields of geological phenomena from a tenth of a second to the age of the earth, across the size range from field outcrops to the whole earth. (These figures and associated discussion, originally developed much earlier for my lecture courses, were first published in my 1962 paper in the *Journal of the Geological Society of India*.)

sense, which can only be studied by synthesis on regional geologic maps and sections. The conclusions of our outcrop-scale observations are abstracted and symbolized on these maps and from them regional structures are deduced. Here we are concerned with folds, faults, nappes, grabens, and plutons. Not rocks or beds but stratigraphic formations are the deformed units.

In tectonics, on a scale of 10 km to 10,000 km, we work on continental and global maps, or the globe itself. Here we are concerned with provincial structures, geosynclines, megashears, and great thrust sheets displaced many tens of kilometers, passing up to continental structures (where the orogen itself is the deformed unit), the primary and secondary polygons, oroclines, sphenochasms, and megashears, and the mid-oceanic ridges and rift systems, thence to the continents and oceans themselves.

In the smaller scales, horizontal and vertical mean little in limiting patterns of motion. Even in structural geology, folds may have vertical deformation of tens of kilometers, although gravity is increasingly important as large structures feel their weight. But in tectonics, gravity dominates. Isostatic equilibrium is the normal state, and departures from it cause horizontal motions toward restoring equilibrium in times that are geologically short. Few structures can have vertical displacements of even 10 km, for by that time the rate of lateral spreading balances upward flow. But there is little restriction on horizontal translations. Megashears occur with translations exceeding 1000 km, but vertical displacement on great faults is restricted to a few kilometers. Orogens are restrained by gravity from being folded substantially about horizontal axes as strata are commonly folded, but gross folding of orogens about vertical axes (oroclines) are common (as recounted in Chapter 9).

On the tectonic scale, all rocks flow. Weight, the driving force, increases with the cube of the linear scale, but the elastic and viscous resistance to deformation only increase according to the square of the linear scale. Ice is a brittle crystalline solid on a small scale, but glaciers flow according to fluid laws, and microscopic study of glacier ice finds crystal fabrics like those of wrought iron and of highly deformed schists and gneisses, which develop these fabrics because they too have flowed long distances. The internal inhomogeneity of the earth (discussed in Chapter 20), differential motions of the earth's shells, and the perturbations of the earth's rotation and revolution, which have no expression on the scale of structural geology, have controlling significance on the tectonic scale, and so also has secular expansion.

Within each scale of size there is also a spectrum of the scale of time, ranging from short-term impulsive phenomena to secular phenomena through geologic time. On Fig. 46 both scales are logarithmic. The maximum size and the maximum time are limited respectively by the size and age of the earth. Most phenomena have fields that slope upward to the right at 45°, and therefore express rates of flow.

At the extreme left is the impact field, including impact craters and astroblemes (Greek ἄστρον, star; βλῆμα, wound from thrown object). In the latter, the kinetic energy of the asteroid is totally destroyed in seconds, first by conversion to plasma of the impacting body and a similar mass of the earth, surrounded by a zone of volatilization, then one of melting, then a zone of shock deformation of minerals, surrounded in turn by a zone of pulverization, where the stress still to be transmitted exceeds the strength of the materials, then finally an envelope of seismic-wave transmission that contains some shattered zones where reflected waves interfere with outgoing waves to reach stresses beyond rock strength.

Necessarily overlapping the impact field is the seismic field, where transmission is entirely elastic, with increasing attenuation as the longest periods reach the relaxation time of the medium. Increasing temperature reduces viscosity and hence relaxation time, which cuts back the seismic field both in wave period and wave length.

Next to the seismic field is a succession of fields stacked in order of their "effective viscosity," first the *nuées ardentes* (incandescent volcanic explosion clouds), with effective viscosities of 10^{-3} poise, turbidity flows, 10^1 to 10^4 poises, through various types of landslides to glaciers at 10^{13} poises, and on to salt and sediment deformation with viscosities through the range 10^{14} to 10^{21}, to an upper practical limit of flow at about 10^{27} poises, the "viscosity" of cold crystalline rocks.

The field of isostasy slopes down to the right at 30° with increasing time, which expresses the fact that the wider the area of the load, the more rapidly is adjustment reached (the 1-on-2 slope is time against length squared). Following the upper edge of the isostatic field, and watching the time and size coordinates, loads involving the earth's figure would reach complete isostatic adjustment in a century (elastic adjustment occurs in much shorter time). In 1000 years, there would be complete isostatic adjustment of loads the size of an ocean or a large continent, and significant adjustment of loads 2000 km in diameter. The recovery following the melting of the ice sheets some 10,000 years ago agrees with these scales. In 100,000 years, isostatic adjustment would be virtually complete for loads 2000 km in diameter, and significant adjustment would have occurred for loads 500 km in diameter. In a million years, adjustment of the latter would be virtually complete, and noticeable adjustment would have occurred for loads 50 km in diameter. For the phenomena I address in this book, gravity truly rules the earth.

CHAPTER 16

Folding

FOLDED STRATA, and orogenesis generally, have been interpreted (almost universally by English-speaking geologists, but less so in Europe and Russia) as implying compressional contraction and that the folded strata formerly occupied a wider area on the surface. A close examination of the geometry of folding shows that this assumption is not necessarily true. This false axiom coupled with the subduction myth to conceive "collision" tectonics.

Similar and Concentric Folding

The left-hand part of Fig. 47 shows the *similar* folding of the beds below. The deformed shape of each bed is exactly similar to the deformed shape of the bed below it and to all the other beds. The beds appear to be stretched, and they vary greatly in thickness. But the thickness of each bed remains constant when measured in the direction of the flow lines. The area of each bed in this cross section and the area of the whole group remain constant, because between each pair of flow lines you have a parallelogram with the same area as the corresponding rectangle in the original beds. In the bottom right of the figure, the beds appear to be undeformed, but the white lines across them indicate the position of the original block to the left after the deformation. In this case, the beds that appear quite undeformed have been deformed to exactly the same degree as the beds at the left, which appear to be strongly deformed. From this diagram it is clear that folding does not disclose the whole deformation, only the deformation transverse to the bedding; the deformation parallel to the bedding is real but unseen. This latent component may even be the greater component, because deformation is easiest along the most

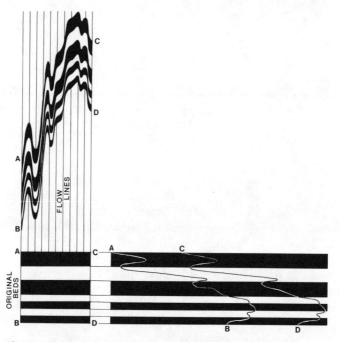

Fig. 47. In "similar" folding, the flow lines of the deformation are transverse to the bedding, and the thickness of the beds measured in the direction of flow is constant. The bottom right diagram, which has exactly the same amount of deformation as the top left diagram, shows that if the flow is parallel to the bedding, the beds appear undeformed.

Fig. 48. Concentric folding. The thickness of each bed remains constant at right angles to the bed. Hence where the beds are curved, the beds are "concentric."

ductile layers whereas deformation transverse to bedding must deform the least ductile beds.

Geologists have recognized two contrasting paradigms of folding, similar folding and concentric folding, as in Fig. 48. In *concentric* folding, the orthogonal thickness of each bed remains constant, and a line

Folding 219

at right angles to any bed is also at right angles to the beds above and below it, which means that the bedding surfaces are mutually concentric.

In similar folding, the thickness of each bed remains constant in the direction of the flow lines, which means that all beds have the same shape, and if collapsed along the flow lines they would all coincide. Because the flow lines are parallel, and the thickness is constant in their direction, the area of each bed between any pair of flow lines remains constant no matter how extreme the deformation, and so does the total area of each bed.

These concentric and similar folding models are extremes, and real folding is usually a compromise between them. In similar folding, the deformation pattern is governed solely by the stress field, ignoring the bedding entirely as though the material was uniform throughout, with successive layers merely colored differently to show the deformation. It is therefore the simplest form of folding physically.

In concentric folding, the physical difference between the beds is so great that almost all the deformation is constrained to slip parallel to the beds, with minimum deformation across the beds. Concentric folding looks simple, and similar folding looks complex, but the reverse is true. Homogeneous materials like rock salt and glacier ice deform with similar folds. Folding in shallow beds, in which the initial great differences in strength between sandstone, mudstone, conglomerate, and limestone still remain, tends to follow the concentric pattern, but deep burial, rising temperature, and pervasive pore fluids progressively reduce the differences in strength, and folding progresses toward the similar model, as illustrated by the progression in rocks from shale, through slate, schist, and gneiss.

Similar Folding Implies Flow, Not Static Compression

When the directon of the flow lines changes in similar folding, very complex patterns result (Fig. 49), but even these are relatively simple to reconstruct, as I elaborated in my anniversary address to the Alberta Society of Petroleum Geologists a quarter of a century ago. The original thickness of a bed was at least as great as its *maximum* orthogonal thickness after deformation. Although the thickness of the bedding in the direction of the flow is consistently maintained, the length along the bed and the surface area of the bed may be enormously increased, as is plain in Fig. 49. Because of this very great increase in surface area, any bed that is a little stiffer than its neighbors stretches into

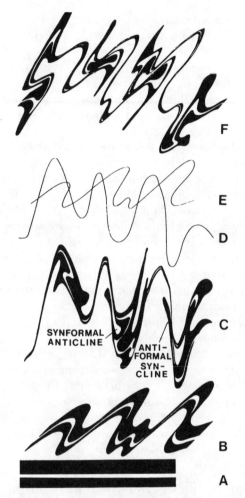

Fig. 49. These strata have been folded twice, with the same trend, but different attitude. A shows the strata before folding. B shows these strata after the first folding according to the pattern E. C shows the refolding of B according to the pattern D. F shows the different result if the D folding preceded the E folding. Note that in similar folding, the thickness of any bed is never greater than, but usually much less than, the original thickness of the bed.

Folding 221

separate lensoid bodies shaped like cow-pats or sausages and called boudins (French *boudin*, pudding or sausage). The highly complex appearance of such deformation, the presence of boudins, and the fact that beds are commonly stretched to paper thinness, gives the impression of gross vicelike compression, and as such deformation is common in the core zones of fold belts, where higher temperatures have homogenized the strength characteristics of the rocks, the notion that such fold belts implied intense horizontal compression and crustal shortening was reinforced.

Fig. 50. Complex folding does not necessarily imply narrowing. If the flow lines diverge (as they do when material is squeezed upward to regions of lower pressure), the folded strata may occupy greater width than before folding. In this diagram the width has doubled, but the area of the folded beds (and volume if the third dimension is added) is exactly the same as the unfolded beds below, because the complex folds can be analyzed as a series of parallelograms, each having twice the width between flow lines and half the thickness along the flow lines.

But this notion is quite false. There is no shortening whatsoever in Figs. 47 and 49. These deformation patterns only indicate strong *flow* in the direction of the flow lines, which coincide with the axial surfaces of all the little folds and crenulations, and with the cleavage and lineation. If the flow lines diverge in the direction of flow (as is common in the upward direction in the cores of fold belts), these fold patterns actually *increase* their horizontal width during the deformation. To make this clear, Fig. 47 has been redrawn in Fig. 50 showing that the identical fold pattern has been produced while the horizontal width has doubled. If the separation of the flow lines doubles, the thickness of the beds in the direction of flow is halved. The area of each bed in the cross section in the upper deformed state in this figure is exactly equal to the area of the same bed in the undeformed state below.

Concentric Folding Implies Décollement at Depth

The concentric folds of Fig. 48 are continued downward in Fig. 51 with strict adherence to the rule that beds maintain their thickness throughout. This illustrates three important rules governing concentric folds: (1) all concentric folds die out at depth; (2) the underlying basement is not shortened; (3) the length along beds through one or a series of folds is constant while the centers of curvature fall outside the section but become substantially shorter as each center of curvature (indicated by numbers) is passed. Accordingly all concentric folds are underlain by a *décollement* or surface of separation (French *décollement*, coming unstuck).

The décollement may be a single surface, as in Fig. 51, where there is a weak bed, or it may be distributed between several weak beds, or in a repetition of strong sandstones and weak mudstones all the mudstone beds may share the adjustment. In the concentric folds of the Jura Mountains, the décollement occurs in yielding beds of salt. In the concentric folds of West Virginia, the décollement zone is a soft black shale above the Oriskany Sandstone, while further south, in the Valley and Ridge country, it is in soft Middle Cambrian shale of the Rome and Conesauga Formations.

Similar and concentric folds may be compared and contrasted, as in Table 3. These models are ideals in which, in the case of similar folds representing deformation of physically homogeneous material, the bedding offers no constraints on the manner of yielding to unbalanced stresses, and in the case of concentric folds the beds are so

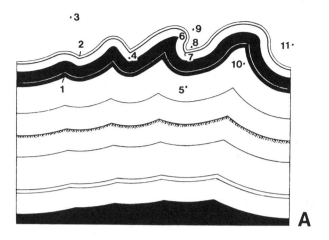
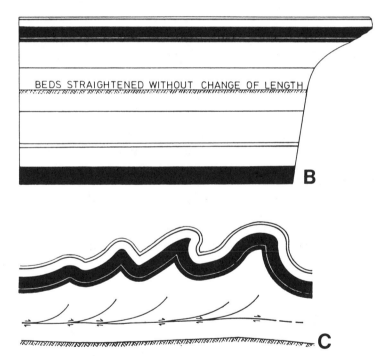

Fig. 51. Because the thickness of each bed measured at right angles to the bedding remains constant, concentric folds consist of a series of tangential circular arcs. In A, the centers of these circular arcs are indicated. Below any such center, the beds become cuspate. B shows the length of each bed in A. The upper beds occupy a shorter length than before folding, but the basement length has not changed. Concentric folding implies yielding beds below (such as salt or weak shale) to allow the upper beds to glide over the basement as they are folded. C shows such a décollement, or surface of detachment.

TABLE 3
Similar and Concentric Folds

Similar folds	Concentric folds
Isotropic deformation	Anisotropic deformation
Shear direction not related to bedding	Shear parallel to bedding
Bedding thickness constant in direction of flow	Bedding thickness constant orthogonally
Length of bed around fold increases greatly	Length of bed around fold constant
Fold persists indefinitely down fold axial surface	Fold dies out down axial surface
No implication of décollement below	Implies décollement below
No implication of crustal shortening	Implies superficial shortening not involving basement
Implies transport in direction of axial surface which proceeds from depth	Superficial transport in direction at right angles to axial surface

different in their physical properties that all slip is parallel to the bedding. Real folding is distributed along the intervening spectrum. But neither category, nor a combination of them, necessarily carries implication of shortening of the underlying crust.

The terms concentric folds and similar folds were defined in 1896 by C. R. van Hise. More recently the terms flexure folds and shear folds have been substituted for concentric and similar folds respectively, and have gained wide currency in America. However, as both types of folds involve shear to a comparable degree (see Fig. 47), and both are flexures semantically and in common English, these terms should be dropped on grounds of priority and precision, and also of ambiguity, because "flexure" was defined in 1876 by John Wesley Powell for the bend between two plateau surfaces, the usage followed by such eminent authors of structural geology textbooks as Bailey Willis, C. M. Nevin, and E. Sherbon Hills.

CHAPTER 17

Diapirs

DIAPIRIC (from Greek διά, through, and πείρω, pierce) means that material from below pierces through overlying rock, often right to the surface. In this sense a volcano would be a diapir, because lava and shattered rock fragments (called tuff) are forced to the surface; but "diapiric" usually refers to intrusion of plastic or solid material, and the injection is not violent but slow, like the flow of a glacier.

Salt Domes

During several geologic periods, beds of salt many hundreds of meters thick were deposited when extensive inland seas, or seas with restricted exchange with the ocean, evaporated, and the salt deposits were later covered by other sediments that formed shale and sandstone. The subsequent intrusion of the salt is best understood by regarding both the salt and the overlying sediment as highly viscous fluids, and that the driving force causing them to flow and form salt domes is the differential weight due to the density difference between the salt and the surrounding sediments. The energy for the intrusion is thus gravitational potential energy.

The density of rock salt is about 2.2 (relative to water) and increases very little by elastic compression under load. The density of the associated sediment increases rapidly as it is buried, from less than 2 at the surface to nearly 2.35 at depths of 4000 m (Fig. 52). Salt is therefore denser than the sediment at shallow depths, but they balance at around 500 m, while at greater depths the sediment becomes progressively more dense than the salt, so that if a salt bed is buried to a depth of 4000 m, it is overlain by 3500 m of sediment denser than the salt, with a further load of 500 m of sediment above that. This is inherently un-

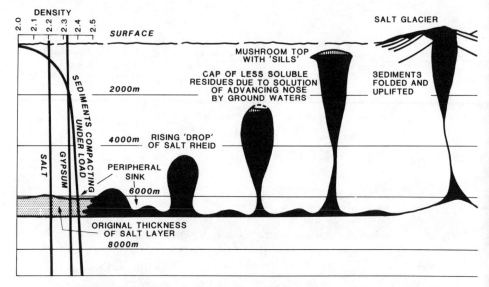

Fig. 52. Crystalline salt rises through the heavier, overlying sandstone and shale with a pattern similar to the rise of a layer of thick oil through water.

stable. It is much the same as having a layer of stiff oil below a layer of water: the center of gravity of the whole system is lowered as the oil rises slowly as drops or columns through the water.

This is exactly what the salt does, forming roughly cylindrical columns from about a kilometer to 5 or 6 km in diameter, which intrude vertically upward for as much as 6 km. Nearly 200 such salt domes are known in the coastal plain of the Gulf of Mexico and offshore, more than 100 in the Emba district around the Caspian Sea, and many others in Rumania, Germany, Iran, and elsewhere.

When such salt domes reach within about 500 m of the surface, the density of the sediment above is less than that of the salt, so there is a tendency for them to spread laterally toward a mushroom shape or to form sills or overhangs, as less work is done in arching the lighter sediments than if the salt continued to rise (Fig. 52). However, the weight of a full column of salt right to the surface is still less than a column of sediment to the same depth, so there is still energy to drive the column upward as well as sideways. In limestone, which is much denser than salt, the column may pierce right through to the surface, and as the salt column weighs less per square meter than a square meter of the limestone, the salt may continue to flow upward until its source at the bottom is emptied. The salt then flows away on the surface as a "salt glacier" (Fig. 53).

When the top of a nascent dome has risen 100 m above the general level of the salt bed, the difference in pressure at the dome compared with the rest of the bed is about 10 tonnes per square meter, but by the time it has risen 3000 m, the difference in pressure between the base of the dome and the surrounding salt bed is more than 200 tonnes per square meter. Hence the dome increases its speed of ascent as it rises, and fractures begin to appear in the surrounding strata and in the beds above the dome but not in the salt itself, because its viscosity is much less.

The leading crown of the dome breaks through successive strata, all of which are saturated with water, some of them with high permeability. These waters continually dissolve the top of the rising salt and leave insoluble impurities, which accumulate as a cap. At the very top is the least soluble impurity, calcite, below that the next least soluble, anhydrite with some calcite, next, gypsum with some anhydrite and calcite, then the ordinary salt with thin bands of gypsum and anhydrite, and occasional grains of calcite.

The flow pattern in a salt dome is very similar to a convection cell (Fig. 54), which circulates for exactly the same reason—a rising column of hotter (and therefore less dense) fluid, and a sinking zone of colder (and therefore more dense) fluid. Above the cap of the dome the circuit is radially outward, so here we find tensional patterns with radial faults and rift troughs (Fig. 55). This radial flow adds to the spreading of the salt described above to produce overhangs and overthrust structures like horizontal nappes overriding steeply overturned strata (Fig. 56). Surrounding the dome is a zone of subsidence, because the salt originally below it has flowed into the rising dome. If the salt bed were originally 1000 m thick, the whole of this could be driven into the dome, and hence the potential subsidence of the sur-

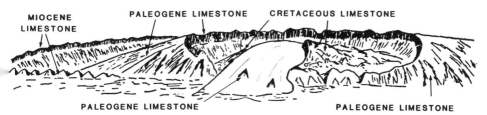

Fig. 53. Salt "glacier" extruding from the Kuh-i-Anguru salt dome, Iran. (From a sketch by Dr. G. M. Lees.)

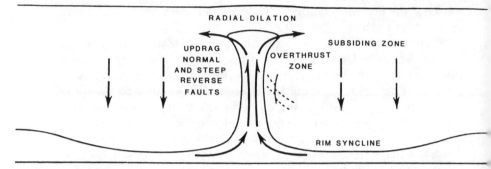

Fig. 54. A salt-dome toroidal circuit is caused by the same kind of density inversion as a thermal-convection toroidal circuit, so the movement pattern is the same.

face is as much as that. However, salt domes usually commence rising while sedimentation is still going on, and the surrounding depression tends to get filled as it grows, so the surface depression does not usually reflect the full amount of subsidence. The resulting "rim syncline" surrounding the dome therefore has more depth in the lower strata than higher up. This downward flow is part of the "convective-cell" circulation pattern, which is completed by the convergent flow of the salt into the dome, and the flow up the column.

The structure of the Heide salt dome is shown in Fig. 56. The overhang at the right has been driven more than a kilometer into the Senonian sediments, and it has dragged the Triassic sediments up from 1000 m or more below, turned them over, and thrust them over the highest Cretaceous sediments. In a mountain belt, orthodox geology would interpret such an overthrust of inverted Early Triassic over the youngest Cretaceous as evidence of crustal compression. Certainly there is local lateral compression, but it is superficial, and a secondary by-product of the salt dome, driven vertically by gravity. There is obviously no crustal shortening whatsoever.

The internal structure of this dome is well known thanks to extensive potash mining. The salt marked with stipples was the first salt to rise. This was then injected by the unpatterned salt beds, and finally the salt shown in black was driven through all the earlier salt, like toothpaste being injected into soft clay. During the whole of this time, all the salt beds were a coarsely crystalline metamorphic rock, with each salt crystal distorting and recrystallizing countless times during the flow. This continuous recrystallization is similar to that which oc-

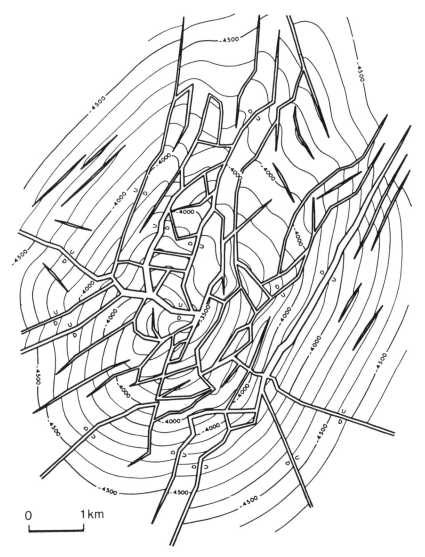

Fig. 55. Radial tensional faulting in the strata above a salt dome. The thin lines are subsurface contours at 100-foot intervals on the Eocene Woodbine Formation above the Hawkins salt dome in Texas. Numbers are subsurface depth; U and D signify upthrown and downthrown sides of faults, respectively. (Redrawn from T. J. Parker and A. N. McDowall.)

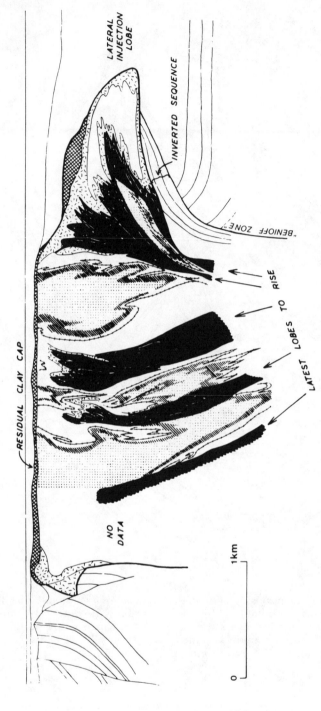

Fig. 56. Internal structure of the Heide salt dome, Germany. (From Dr. Alfred Bentz.)

Diapirs

curs in a flowing glacier (which is also a metamorphic rock) and in the iron crystals when a steel billet is drawn into a wire kilometers long. This is also precisely how gneiss and schist are injected into the cores of mountain belts.

In plan, the folding in a salt dome is extraordinarily complex (Fig. 57), but in the walls, scarcely any folds or contortions can be seen, because the direction of flow is vertical. For an analogy, stack a lot of very tall dunce's hats to form a single high column; squeeze them in and out so that they are strongly crenulated by vertical ribbing; now cut them across horizontally, and you have a simulation of the very complex plan of a salt dome; cut them vertically, and the pattern is simple with the cut edges of the hats running up in parallel lines, each inside its neighbor. A map like that of Fig. 57 could be matched in many places in the deeply eroded cores of any ancient orogen. There is a dominant east-northeast trend in the southern part of the map, veering more northeasterly in the southeast corner. In the northern part of the map the trend is north-northwest. This clearly suggests an area of overprinted (that is, refolded) folding with intense compression from the north-northwest and northeast. The map on the right shows the surroundings of this map on a smaller scale, which confirms these two conflicting trends nearly at right angles to each other. In the northeast corner a northerly trend starts to appear, which dominates the area just to the east of this map. The dips of the strata are consistently steep. In plan there is intense crenulation on all scales from a few millimeters up to the limit of the exposures. The crestal traces of the crenulations on all scales and the mineral lineations vary little from the vertical. These are the flow lines.

Fig. 57. Plan of folding of strata in the Grand Saline salt dome, Texas. The regional setting of the large rectangle is shown at right. The top is north. (After Professor William R. Muehlberger.)

There is no doubt that in a field of gneiss and schist we might conclude that these were strongly overprinted structures that have suffered intense horizontal compression in not less than three orogenies. But how wrong we would be! The horizontal compression in this area is nil. The direction of flow is along the lineations and fold axes, normal to the page, not transverse to them in the plane of the page. This map is part of a level plan in a mine in the Grand Saline salt dome, Texas, mapped by Professor William R. Muehlberger.

Gneiss Domes

Domes of crystalline gneiss mantled by schist are common in orogenic belts of all ages. They pierce upward much in the same way as salt domes, and they commonly have mushroom tops (Fig. 58). In old mountain belts like the Caledonides of Norway, they are so deeply eroded that the complete shape of structures can be seen. Recently, however, Professor C. D. Ollier of the University of New England and Dr. C. F. Pain of Sydney described a group of such gneiss domes in southeast Papua that are actively rising now and have risen several kilometers above the surface in recent times, shouldering aside the surrounding rocks (Fig. 59). They are 2000 to 3000 m high and tens of kilometers across, more elliptical than circular. The most perfect example is Goodenough Island. (The figure number originally in this quotation has been changed here to Fig. 59.)

Goodenough is one of the most mountainous islands in the world: the highest point is over 2500 m, yet the island is only 35 km by 25 km, and the high part of the island is a dissected dome only 20 km by 16 km, bounded by

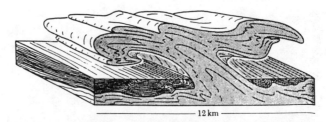

Fig. 58. Reconstruction of the mushroom-shaped diapiric core of migmatite in the early Paleozoic (Caledonide) orogen in Greenland after removal of the overlying eugeosynclinal strata. Migmatite was originally sedimentary strata, but was permeated by superheated water above the critical temperature so that feldspars, quartz, and other minerals crystallized in it, producing an intermediate stage between sedimentary rock and granite. (After J. Haller.)

Diapirs

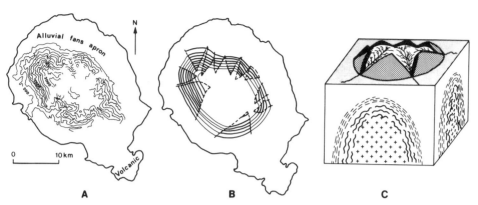

Fig. 59. Goodenough Island gneiss-dome diapir, which is still rising now. (From C. D. Ollier and C. F. Pain.) See text for details.

alluvial fans and a volcanic area to the south. Figure 59-A is a contour map of the island, and like Figure 59-B, it brings out the dome-like nature of the center of the island. Drainage is roughly radial, with steep-sided valleys debouching on to the outer plains and depositing huge fans. The valleys intersect the dome surface and leave triangular facets of little-dissected remnants of the original dome surface. These are analogous to "planezes", triangular facets of original volcano surface on a dissected volcano, or to "flatirons", the triangular facets formed on dipping sedimentary rocks when V-shaped valleys cross them. We shall refer to them as "dome facets" or simply "facets". Their slopes are rather even at gradients of about 20°. Some individual facets are over 2000 m high. The generalized contour map of the dome is based on the contours on the facets which are easily recognised in the field and on air photos as well as on contour maps.

Geologically most of the dome consists of quartzo-feldspathic gneiss with minor amounts of amphibolite and calcic gneiss, and small intrusions of granodiorite in the center. All the metamorphic rocks are of amphibolite facies. The foliation of the gneiss near the dome surface is parallel to that surface, both inside the dome and in the outer rock. Strike directions on the major gneissosity planes therefore describe circles concentric to the dome outcrop, and there can be little doubt that the two are causally related. We have found tight folding, shearing, and boudinage close to the dome surface, which implies very considerable deformation and shows that the gneissosity is not simply parallel to the original bedding.

The dome has been driven upward by the buoyant force of granite below, as indicated in Fig. 59-C. As with the stretched wire, the glacial ice, and the dome salt described above, the quartz and feldspar and other minerals of the gneiss have been continuously recrystallizing during the solid-state flow of the intrusion.

Glacier Analog

It is interesting to compare the movement pattern of such a dome with that of the great Malaspina Glacier of Alaska where it flows out over the coastal plain (Fig. 60). The width of the glacial lobe is about the same as that of the Goodenough dome, and the foliation pattern of the dome matches that of the glacial lobe.

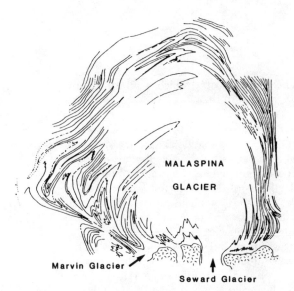

Fig. 60. When the Seward and Marvin glaciers flow out of the mountains onto the southeastern Alaskan coastal plain, they unite as the Malaspina Glacier. This map shows the pattern of flow developed by the moraine ridges. (After R. P. Sharp.)

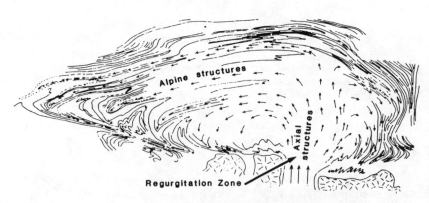

Fig. 61. The Malaspina Glacier flow pattern is shown compressed 3 to 1 as though its outflow was resisted, to simulate a diapir being compressed by its weight. Compare this with the actual pattern of the Alps.

The Malaspina flow pattern also resembles that of the Alps, but as the scale of the Alps is five times as large, and as gravity severely flattens upward movement, Fig. 61 shows the Malaspina lobe foreshortened by a factor of 3 for comparison with the flow pattern of the Alps. We now have a relatively narrow axial zone of "roots," whence the "nappes" (which means sheets in French) rise steeply out of the "pre-orogenic trough." One by one they turn over and flow outward to the "northwest" as great nappes, most of them overriding above, a few driving into the backs of others ahead of them. The outermost structures of the "Pre-Alps" are the farthest traveled. Some lobes, fewer in number, turn over toward the "southeast."

If not true in detail, the Malaspina pattern certainly contains the fundamentals of the Alpine structures. Yet orthodox geologists tell us that the Alps represent great shortening, and that the overthrust belt has been compressed to one-eighth of its original width. But we know that the Malaspina lobe produced similar geometrical and structural patterns while it was being dilated to 30 times its width! Can there be any question that the structures of the Alps could be produced by the upward extrusion of the contents of the geosyncline, spreading laterally where the pile lacked lateral support? The jaws of the extruding zone may have remained fixed while the material was squeezed out, or they could themselves have dilated during the process. In that case, the crust would have widened during the formation of the Alps.

Tectonic Diapirs

Apart from demonstrable oroclines where an orogen has been bent, orogens are so frequently composed of circular arcs that a general cause is implied. The Italian geotectonicist Professor Forese Wezel of Urbino has called them krikogens (circle generators). Their common characteristics are: (1) a sector angle between 50° and 80°, and radius between 1000 and 2000 km; (2) convexity commonly eastward, or less commonly equatorward; (3) location commonly in a zone between continent and ocean; (4) high heat flow from the mantle and a backbone of andesitic volcanoes; (5) recent uplift of several kilometers with folding of the strata; (6) a trench on the convex side, with a Benioff seismic zone; and (7) a stretching area within the arc, often to the extent of appearance of a small sea basin, perhaps with residual horsts of prekrikogen rocks.

These characteristics are consistent with interpretation as diapirs. The radius is determined by the depth of generation of diapiric material and the effective thickness of the tectonic crust to be pierced. The vulcanism and high heat flux indicate ascent of molten material and general upward motion of the diapir, which brings up the isotherms. An extensional zone within the arc is normal above a diapir (Figs. 52–54). The axis of the diapir is under the extensional area. The ultimate result is to sweep the central area free of continental rock exposing an oceanic core. The Benioff zone is the boundary of the diapir against the surrounding crust. (Compare this with the diapir boundary marked "Benioff zone" in Fig. 56.) An ideal diapir would be circular and symmetrical, but the vertical motion combines with regional dynamics. None of the preceding examples (the Heide salt dome of Fig. 56, the salt glacier of Fig. 56, the outflow of the Malaspina glacier, or the gneiss domes) are symmetrical or complete a circle. Commonly the eastward rotation of the earth is the helm that steers tectonic processes. The Tyrrhenian Sea, Po basin, Pannonian basin, Aegean Sea, and Black Sea are examples of such krikogens, where rising tectonic diapirs have spread the bordering trends and brought mantle material to the surface. The so-called "back-arc basins" along the Pacific margin of Asia are also tectonic diapirs.

CHAPTER 18

A Simple Model of an Orogen

THE STAGES in the development of an orogen are depicted in Fig. 62. The initiation of primary stretching in continental crust leads first to necking or thinning, until the top and bottom surfaces of the lithosphere converge toward zero at some 5 km below sea level. By the time the continental crust has thinned to zero, the mantle below it has already risen 30 km. Thus, although the surface of the continental crust subsides steadily, the bottom and the mantle diapir below it steadily rise, and the mantle continues to do so throughout orogenesis.

Orthodox compressional theory agrees that during the geosynclinal stage of orogenesis the crust must be stretching, because otherwise there is no possibility of maintaining even approximate isostatic balance through the millions of years involved in this stage. The fact that after the melting of the ice sheets of the last ice age, isostatic balance has been nearly completely restored in only 10,000 years in the cold inactive crust of Canada and Scandinavia, implies that in the much hotter profile of an orogenic zone, isostatic balance would be closely maintained during this extensional stage. But orthodox tectonics reverses from crustal extension to gross shortening. By contrast, in the expansion model, extension persists through all stages, and the subcrustal diapiric motion is upward at all times.

At the stage when the continental crust has thinned to zero (middle diagram of Fig. 62) there are two contrasting sites for deposition of sediment. In the central zone is a deep *eugeosyncline* with an active and unstable floor, many faults and rifted troughs, and volcanic magma coming up from the hot rising diapir below. Sediment accumulates rapidly without much reworking by wave action, which otherwise breaks up rock fragments and minerals not strong mechanically or easily weathered. Earthquake jolts trigger flows from the slopes of

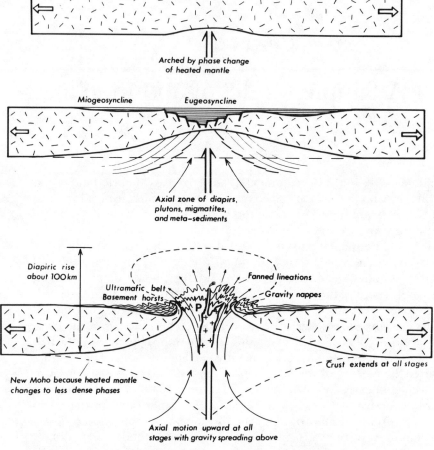

Fig. 62. The development of a geosyncline and orogen involves continuous crustal extension and continuous diapiric rise of about 100 km.

sand- and silt-laden water (turbidity flows), which settle out with characteristic patterns. Whole slabs of sediments, several hundred meters thick and several square kilometers in area, creep slowly downslope, gliding on deeper sand layers that have become quicksand because of excess water pressure between the grains, and sometimes become contorted in the process. I discussed such phenomena in the 1963 International Symposium on Syntaphral Tectonics at Hobart, where John Elliston described in detail several field occurrences and the processes by which these sediments became rock.

These eugeosynclinal sediments contrast with sediments in the *miogeosyncline,* a gently sagging zone at the side of the orogen proper (Fig. 62, middle diagram). The floor of the miogeosyncline subsides because the continental crust below it is thinning, and isostatic balance therefore requires that the surface fall below sea level, deepening in the direction of thinning, that is, toward the main trough. The floor is much more stable than that below the eugeosyncline; temperatures are lower, with no vulcanism; seas are shallower, so wave action winnows out rock fragments and weak mineral grains; and marine life is abundant, so limestone is common.

A ridge of basement commonly rises as a rim separating the miogeosyncline from the eugeosyncline. This is because the weight at the eugeosyncline from the surface right on down is less than normal, and although isostasy is in the process of correcting this by the upward flow of the diapir, the upward drive is transferred in part laterally by the strength of the crust, so the crust adjacent to the eugeosyncline is uplifted. (Compare this with the raised rims of rift valleys in Fig. 19.) This basement ridge separates two distinct facies. On the eugeosynclinal side, not only are there contemporary volcanics, and injections of granite during the subsequent folding, but also serpentine belts and ophiolites, rock types derived directly from the mantle, which become intimately interlarded with the eugeosynclinal sediments.

Folding and Thrusting

After the stretching of the crust has thinned it to zero, the continuing and accelerating ascent of the deep diapir begins to drive out the new sediments, regurgitating the geosynclinal gut, which then spreads laterally at the surface. Consider the point marked P in Fig. 62, in the middle of the rising orogen at the level of the former land surface. It is being driven upward by the rising diapir below, but pressing down on it is the weight of the pile above. The two vertical forces squeeze it to spread sideways. The rate of spreading depends on the size of the overburden load and the viscosity of the materials. If this rate is less than the rate of the rising diapir, the orogen must rise higher, increasing the overburden load at P (at the general surface level away from the orogen). The height of the orogenic zone continues to rise in this way until the sideways flow matches the rate at which the diapir is pushing up from below. The surface of the orogenic zone then ceases to rise, but the orogen continues to spread laterally, like the Malaspina glacier analog (Fig. 61), and will continue to spread as long as the diapir rises, piling nappe on nappe.

Within the orogenic zone, the lineations and thrust surfaces are all in the direction of the flow, the inner zones always overriding their flanking neighbors, very steeply near the center, but becoming flatter and flatter outwards, as nappes override the miogeosyncline. The soles of these nappes may even come to slope downward like the sole of the salt glacier in Fig. 53, but just as in that example, if the sole is traced back to its source it originates from the steep upward drive from the diapir. Observe, in the bottom diagram of Fig. 62, that the thinned-out edge of the original continental crust is turned up and overturned to overthrust the miogeosyncline, where its driving pressure and the pressure from higher nappes overriding it push the miogeosynclinal sediments forward, resulting there in more thrust sheets and trains of concentric folds.

The Myth of Alpine Foreshortening

Orthodox geologists, who believe that the stacking of nappes on nappes, each with tens of kilometers of travel over its sole, proves that the Alps were formed by several hundred kilometers of crustal shortening as the two sides were squeezed together, might ask, where is all the material coming from to account for the observed stacking of nappes? Let us look at actual numbers involved.

According to the expanding-earth model, the volume of the interior is increasing. The outer crust (lithosphere) is relatively cold and brittle, so it ruptures into polygonal areas a few thousand kilometers across (as described in Chapter 20) to accommodate the increased volume. These ruptures include the mid-ocean rift zones where the seafloor crust is growing, but the orogenic belts are also part of this spreading system. In Fig. 20, the orogenic belt through Europe and southwest Asia is necessary to complete the African polygon, the Himalayan orogenic zone is necessary to complete the Indian polygon, and the Indonesian orogenic belt completes the Australian polygon. All of these polygon boundaries are zones where material from the interior extrudes to increase the surface area of the globe to fit the new volume.

The root zone of an orogenic belt (Fig. 62) would be something like 100 km wide; to be conservative, let us assume only 50. The vertical rise of the driving mantle column would also be something like 100 km, but again, to be conservative, let us assume only 50. So the rising ram

A Simple Model of an Orogen

of the orogenic axial zone would be 2500 square kilometers in cross section. If this extruded at the surface (like the salt glacier of Fig. 53) as a sheet 1000 m thick, the sheet would flow out for 2500 km. (The overthrust sheets in orogenic belts vary in thickness, but 1000 m would be about the average.) Instead of being just one sheet, the sheet buckles or fractures after flowing less than 100 km because of bottom friction, and the rest of the sheet overrides the first part, and this continues so that nappe stacks on nappe, or is thrust into or below earlier nappes. But the cumulative sum of travel would still add up to the 2500 km. This is three or four times the total nappe overthrusting observed in the Alps, which according to orthodox dogma implies crustal compression by at most 700 km. How wrong can they be? Far from crustal shortening of several hundred kilometers, the Alps probably represent crustal widening of a few tens of kilometers.

The outflow of nappes is relatively fast, perhaps one-thousandth of the speed of glaciers. Nappes could advance 500 km in a single geologic epoch; the speed would vary greatly according to the excess water pressure at the sole and the kind of material there—shale, evaporites, limestone, and so forth.

In the folded frontal belt, all the manifestations of horizontal compressional tectonics develop. The miogeosynclinal zone *is* foreshortened. The microscopic fabric of minerals *do* conform to compressional patterns. Folds *are* flattened transverse to the axial surfaces of the folds. But none of these phenomena involve shortening in the basement, nor narrowing of the orogenic belt as a whole, which actually widens during orogenesis.

The Orogenic Root

The boundary between the earth's crust and the mantle is defined by a sudden jump in the velocity of seismic compressional waves from about 7 km/s in the lower crust to about 8 km/s at the top of the mantle. This boundary is formally called the Mohorovičić discontinuity after its discoverer, the Croatian seismologist Andrija Mohorovičić (1857–1936), but commonly contracted to "Moho." In the top diagram of Fig. 62, the Moho is the base of the continental crust. In the bottom diagram, it is shown as a dashed line plunging down under the orogen to a depth of several hundred kilometers. The material in the rhomboid region above this new Moho was originally below the Moho, but it has risen a hundred kilometers or so with the diapir. (As a guide to the amount of vertical rise, a thin dashed line completes a broad arch to connect the back-turned edges of the base of the continental crust.)

This risen material is still relatively hot but under very greatly reduced pressure, so it undergoes a phase change, usually from dense eclogite (or equivalent material) with a seismic velocity above 8.0 to less dense gabbro with a seismic velocity well below 8.0 (or other similar phase change). A considerable volume of water in the gas phase (that is, above its critical temperature) is also introduced that causes serpentinization, reducing the rock's density and seismic velocity. So a seismologist reports that there is a deep "root" under the orogen. Likewise, a geophysicist measuring the gravity field finds densities of about 3.0 extending down deep under the orogen instead of the

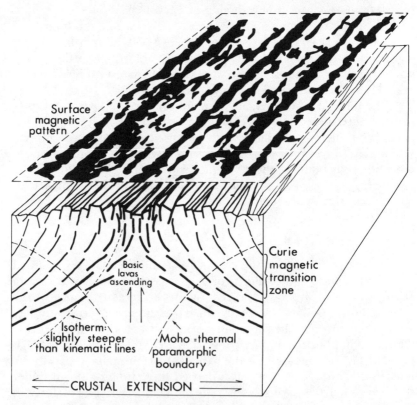

Fig. 63. Except for the absence of continental crust and abundant sediments, an oceanic spreading ridge is similar to an embryonic orogen. The diapiric root is genetically identical. The black stripes are positive magnetic anomalies recorded by ships' traverses.

A Simple Model of an Orogen

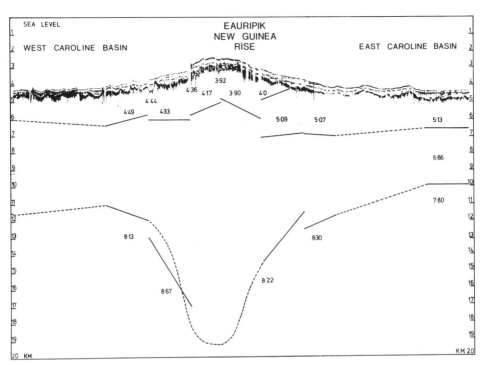

Fig. 64. The diapiric root of a second-order polygon ridge (see Chapter 20) on land or at sea is identical with that of a spreading ridge or orogen, but of less degree. The numbers are velocities of compressional seismic waves in kilometers per second. The straight lines are seismic-refraction velocity boundaries. The surface structure is from seismic-reflection continuous profiles. The lowest boundary (partly dashed) is the Moho. (After Den et al.)

higher densities of the mantle, so he confirms that the orogen has a deep root.

Those who believe that the orogen is the locus of intense crustal compression interpret the depression of the Moho as evidence that the crust has been squeezed together and doubled in thickness to form the root. But let us compare orogens with the mid-ocean spreading ridges (Fig. 63) and with the ridges of the second-order polygons (Fig. 64), which are introduced in Chapter 20. Under them the Moho (as determined by the compressional seismic velocity) plunges down exactly as in Fig. 62, and so does the density distribution (as determined by the gravity profile). There is no continental crust there to squeeze down to form a root, and in any case everybody agrees that these are zones of extension. In the subsurface, orogens and spread-

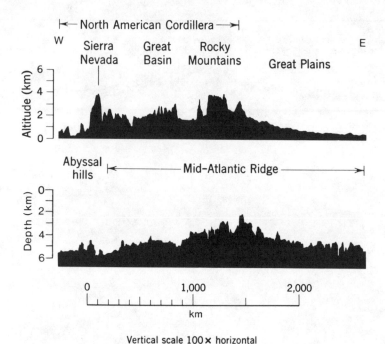

Fig. 65. Comparison of the topographic profiles of a mid-ocean spreading ridge and an orogen. (After Shepherd.)

ing ridges are identical seismically and gravity-wise. Both link up into the extensional mosaic of the earth's expanding surface. They differ because orogens incorporate a good deal of continental crust and a large volume of sedimentary material. Because the underlying distributions of mass below an orogen and an oceanic spreading rift are identical, their surface expressions are also similar (Fig. 65).

The genetic similarity of spreading ridges and orogens has been recognized by geologists not blinded by the compressional tectonics creed. Thus the eminent tectonicist L. P. Zonenshayn at the 1972 International Geological Congress crystallized the wisdom of many Russian geologists in stating that geosynclines, fundamentally, are the sites where initially new oceanic crust is created. This is the antithesis of the subduction concept. He concluded:

The origin of a geosyncline is accompanied by tension, by moving apart of lithosphere plates and by creation of a new oceanic floor. In some regions (South Mongolia, for example), one can restore ancient sea-floor spreading. Eugeosynclines appear to be analogous in their initial stage to ocean ridges,

A Simple Model of an Orogen   245

and in the mature stage to island arcs. On the whole, the evolution proceeds, possibly from conditions resembling mid-ocean ridges to an island arc and further to an orogenic zone. . . . Geosynclines (and eugeosynclines especially) are deep-rooted structures in which the energy and substance of the upper mantle rise up toward the earth's surface. In this process a new crust is created, including "basaltic" and "granitic" layers.

The granitic cores of orogens have commonly been assumed to extend down to great depths; indeed, the name bathylith or batholith (Greek βαθύς, deep; λίθος, stone) implies this. However, evidence has been increasing that true granite forms only the hood of such plutons. Recently, Donald C. Ross of the U.S. Geological Survey inferred, from a study of the most deeply eroded parts of the Sierra Nevada bathylith and the fragments brought up in the rising magma, a downward sequence from granite to 10 km depth, through tonalite (plagioclase granite). Beneath this, sillimanite replaces andalusite. The next underlying layers are migmatite (interlarded granite and gneiss, from Greek μῖγμα, mixed), then hornblende granulite to 25 km, then two-pyroxene granulite at 30 km. Such a sequence is a logical consequence of the paradigm of Fig. 62.

The model of an orogen is the simplest case. It is bilaterally symmetrical; but orogens commonly develop at the margin of continental crust against oceanic crust, with sediment supplied from only one side, so the resulting orogen develops asymmetrically, with nearly all the thrusting directed to one side. Furthermore, as will be detailed later, the rotation of the earth leads to asymmetry. There is also a progression of the rate of orogenesis through geologic time, so that ancient orogens may differ in some characters from later ones. The model considers only one orogenic cycle of sedimentation and folding, whereas several such cycles may follow each other. Although each orogenic zone has its own individuality, the basic principles outlined above apply to all. Bearing this in mind, let us now compare three orogens, the Appalachians, the Alps, and the Himalaya, with the basic model.

The Appalachians

In the mid-west of the United States, the pre-Appalachian basement crops out, as in the bottom diagram of Fig. 62. Proceeding east we come to the flat-lying strata of the miogeosyncline in the Allegheny synclinorium, which thicken progressively until we reach the concentric folds, then the nappes of the Valley and Ridge zone, which are thrust forward on flat or gently dipping thrust surfaces. Next we find the basement rocks of the Blue Ridge, which are thrust over the miogeosynclinal strata, just as in the model. Beyond that are the complexly folded rocks of the eugeosynclinal facies, injected by granites and ophiolitic rocks. Next we come to the Piedmont, where the Appalachian rocks disappear under the younger cover of Mesozoic rocks.

The surface pattern of the Appalachians is consistent with diapirism but inconsistent with the vice model of compression between converging continents. The front against the miogeosyncline is bowed into a series of arcs. The overthrusting is most intense where the arcs bow outward (westward), and much less intense where the arcs bow inward toward the crystalline core. This is what should be expected from a chain of diapirs, but in the compression model, the overthrusting should be most intense where the craton advanced farthest into the orogenic zone.

Dr. Philip B. King, formerly of the U.S. Geological Survey and author of several authoritative books, reviews, and regional maps of the geology of North America, depicted four independent sediment cones spreading northwestward from the Appalachian axial zone (Fig. 66). In the Middle Ordovician, a tumor rose actively in North Carolina, and its rapid erosion shed a fan of marine clastic sediments 2400 m thick over Kentucky, while less than 1000 m accumulated else-

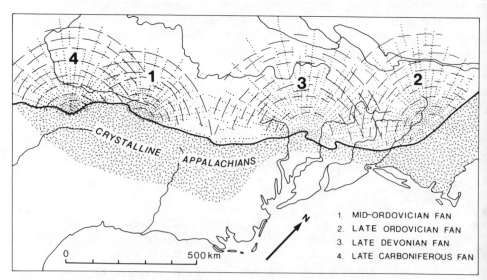

Fig. 66. Paleozoic diapirs and clastic wedges of the Appalachian orogen.

where along the Appalachian front. In the Late Ordovician, an active tumor regurgitated in Maine and spread a thick fan of sediment across Vermont and far into Quebec. In the Late Devonian, it was New York's turn, and a clastic fan up to 2700 m thick spread across West Virginia, Pennsylvania, and upper New York State while only 600 m was deposited in Tennessee and about 30 m in Alabama. But Alabama had its turn in the Late Carboniferous, when a regurgitating tumor rose in Georgia to spread a fan of sediments up to 3000 m thick.

This succession of events is logical in the diapiric model, but in the convergent-continents model a block of the foreland a couple of hundred kilometers wide would have to close on the Appalachians in the Middle Ordovician toward Kentucky, another would have to converge on Maine in the Late Ordovician, still another against New York in the Late Devonian, and another in the far south in Late Carboniferous time.

Even though the diapiric process may have been initiated more or less equally all along the trough, such an equality would be transient: any slight inequality that raises the isotherms is favored by feedback, because the rate of yield under a given load increases exponentially with absolute temperature. Hence diapirism always tends toward rounded tumors (called krikogens by Professor F. C. Wezel, of Urbino, from Greek κρίκος, a ring).

Observe that the excess thickness in the sedimentary cone around a diapir implies not only more rapid supply of sediment, but also more rapid subsidence of the floor. Three factors may contribute to this: (1) the process described earlier that produces the sequence of miogeosyncline, basement ridge, and eugeosyncline is intensified locally by the more active tumor, so the miogeosyncline deepens there; (2) the flow of material from below toward the diapir may cause subsidence corresponding to the rim syncline around a salt dome (Fig. 54); (3) the additional load of sediments in the cone adds isostatically to the subsidence.

The Alps

The Vosges, Black Forest, and Bohemian massifs represent the pre-Alpine basement. Spreading onto them are the flat-lying strata of the Permian and Mesozoic in a miogeosynclinal facies. Farther to the southeast, these rocks start to buckle in the trains of concentric folds of the Jura Mountains, which are underlain by flat-lying thrust surfaces as in the Valley and Ridge zone in front of the Appalachians.

The Swiss plain, with Lakes Geneva and Neuchâtel, intervenes between the Jura and the Alps. This is a sagging zone caused by the continuation of the regional extension represented by the Alps themselves. The van of the Alps are the Helvetic nappes, a stack of flat overthrusts of limestone and shale characteristic of the miogeosyncline facies; then we come to the upturned pre-Alpine basement, represented by Mont Blanc, the Aiguilles Rouges, and the Aar massif, which, as expected, signals a complete change of facies to the *schistes lustrés*, then the zone of roots, and the ophiolites of the Ivrea zone. All this is consistent with the model.

The Himalaya

Figure 67 shows two profiles across the Himalaya, the first by Augusto Gansser, the leading authority on Himalayan tectonics, and the second interpreting the same surface information in accordance with the orogenic model of Fig. 62. As in the model, we start from the plain

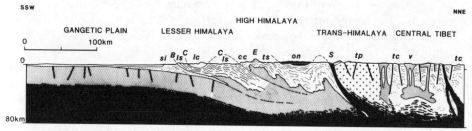

B BOUNDARY THRUST	sl SIWALIKS	ts TIBETAN SEDIMENTS
C CENTRAL THRUST	ls LESSER HIMALAYA SEDIMENTS	on OPHIOLITE NAPPE
S INDUS SUTURE	lc LESSER HIMALAYA KLIPPE	tp TRANS-HIMALAYA PLUTONS
E MT EVEREST	cc CENTRAL CRYSTALLINES	v VOLCANICS
	tc TRANS-HIMALAYA CRYSTALLINES	

Fig. 67. Comparison of compressional model and diapiric extension model of Himalayan tectonics.

of the Indo–Gangetic lowland where the Mesozoic and Tertiary strata of the miogeosyncline thin out onto the pre-Himalaya basement rocks and thicken toward the synclinorium of the miogeosyncline. The first folds across the Ganges Valley are the Siwaliks (so named because each dissected hill is said to resemble the wig of Siva, the great god of the Hindus). These are concentric folds in the youngest strata of the miogeosyncline, as always occur in this position. Continuing through the miogeosynclinal strata of the Lesser Himalaya, we quickly reach the "Great Boundary Thrust," where the mountains of the Lesser Himalaya are thrust forward. Thence we meet a succession of thrusts until we come to the "Great Central Thrust," where begins the High Himalaya.

The Main Central Thrust is where the pre-Himalayan basement first turns up to reappear at the surface, and is carried forward as a great nappe. Most of this has already been eroded away, so that isolated remnants of it lie on the miogeosynclinal sediments. Such an isolated piece of a nappe is called a klippe (merely the German word for cliff). A peculiar thing about these klippen is that they are upside down—they are less strongly metamorphosed at the base than they are higher up. This is as should be expected, because the dashed line of the thrust surface connects back to the thinned-out upturned basement, which is overturned so that the under surface is above the upper surface.

Because the Himalaya is such a great orogen, we remain in the miogeosyncline until the next major thrust is met at the Indus suture, where we change to the eugeosynclinal sediments. Mount Everest consists of Mesozoic limestone dipping north-northeast toward the deepest part of the miogeosyncline, which contains a conformable sequence of strata all the way from the Cambrian to the Eocene. Ophiolites, which belong to the eugeosynclinal facies, do not occur in this zone. Décollements are frequent, particularly near the crystalline floor, with much back-folding as in the calcareous Helvetic Alps. Deep rift troughs developed near the Indus suture zone during the early Tertiary and rapidly filled with sediment from the eugeosynclinal zone.

At the Indus suture we first find ophiolites and serpentinites, driven up in the rising diapir. On reaching the spreading zone, these are pushed laterally as a great nappe thrust for 100 km southwestward over the miogeosynclinal strata. The Indus suture has the long straight outcrop line characteristic of megashears, that is, where a whole continent has been displaced horizontally rather than pushed up and overthrust. Indeed this suture is the main megashear of the Tethyan torsion, which moved the northern hemisphere continents more than 1000 km westward with respect to the southern continents (of which more in Part Five). Hence Tibet formerly lay immediately north of Australia, and Afghanistan and northern Iran formerly lay north of the Himalaya.

In the zone southwest of the Indus suture, we find large exotic blocks of limestone that are quite foreign to that region; their source is therefore to be sought in Afghanistan. This long horizontal tearing along the Indus suture complicates the structure there, producing steeply dipping slices of various rocks, and frequent ophiolitic mélanges. *Mélange* is the French word for a mixture, and that is what the mélanges are here, a hodgepodge of blocks, large and small, some as big as city blocks, with ophiolite a common constituent. Such mélanges are a common feature of megashears; indeed, whenever mélanges occur along a long straight line of valleys, a megashear should be suspected.

Beyond the Indus suture we find a complete change of facies to the eugeosynclinal sediment of central Tibet, intruded by large bodies of granite, tonalite, and gabbro, and extensive Tertiary volcanics. Himalaya is a very young orogen that is still rising diapirically and thrusting out its nappes. It is nearly 400 million years younger than the Appalachian orogen. After the lapse of that amount of time, erosion will have cut 10 km or more off the highest parts of the Himalaya, reducing it to a peneplain, which will have arched up again to maintain isostatic balance because of its remaining less dense roots, and this second-generation epeirogenic Himalaya will have been dissected by new rivers. Himalaya will then resemble more closely the present Appalachians. I say "resemble" because the tempo of orogenesis has accelerated significantly, and late Tertiary orogenesis differs in degree from late Paleozoic orogenesis, although conceived in and delivered from similar orogenetic wombs.

CHAPTER 19

The Benioff Zone

THE BENIOFF ZONE is a sheet of earthquake foci that dips downward under an orogen at about 50° to depths of about 300 km, starting from an oceanic trench. Andesitic volcanoes commonly occur above where the Benioff zone reaches a depth of 120 km. Further earthquakes may occur down to depths of as much as 700 km, but these deep-focus earthquakes, although clearly associated with the normal Benioff zone, appear to be somewhat independent of it: they only occur in the most active regions, so long stretches of Benioff zones do not have them. A discontinuity also occurs down the dip of the zone; even where deep-focus earthquakes are present, a discontinuity or gap occurs below the normal Benioff earthquakes and the cluster of deep earthquakes. Incongruence may also be seen in plan view. For example, deep-focus earthquakes occur where expected down dip from the Bonin–Marianas trench, but the belt is less arcuate than the normal part of the Benioff zone, and continues on to cut across the orogenic arc and the trench at the southern end near Guam, and in the north continues roughly at right angles across the Honshu arc and trench and across the Sea of Japan, and after a short break reaches the Asian mainland near Vladivostok. This independence is strikingly apparent when the earthquake distribution is studied stereoscopically from computer-generated plots. Many of the deep shocks show transcurrent movement along the belt, which may account for its straightness. In contrast, the most active belt of deep-focus earthquakes anywhere conforms faithfully to the hooked pattern of the Tonga arc, although even there a pronounced gap is present between the normal Benioff zone of shallow and intermediate foci down to 300 km and the very numerous deep foci.

The expansion and subduction models agree that this is a zone of shearing fracture, with the orogen side moving *relatively* upward and over the oceanic side. This relative motion is confirmed by the sense of first motion deduced from the earthquakes that propagate from the fractures. But the subduction model claims that the oceanic lithosphere is thrusting downward beneath the orogen and has done so for thousands of kilometers. In contrast, in the expansion model, the oceanic crust is stationary and the Benioff zone is the boundary of the upward-thrusting diapir, which flares in a bell shape as it rises. The total upward motion in the center of the orogen is a hundred or so kilometers, not thousands, and the motion at the Benioff zone is only tens of kilometers at most. The question is, which side moves, the oceanic side or the orogenic side?

Continuous seismic profiling shows relatively thin, regular sediments on the oceanic side, monotonously undisturbed for thousands of kilometers, right up to the trench. But as the trench is crossed, tectonic violence of all kinds erupts: seismicity, thrusting, and gross slumping toward the trench repeatedly triggered as the continuous oversteepening of the rising orogen reaches slope instability. Heat outflow from the interior is consistently low on the oceanic side, but as the trench is crossed it more than doubles, and even increases locally up to tenfold. This situation is incompatible with the subduction model, in which the overall trench−orogenic arc system is where cold oceanic lithosphere is descending deep into the mantle. This requires that the total heat flux in the region should be much *less* than average. By contrast, the expansion model asserts that the whole orogenic zone is where hotter mantle material is forced upward and outward by the expanding interior; this requires that the total heat flux in the region should be significantly *more* than average, which is precisely what is found.

Everything in the orogenic zone moves *up*. The isotherms are tens of kilometers higher than average. In the axial zone, volcanoes bring up lavas from the partially melted mantle; magmas rise to crystallize in plutons; crystalline metamorphic rocks originally deep below the floor of the geosyncline are forced up a dozen kilometers or more, to crop out high in the mountains; serpentinites and peridotites, originally even deeper below these crystallines, do likewise; gneiss domes intrude diapirically upward into the geosynclinal strata. These diapirs pierce upward in the viscosity pecking order displayed in Fig. 46— gneiss domes, migmatite diapirs, magmatic plutons, volcanic lavas, and *nuées ardentes*. Overthrusts are numerous, always with the orogen side riding up over the trench side, on very steep thrust surfaces near the axis of the orogen but on progressively flatter slopes farther from the axial zone.

The Benioff Zone

What forces all these bodies up? Not buoyancy, because except for the magmas they are denser than the rocks they penetrate, and they are driven to higher altitudes than their density would warrant. They are forced up by the ascending mantle below them, which changes phase to less dense forms as higher temperatures are brought to levels of lower confining pressure.

From the axial zone right out to the Benioff zone is a single complex diapiric unit. The Benioff zone is the boundary of the ascending diapir against the stationary oceanic lithosphere. The boundary of a salt diapir is sharp because of the difference in the "viscosity" of salt and intruded strata, but the boundary of an orogenic diapir is gradational, because the resistance to deformation ("viscosity") is not a discontinuous step but diminishes with temperature. The axial zone ascends fastest because the temperature is highest there, and this is maintained because the ascent itself continues to bring up hotter material. The pressure driving the diapir is great enough to fracture rock anywhere in the orogen, but in the central zone the rock yields by flow at lower stress-difference than the fracture stress; at lower temperature high in the orogen, thrust fractures start to appear even in the central zone. Laterally from the axial zone, stress-difference has to build up progressively higher to force flow as fast as the diapir requires, and eventually the stress-difference reaches fracture level before relief by flow. The resulting zone of fracture is the Benioff zone. It is the ultimate boundary of the diapir. The distribution of earthquake foci reflects this pattern. At shallow depths, earthquake fractures occur right across the orogen, but contract to the narrow Benioff zone progressively with depth.

Imagine a cylinder maintained at red heat pressed against a mild steel plate by a hydraulic ram. The steel in contact with the hot cylinder would slowly yield, but cracks would appear in the cold peripheral ring—the Benioff zone. Compare this with a bar of hot toffee, clamped horizontally in a vice, and with a heavy weight hanging on the projecting end. The toffee bar bends down immediately. Repeat this with a similar bar, not quite so hot, and hang on the same weight. Again the bar bends, but more slowly. Repeat with a still cooler bar; again the bar bends down, but still more slowly. Repeat with a cold bar and bending is scarcely noticeable before it snaps off, if the weight is greater than the strength. This is what happens at the Benioff zone, the outer boundary of the orogenic diapir.

Where the upward driving force is great, and the rate of ascent of the diapir relatively fast, the Benioff zone of fractures goes deeper to higher temperatures. (A bar of warm toffee can be broken if bent rapidly, so that it breaks before it can bend.) Where the diapir driving force is less, the rate of upward movement can be accommodated by flow in relatively cooler rocks, so the Benioff fractures only go down a couple of hundred kilometers, even though the ascending diapir originates much more deeply than this.

The rocks just outside (beneath) the Benioff zone are held in a state of elastic strain, at a stress level just below what would break them. They are the constraining wall of the diapir. Shear fracture could not occur without such a constraint, because an unconstrained shear stress produces rotation, not shear fracture as in the Benioff zone. Experiments on rocks held under shear stress below fracture level show that they transmit sound and seismic waves 5 percent faster than when not stressed. The seismic velocity in the zone underneath the Benioff zone has been found to be higher than normal. This has been interpreted as evidence for a cold descending lithosphere slab, whereas such anomalous velocities are inevitable in the constraining rocks of the ascending diapir, just below the threshold of fracture.

In a salt dome, the outer boundary surface between the rising diapir and the strata that are only dragged upward, the surface on which the major stratigraphic discontinuity occurs, is the equivalent of the Benioff zone (Fig. 56).

Seismograms from a number of observatories of the same earthquake enable determination of the direction of motion at the point of rupture and differentiation between transcurrent, extensional, and shear failure. Except for shallow earthquakes, ruptures in the Benioff zone are shear failures. This has been interpreted as indicating crustal compression, incompatible with a generally extensional regime. However, the bell-shaped boundary of a diapir must have this character, with the diapir side riding up over the static side. However, it has already been pointed out (Fig. 19 and associated text) that even in an extensional regime, tensional ruptures are only possible at shallow depths where the weight of the overburden is less than the shear strength of the rock, because at any greater depth shear rupture must occur before the regional extension can reduce any stress to zero to induce tensional failure.

Professor William Tanner, of Florida State University, has demonstrated both empirically and experimentally that diapirs only occur in extensional regions.

The orogenic arcs of East Asia, particularly the Japanese arcs, which are the type areas on which the Benioff concept was founded, differ from the paradigm of Fig. 62 in three fundamental ways. Firstly, they

The Benioff Zone

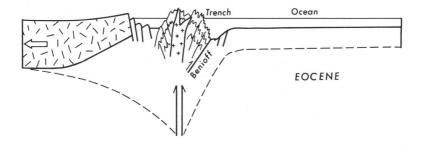

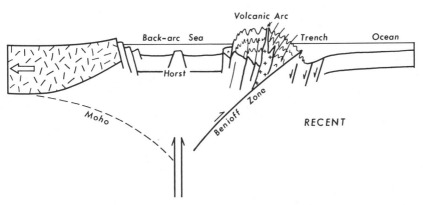

Fig. 68. Development of East Asian back-arc sea by asymmetric scouring of the axial zone by mantle diapir.

are polycyclic, with a Tertiary orogenic cycle overprinted on a late Mesozoic cycle, along a zone which had been active far back in time to the Proterozoic birth of the Pacific. Secondly, the Mesozoic and Tertiary orogens were intrinsically asymmetric, in that the western flank was continental lithosphere, whereas the eastern side was oceanic. Thirdly, the rate of orogenic diapirism became exceptionally fast (Fig. 68).

The orogenesis model of Fig. 62 starts with normal continental lithosphere, which is progressively thinned to zero and replaced by the orogenic diapir. If the rate of ascent of mantle-derived material increases with further more rapid extension, the orogenic complex may be swept aside, leaving only this simatic material above the axial zone. The result is a nascent ocean basin with a central spreading ridge.

The bottom of the Benioff surface where it can be traced deeply enough lies under the small tensional seas (which 30 years ago I called "disjunctive seas" and recently were called by plate tectonicists "back-arc basins"), for example, the Sea of Japan. Professor Forese Wezel of Urbino has emphasized that the axis of the Cretaceous orogen lies here, but as the diapir continued to bring up material derived from the mantle, a stage was reached where the central zone of the orogen consisted only of mantle material, so that a new sea was born with a floor of mantle-derived (oceanic) crust. The earlier "continental" parts of the orogen were largely swept aside during the Paleogene like surface scum over a convection cell, although residual horst-slices of it remain, as in the ridges in the center of the Sea of Japan. If the African rift valleys were expanded by such a process until oceanic floor widened between the outer rims, the Ruwenzori horst would be left as such a ridge in the center of the new sea.

Wezel has pointed out that Eocene to Oligocene tensional faulting dominates the coasts on both sides of these disjunctive seas, with Miocene molassic sediments unconformably overlying them. Because the Sea of Japan marks the swept central area of the diapir, the heat flow from the mantle is much higher there than normal, commonly in excess of 2.4 heat-flow units.

Recently a French oceanographic team led by Guy Pautot on the *R.V. Jean Charcot* has identified the extinct spreading axis for 500 km in the South China Sea by the inward-facing fault scarps (which trend N 50° E ± 10°) and the outward tilting of the blocks.

As pointed out in connection with the Appalachian clastic fans (Fig. 66), an orogen does not grow uniformly along its length, but tends to form diapiric foci some 600 or 700 km apart. If conditions were symmetrical, funnel-shaped diapirs would form ring-shaped orogens at the surface (Wezel's krikogens). But the intrinsic asymmetry of the lithosphere has resulted in the tendency for the continental lithosphere to move westward with respect to the oceanic lithosphere. Hence, the spreading ridges did not grow symmetrically, because crustal increments were inserted on the west side of the spreading ridge, rather than equally on both sides, as in the standard plate-tectonic model. The result is a line of basins, with rifted continental crust on the western side, a basin floor of oceanic crust which has grown from west to east, and an orogenic arc on the eastern side.

Diapiric salt extrusion should ideally be symmetrical also, but the salt glacier of Fig. 53 only flows one way. Similarly, the spreading lobe of the Heide salt dome (Fig. 56) spreads only in one direction. We shall see later that there are more compelling fundamental causes of asymmetry related to the earth's rotation. With the exception of the

Aleutians and Sunda arcs, which are convex equatorward for similar reasons, all Benioff arcs, including the Antilles and Scotia arcs, are convex eastward, and orogenic overthrusting tends to be eastward, not only in the East Asian orogens, but also in the Cordilleran orogen.

Intra-continental spreading, such as the African rift valleys, continuing as the Red Sea, and the south Atlantic and south Pacific, are all essentially symmetrical because in these cases the controlling conditions were symmetrical: initial rift in cratonic continent, through to continent facing continent across oceanic crust with a median spreading zone.

From the foregoing it is clear that orogenic zones are genetically identical with the misnamed "mid-oceanic spreading ridges," which are not always mid-oceanic (I will show later that they never were mid-oceanic in the North Pacific), and when joined with the orogenic belts are seen to be really circumcontinental.

The question is often asked, why do orogenic zones mainly occur at margins of continents? This concept itself is inherited from earlier times before the mobility of continents had been recognized. When Pangaea is restored, the great Tethyan orogen cuts the Pangaean megacontinent in half, and the Caledonian–Appalachian orogen developed within the contemporary continent to separate America and Africa–Europe. We will see later that the Cordilleran orogen was born as an intracontinental orogen to separate east Asia from North America, and Australasia–Antarctica from South America (see Fig. 97 in Chapter 22).

Once an orogenic zone has developed, it remains a hotter weaker zone right through the mantle, and subsequent expansion of the earth tends to be concentrated there. Thus the Mesozoic North Atlantic mainly followed the recently active Caledonian–Appalachian orogenic zone, although rotated from it by a transcurrent phenomenon which will be discussed later. The Cordilleran orogen was born about a thousand million years ago as a major rift zone separating Asia and North America and has continued to widen since, developing the Pacific Ocean. The Tethyan orogen has remained a weak zone, along which gross sinistral displacement has occurred (to be discussed later), but also along which the Mediterranean and Caribbean seas have opened.

Another question asked is why does the Atlantic have no current orogen if orogens and oceanic spreading ridges are genetically the same? The same question could be asked for the Southern Ocean between Australia and Antarctica. The answer lies in the rate of diapiric spreading. We will see later that the rate of Earth expansion rapidly increased during the late Mesozoic and Tertiary. With such rapid expansion, the spreading zone is soon carried beyond reach of residual continental crust and rapid sedimentation, so that although the mantle diapirism proceeds exactly as before except that the rate is faster, the summit of the ridge rarely reaches sea level, so neither a nearby continent nor self-cannibalism contributes sediments. Vast outpourings of lava (all submarine), seismicity (greatly reduced because of the high temperature right to the surface), a typical seismic and gravity root (Fig. 64), deposition of base-metal sulfide ores, and topographic relief on a similar scale as on a continental orogen (Fig. 65) all occur. But there are no granites or andesites, which are typical of orogens adjacent to continents, because these require the presence of sediments and remnant continental rocks for their generation. The absence of a Benioff zone, or any earthquakes of intermediate focal depth, and shallow earthquakes mainly confined to transverse fractures, would seem to indicate that the expansion is largely filled by basaltic lavas.

PART FIVE
TECTONICS OF THE WHOLE EARTH

CHAPTER 20

Global Extension

ONE OF WEGENER'S criticisms of the contraction theory was that if the fold ranges of the earth's crust were due to cooling shrinkage of the interior, the pattern should be uniformly distributed, like the shrinkage of a shriveling apple, which it wasn't. The same argument should apply in reverse to the pattern of Earth expansion, when the brittle crust responds to the expanding interior (Fig. 69). Indeed, the global expansion pattern is uniform.

The Hierarchy of Expansion

The earth's surface is made up of eight continental polygons, each a few thousand kilometers across, that meet each other along tectonically active zones (Fig. 20). The plate theory calls them plates, about 100 km thick. I believe them to be polygonal prisms, extending down for the full depth of the mantle, 3000 km. I ask, what is the primary inhomogeneity of the earth? Surely, a fluid core of nearly half its radius, overlain by a crystalline mantle. Equally surely we should expect that this primary inhomogeneity would find the most obvious expression on the surface, as indeed it does.

The primary polygons, 3000 km thick, are a few thousand kilometers across. If the crystalline mantle were only 1000 km thick, we would expect the surface to express this with 20 or so primary polygonal prisms. Observe on Fig. 20 that each primary polygon (including the Eo-Pacific, formed more than 100 million years ago, called by some the "Darwin plate," which has acted as a continent) consists of a continent surrounded by its accreted oceanic crust that has been added to it mostly during the last 100 million years. If that growth is removed, the polygons are reduced to about half their size. This sug-

Fig. 69. Uniform distribution of crustal extension.

gests that the crystalline mantle was then only 1500 km or so thick, which agrees with the rate of expansion implied by other data.

I next ask, what is the second-ranking inhomogeneity of the earth? Surely the presence of the asthenosphere, a tectonically weaker, more yielding zone that separates the 100 km or so of stronger lithosphere above from the deeper mantle below. Surely we should expect the presence of this inhomogeneity also to be expressed at the surface, as the next most prominent feature after the primary polygons, and of course this is so (Fig. 70). The primary polygons are patterned by second-order polygonal basins and separating swells. A critic once objected that such basins and swells as in Fig. 70 are confined to the continents. Are they? Figure 71 shows that this pattern extends over continental and oceanic lithosphere throughout the earth's surface without any significant difference.

Looked at globally, by far the greatest share of the heat flowing out of the earth emerges at the actively spreading ridges between the primary polygons, coming up all the way from the core. Away from the spreading ridges, the rate of heat flux diminishes. Likewise the second-order polygons: the highest heat flux is along the swells. Indeed, that is precisely why they are swells. Most mineral substances exist in different forms according to temperature and pressure. Thus silica exists as quartz under shallow conditions, as coesite deeper, and as much denser stishovite at still greater depths. Carbon crystallizes as graphite at the surface and as denser diamond at high pressure. The rock called gabbro, made of feldspar and augite at moderate depths,

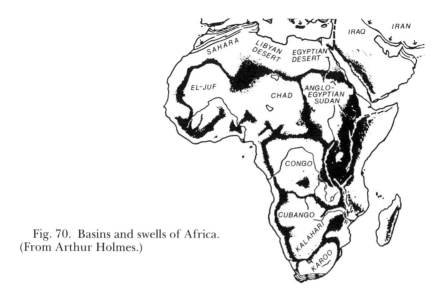

Fig. 70. Basins and swells of Africa. (From Arthur Holmes.)

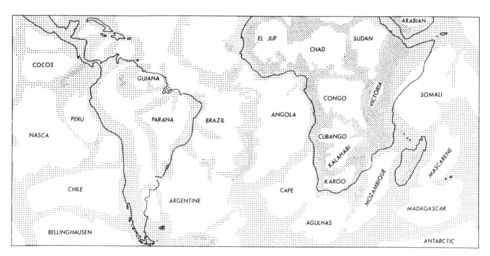

Fig. 71. The second-order pattern of basins and swells extends similarly over continental and oceanic crust.

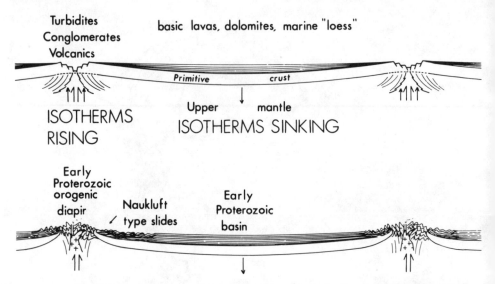

Fig. 72. Early Proterozoic pattern of passive basins bordered by active diapiric swells, which may develop into rifts through the crust, then into orogens.

changes to a much denser mix of garnet and jadeite known as eclogite. Temperature has the opposite effect to pressure. So if two places have rocks of the same composition below them, but one is hotter than the other, the depth at which each of the density changes occurs will be deeper at the hotter site, which means that it swells upward like rising dough. Thus the swells rise higher because the temperature below them is higher, and the basin floors remain low because the temperature below them is lower (Fig. 72).

The boundaries of the primary polygons have by far the greatest share of the world's earthquakes. But the second-order swells are also seismically active, which shows up when the distribution of the thousands of very small earthquakes is plotted. In some cases the fault fracturing associated with swells develops into major rift valleys, like the rifts that frame the Lake Victoria basin in Africa.

As the earth expands, the first adjustment to the decreasing surface curvature occurs at the primary spreading ridges, but if this were all, the curvature within the primary polygons would remain too much. The lithosphere is not strong enough to support a major departure from isostasy, so the surface of the primary polygons adjusts to average out the new curvature by a fracture pattern, the dimensions of which are determined by the thickness of crust down to the asthenosphere. This is the mechanism behind the second-rank basins and swells. Even so, the unbalanced weight to be borne by the rocks would still exceed their strength, so adjustment continues on down the hierarchy of extension to the third rank, the fourth, and so on.

The thicker lines on Fig. 73 outline polygons some tens of kilometers across, which Japanese workers have found to jostle independently during earthquakes, and which are the locus of low-level seismic activity. Within them are a still lower rank of polygons, 5 km or so across, which tilt independently and have a still lower level of seismicity. The hierarchy reaches down to the master joints, a few hundred meters apart, and within them in turn the systematic joints, by which all rocks are broken, and which allow the final adjustment to the changing curvature of the earth's surface and to any other stresses the lithosphere may suffer. All regions of the earth's surface (except the youngest sediments, which still yield rather than fracture) acquire two sets of joints nearly at right angles and with their intersections nearly vertical. These are the epeirogenic (continent-forming) joints, which adjust to the decreasing curvature of the earth's surface.

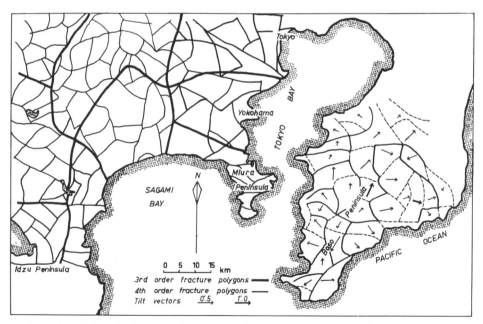

Fig. 73. Tilt blocks in central Honshu (from Miyabe, 1931). These are third- and fourth-order polygons in the hierarchy of crustal extension.

Relative Movement of Primary Continental Prisms

According to the plate-tectonic theory, the continental units are "plates," relatively thin (100 km or so) compared with their surface area, and they move horizontally large distances relative to each other, detached from the underlying mantle. In contrast, the expansion theory regards the continental units as polygonal prisms some 3000 km thick, nearly as thick as they are wide, and continents still rest on the same mantle as they did at the outset; their separation is due to the growth of new oceanic crust between them as the earth has expanded. Which is right?

Klaus Vogel, an engineer in the German Democratic Republic, is one of a group of people in many countries who have fitted all the continents together on a globe nearly half the size of the equivalent earth, assuming that continental crust originally enclosed the whole earth but was broken and dispersed as the earth expanded to its present size. But Vogel went one step further. He enclosed his reconstructed globe inside a transparent outer globe, to show in one model the primitive in relation to the present earth (Figs. 74, 75). Vogel found what Schmidt and Embleton found much later: the separation of the continents occurred when they moved *radially* outward as the earth expanded.

Drs. P. W. Schmidt and B. J. J. Embleton, two Australian paleomagneticians, were investigating the "polar-wander paths" of different continents during the Proterozoic Eon. By moving continents around (together with their polar-wander paths) until the polar curves coincided, they expected to identify the relative positions of the continents during that time period. To their astonishment (they believed in the plate theory), they found that the polar-wander paths for North America, Greenland, Africa, and Australia coincided (within the error limits of the technique) with these continents *in their present widely separated positions on the globe*! This suggested that their relative angular positions with respect to the earth's center now were about the same as they were more than a billion years ago, their present separation being due to their movement out radially as the earth expanded. Schmidt and Embleton reported this unexpected result in the 1981 *Journal of Geophysics*: "A geotectonic paradox: Has the earth expanded?"

This, of course, is the meaning of Dr. Parkinson's analysis of the NASA data (p. 169) that the increases of chord length between Europe and North America and between Australia and South America

Global Extension 267

Fig. 74. Vogel's primitive globe inside a transparent globe of the present earth, showing that the present separation of the continents is mainly due to their movement radially outward as the earth expanded. This should be compared with Perry's computer-generated calculations in Fig. 76.

are entirely accounted for by the 2.4 ± 0.8 cm per year radial extension indicated by the increase in the Australia to North America chord, without any "plate" motion whatever.

Dr. Ken Perry of Wyoming has demonstrated with geometrical precision what Vogel found with his globes. He set up a program based on matrix algebra and a hidden-line algorithm, so that continents may be moved out radially from the center of the earth, using one center of similarity and one rotation pole, and plotted on any desired projection (Fig. 76). His program generated successive positions of spreading ridges, fracture zones, and magnetic anomaly lineations, and

Fig. 75. Vogel (right) demonstrating his globe-in-globe to the author, Werdau, GDR, January 1979.

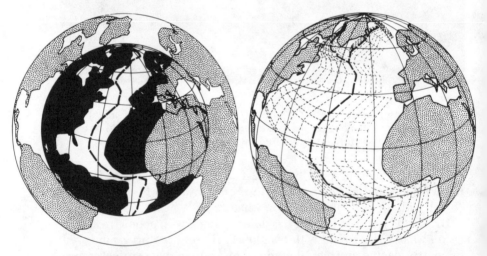

Fig. 76. Dr. Kenneth Perry's computer reconstruction of the opening of the Atlantic Ocean. Figure 74 was made by Vogel in his Werdau home; this figure was made by Perry in his Wyoming home by programming his computer to move the continents radially outward from their positions in Pangaea. Perry's inner globe has already been expanded by 25 percent. The right-hand globe shows the growth strips and present spreading ridge generated by the computer program.

from these he calculated the amount of radial expansion implied by each anomaly. Perry also demonstrated by direct computation that this geometry is compatible *only* with a radially expanding earth.

Mantle-Welded Continents

One of the surprises of the early 1960's was that the rate of heat flow through continents is statistically the same as through the ocean floors. Dr. V. V. Beloussov, of the Russian Academy, and Dr. Uwe Walzer, of the East German Academy, have each pointed out that this general equality of heat flux between continents and oceans contradicts plate tectonics. The radiogenic heat yield of continental rocks is known to exceed that of oceanic rocks by an order of magnitude. If continental lithosphere moved over passive mantle on a yielding asthenosphere, the heat yield below the asthenosphere should be generally uniform, so continental heat flux should be significantly greater than oceanic. On the other hand, if the continents derived their radiogenic elements by differentiation of the mantle fixed below them, the equilibrium heat flux would be more or less constant everywhere irrespective of the degree of differentiation. More recent work has intensified this paradox because the mean continental and oceanic heat flux has now been reported to be 60 and 91 milliwatts per square meter respectively. The difference is predicted by the expansion model, because most of the oceanic crust has risen some 30 km during the last 100 million years, bringing its higher temperature with it. This excess heat, which has a long half-life, is still dissipating at an exponentially declining rate and increases the normal heat flux.

Adam Dziewonski and John Woodhouse, two Harvard seismologists, recently investigated the transmission of large earthquakes, each of which had been recorded at 30 or more observatories, in order to measure the temperature distribution below the surface. They prepared global maps of the temperature at depths of approximately 100 km and 340 km. It was not surprising to find that higher temperatures continued down below the oceanic spreading ridges, including the new crust generated during the last 100 million years. But they did not expect that the colder temperatures below the continental nuclei would persist down through the 100 and 340 km maps, and even to the greatest depths studied, more than 500 km. This is exactly the prediction of the expansion theory, but is quite contrary to the plate theory, which requires the detachment of the "plates" from the mantle and free relative movement on a large scale at the asthenosphere. Even the older parts of the Pacific Ocean floor (the "Darwin plate" formed more than 100 million years ago, which thereafter acted like a continent) remain attached to the underlying mantle. Dr. Woodhouse commented: "That's still a great difficulty. Now it has to be thought about." The "great difficulty" is only for the plate concept.

Paul D. Lowman, Jr., of the Goddard Space Flight Center, has confirmed and emphasized the mechanical obstacles to the movement of continents with respect to their underlying mantle:

Recent studies indicate three problems with the concept of continental drift as an incidental corollary of plate movement: (1) Slab pull can not drive plates with continental leading edges, (2) There is no low-velocity zone under shields, and (3) continents have "roots" 400 to 700 km deep. These problems imply that if continental drift occurs, it must use mechanisms not now understood, or that it may not occur at all, plate movement being confined to the ocean basins.

Lowman is right, the motion *is* confined to the ocean basins on an expanding earth, with new crust inserted there while the separating continents ride passively on their own mantle, as in Vogel's model (Fig. 74) and Perry's computer reconstruction (Fig. 76). In retrospect, this validates Jeffreys's intuition that the horizontal sliding of plates was physically impossible; his error came from not realizing that the earth is expanding.

Vogel's globes, Perry's geometrical analysis, Schmidt and Embleton's polar-wander paths, the equality of continental and oceanic heat flux, and Dziewonski and Woodhouse's seismology-temperature distribution—five wholly independent techniques—all indicate that the separation of the continents has been caused by radial outward movement during Earth expansion.

CHAPTER 21

Global Torsions

THE CONCLUSION of the last chapter, that the continental blocks had moved radially outward by more than 2000 kilometers, has inescapable implications. Quite apart from any effect on the overall rate of rotation of the earth, a continental block like Africa that straddles the equator would have to increase its relative eastward speed by several hundred kilometers per hour to keep up with the daily rotation, in contrast with a polar continental block like Antarctica, which does not have to move any faster to get around. A block like South America would have to increase the eastward speed of its northern parts much more than its southern parts. The changes are very slow—only a centimeter or so per year—and we have already seen that for motions of this kind of speed, the whole of the crystalline mantle flows like a glacier. Clearly we should expect to find that substantial shearing motions had resulted. When a shear goes right round the globe, it becomes a torsion.

Theoretically, on a rotating gravitating celestial body, part fluid and part rheid (that is, subject to solid-state fluid flow on long time scales), various large-scale rotations could be expected, but rigorous mathematical analysis of them has not been mastered. Dr. W. S. Jardetzky, who, during the previous three decades, had written a series of relevant papers in the journals of the Serbian Royal Academy, the Vienna Academy of Science, the American Geophysical Union, and in *Science*, was one of the few mathematicians who might achieve this. Happily, Dr. Jardetzky approached me after my debate with Professor Bucher at Columbia University in 1960, and on returning to the University of Tasmania, I arranged a post-retirement fellowship for him to work with me on these problems. But sadly, he died just before taking it up. So I reverted to looking empirically for evidence of any such motions.

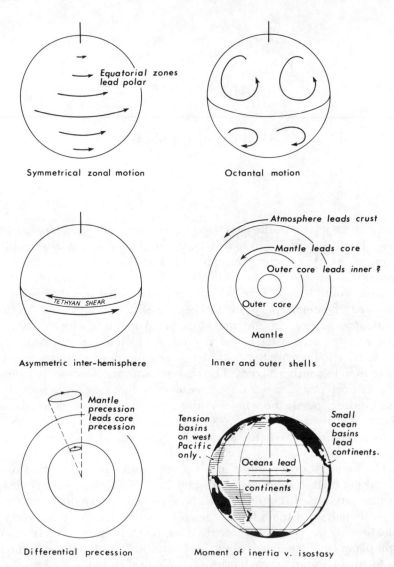

Fig. 77. Rotation modes within the earth.

Kinds of motions to be sought are (1) interhemisphere; (2) differential motion of continental and oceanic lithosphere; (3) octantal motions, as in the top right terrella of Fig. 77; (4) interzonal motions, such as equatorial zones relative to circumpolar zones; (5) differential motions between shells, such as between core and mantle, mantle and lithosphere, lithosphere and hydrosphere, and hydrosphere and atmosphere; (6) differential precession of shells; and (7) nonaxial rotation of the whole earth (polar wander) and perhaps others (Fig. 77).

1. Interhemisphere rotation has indeed occurred, and is the main theme of this chapter.

2. Gravity ensures that all regions of the lithosphere approach buoyant equilibrium, that less dense (continental) regions rise higher with respect to the geoid than denser (oceanic) regions do. Hence, because continents contribute more to the moment of inertia than oceans, continents tend to creep west while oceans tend to creep east. The consequences are discussed on pp. 285–86.

3. Octantal rotations have been proposed by several writers (Sakuhei Fujiwara, Pierre St. Amand, Hugo Benioff, Biq Chingchang, and others). Any block moving equatorward increases its distance from the earth's axis of rotation and so must increase its eastward speed or be left behind (that is, shift westward). This Coriolis effect is greatest in the high latitudes. Any block moving radially outward also has to increase its velocity eastward or lag westward, but as this effect is maximum at the equator, octantal rotations would ensue in the same sense as the Coriolis effect. However, most of the empirical rotations proposed by the above authors are not in the Coriolis sense, which implies a different cause (perhaps inertial disturbance by thermal bulges from asymmetric heat flux).

Ascending diapiric material, molten or solid, increases its distance from the center of the earth as it rises, so should tend to be left behind in the rotation compared with its new environment, that is, deflected westward. Hence a spreading zone might be expected to tend to add new crust more easily along its west side than the east, which would be synergistic with the inertial tendency described in the last paragraphs. As a result, active diapiric orogens would tend to add new crust on their western side, and with each new addition widen the new crust between the active orogen and the parent continent, as we see along west Pacific Asian coasts; in contrast, along the east Pacific coasts, successive orogenic axes through the Paleozoic, Mesozoic, and Tertiary have scarcely migrated relative to each other, and new crust has been inserted on their western flanks. Similarly, material being moved toward the equator would be deflected westward, while material moving poleward would be deflected eastward, which might show up as westward motion of the equatorial zone and eastward motion of the subpolar zone.

4. I have not observed interzonal motions such as those just sug-

gested or other zonal motions, although several authors (Jardetzky, M. Bogolepow, W. F. Tanner, W. N. Gilliland, T. H. Nelson and P. G. Temple, R. Dearnley, and G. E. Thomas, and others) have suggested them. Perhaps I am blinkered by my own creed. I expect the reason is that the primary expansion has broken the mantle and lithosphere into the primary polygons of Fig. 20 and the secondary polygons of Fig. 71, so that the effect of zonal motion is to rotate these blocks. But this does suggest streaming in the core and perhaps the mantle. Jardetzky has written that "there is no trace of mechanical equilibrium in any celestial body," and I must agree that on the time scale of geotectonics, Earth is wholly fluid except for an eggshell-thin crust. Certainly, eastward motion of the equatorial zone relative to the polar zones occurs on Sun, Jupiter, and Saturn. However, these motions, and those cited by Jardetzky, are all in the gaseous envelope, and such equatorial zonal motion does occur in Earth's atmosphere too, causing strong jet streams.

5 and 6. The evidence that the mantle overruns the core and the atmosphere overruns the lithosphere, and also the differential precession of core and mantle, were presented in Chapter 2 in connection with the earth's magnetic field.

7. U.S. Senator Estes Kefauver was said to have expressed apprehension that atomic explosions might tilt the earth's axis. Such anxiety can be allayed, because a million bombs favorably placed to be most effective and each releasing energy of 10^{11} megajoules might move the axis one centimeter! Nevertheless, there is abundant evidence that the poles have wandered and are wandering now. Apart from oscillatory wobbles, the north pole of rotation has migrated about 10 meters away from North America and toward Siberia during the last century. That may sound trivial, but if it were consistently maintained, Singapore would be at the north pole some 90 million years hence. This is not so long on the geological time scale. Paleomagnetic data leave no doubt that polar wander of this kind of rate has occurred commonly in the past.

North America has rotated through 90° with respect to the north pole during the last 400 million years (Fig. 78), and I suspect that this is due to a rotation of the whole lithosphere around Central America, of which more later. This could represent rotation of the lithosphere with respect to the mantle (which is excluded by the permanence of the primary polygons as I have maintained earlier in this chapter), or rotation of the mantle with respect to the core, or rotation of the whole earth. Earth must maintain her rotational momentum axis in

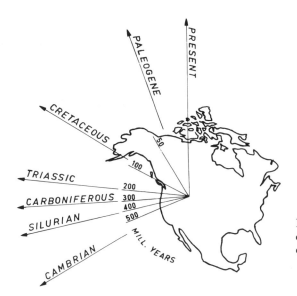

Fig. 78. Rotation of North America, as indicated by paleomagnetic data.

relation to the fixed stars, except as altered by extraterrestrial torque, such as the differential attraction of Sun and Moon on an asymmetric bulge. The axially asymmetric distribution of partial melting in the mantle to yield the vast volume of Jurassic basalts and dolerite must have caused a major tumor in Pangaea, which would have disturbed the moment of inertia notwithstanding isostatic equilibrium. This would have caused both axial wobble and slow migration of the pole until the earth rotated about the new axis of maximum moment of inertia. Likewise the greater growth of new ocean floor in the southern hemisphere, centered near the Falkland Islands (as described later in this chapter), would also have a similar effect.

Impact by an asteroid could give an external torque required to shift the earth's momentum axis permanently, but not by much. If Ceres, the largest asteroid with a diameter of 1025 km and a mass of 10^{19} metric tons, collided tangentially with the earth, the impact might change the axis by about 10°, and the next two largest, Vesta (555 km) and Pallas (538 km), could only manage a couple of degrees. The actual shift would in fact be much less than this, because most of the kinetic energy would be dissipated impulsively as heat, volatilizing and ionizing the asteroid and a comparable local mass of the earth to a plasma. So asteroid impacts are not the cause of the observed large angles of polar wander.

Before proceeding further, let us crystallize the meaning of some

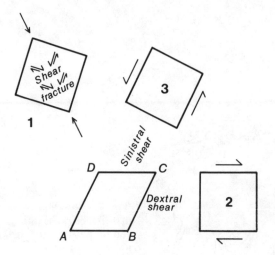

Fig. 79. The rhombus ABCD may have been deformed from a square by any one of these types of stress.

terms. In Fig. 79, the rhombus ABCD could have been formed by squashing square 1, or by a clockwise shearing of square 2, or by a counterclockwise shearing of square 3. When a block of concrete is being tested by compression like block 1 until it fails, both clockwise 2 and counterclockwise 3 shear fractures may develop together in it. Shears of this type are therefore called "conjugate" shears, meaning that they are "married together," and form together in response to the same stress. Instead of clockwise and counterclockwise, they are more commonly called "dextral" or right-handed and "sinistral" or left-handed (Latin *dexter, sinister,* right, left). Conjugate *stresses* are at right angles to each other, but resulting conjugate *fractures* intersect at about 60° because internal friction has to be overcome before fracture can occur.

Half a century ago, I realized that the earth had suffered sinistral shearing on a global scale, but saying so then branded me as a rat-bag because at that time transcurrent faults of even tens of kilometers were outside orthodox thinking, still less offsets of hundreds or thousands of kilometers. It was 1946 before Prof. W. Q. Kennedy demonstrated that 100 km of sinistral shear had occurred along the Scottish Great Glen, and 1953 before Mason L. Hill and Thomas W. Dibblee reported that 300 km of dextral displacement had occurred along the San Andreas shear and that the Garlock and Big Pine faults were its sinistral conjugates. It was not until the 1960's that such movements on the scale I had indicated were widely accepted.

In this chapter I will show that the earth has been subjected to two pan-global torsions, along two great-circle belts at right angles to each

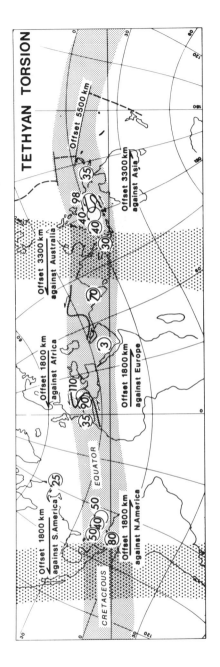
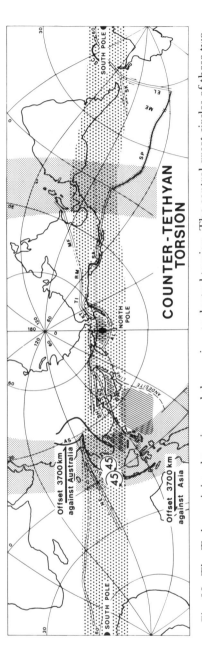

Fig. 80. The Tethyan sinistral torsion and the conjugate dextral torsion. The central great circles of these two Mercator projections are at right angles to each other.

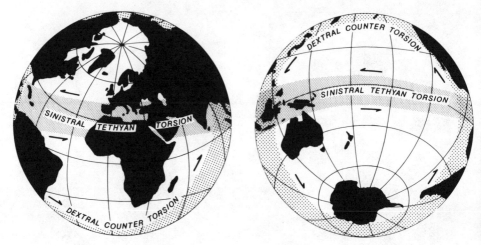

Fig. 81. Equal-area projections of the "land hemisphere" and "ocean hemisphere," showing the Tethyan sinistral torsion and the conjugate dextral torsion, mutually at right angles.

other, the sinistral Tethyan torsion, and the circum-Pacific dextral conjugate torsion. The Tethyan torsion was equatorial, and the circum-Pacific torsion was polar and separated the "land hemisphere" from the "oceanic hemisphere." They cross at right angles to produce the contused areas of the East and West Indies (Figs. 80, 81).

Tethyan Torsion

In 1938, after four years of fieldwork in New Guinea, I wrote in my doctoral thesis: "New Guinea has been sheared westward under a colossal shear system on a scale grander than has been demonstrated anywhere else on the globe. . . . The stresses which are responsible for this great westerly displacement are of continental dimensions. They are probably related to the main architectural pattern of the globe."

In the 1956 Hobart symposium, I traced this zone of sinistral torsion right around the earth. When the Americas are fitted against Africa as in Bullard's computer fit (Fig. 29), North America has to be moved 1800 km farther than South America, indicating an 1800-km sinistral offset through central America. When North America is fitted back against Africa, again according to the computer fit, Europe has to be pushed westward 1800 km with respect to Africa. Some 700 km of this shows up in unwinding the Mediterranean oroclines

(Fig. 14), to which must be added the offset across the Moroccan shear zone south of the Riff. When Pangaea is reassembled with the artifact gape closed, Australia has a 3000-km sinistral offset against Asia, as shown in detail by comparing the relative positions of Australia and China in Figs. 94 and 96 in Chapter 22. The "andesite line," indicated by the broken line in the western Pacific in Fig. 80, is offset sinistrally 5500 km from New Guinea to Samoa. Recently, C. J. Pigram of the Australian Bureau of Mineral Resources, and J. B. Supandjono, of the Indonesian Geological Research and Development Center, inferred displacement of the Sula terranes from eastern New Guinea to Sulawesi:

This examination shows that the stratigraphy of the Sula Platform does not correlate with any of the proposed sites of the origin for the Sula Platform in western Irian Jaya. However we find excellent correlation between the pre-Cretaceous stratigraphy of the Sula Platform and that part of the Australian craton found in Papua New Guinea between long. 141 and 145° E, which suggests that the Sula Platform was detached from a site 1200 km farther east than any previously proposed site of origin and implies a total displacement of more than 2500 km.

Right around the Tethyan girdle, blocks broken from the continents and local orogenic belts are *rotated sinistrally* by the torsional drag (see Table 2 on p. 113). Included are Spain (35°), Corsica-Sardinia (90°), Italy (110°), India (70°), Seram (100°), New Guinea (35°), Mesozoic Mexico (130°), Colombian orogens (40–60°), Greater Antilles (35–45°), and Newfoundland (25°). Each of these rotations was published by me and later confirmed by paleomagnetic measurements. (In the Malaysia–Indonesia archipelago, there are also counter-rotations from the conjugate circum-Pacific torsion, which will be discussed next.)

When an orogenic belt is dragged by such torsion, it forms an S-shape when dragged sinistrally and a Z-shape when dragged dextrally. Several sinistrally dragged oroclines occur around the Tethyan torsion zone: the Atlas of North Africa through Sicily and Italy and the Alps (Fig. 14); from the Zagros of Iran, through Baluchistan, Kashmir, and the Himalayas (Fig. 12); Sumatra, Java, the Banda loop, southeast and northeast Sulawesi, the Sunda spur to New Guinea; the loop from northern Venezuela, Trinidad, the Lesser Antilles, and the Greater Antilles is the southern half of such an S, the upper half being represented by a great sinistral megashear through the Cayman trench, and the Motagua and Clipperton shear zones. (In contrast, the Himalaya, Assam, Thailand, Malaysia, Sumatra double drag is Z-shaped, and belongs to the conjugate circum-Pacific torsion.)

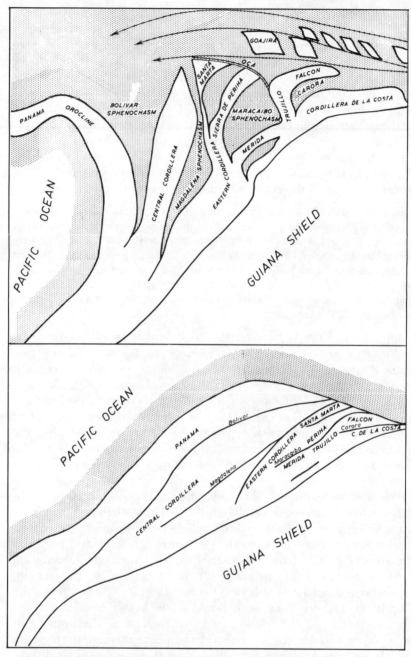

Fig. 82. Splaying of trends in Venezuela and Colombia. The bottom diagram reconstructs the position in the Early Cretaceous, before the drag of the Tethyan torsion.

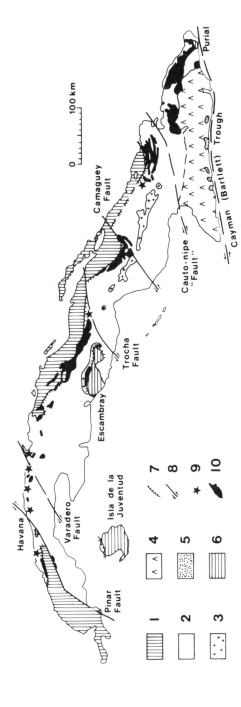

Fig. 83. Principal features of Cuban geology, excluding most Cenozoic sediments. (From Dr. M. A. Iturralde-Vinent.) 1, northern continental margin suite; 2, central eugeosynclinal suite on oceanic crust; 3, Late Cretaceous granitoids; 4, Paleogene volcanic center; 5, late Eocene granitoids; 6, southern metamorphic suite; 7, line of thrusts; 8, Paleogene wrench faults; 9, selected deep wells that found suite 1 underlying suite 2; 10, outcrops of the ophiolite associations.

The Tethyan torsion shows up in the gross structure everywhere. Figure 82 shows how the complex pattern of deep sediment-filled troughs and the bends in the grain of the pre-Mesozoic basement through northwestern Venezuela and Colombia are simply reconstructed to the condition of 150 million years ago by reversing the sinistral drag that spread the region like the fingers of a hand. Sinistral torsion dominates the structural pattern of Puerto Rico, with northwest-trending folds and thrusts, northerly tensional faults, and east-west sinistral transcurrent faults. Figure 83, prepared by Manuel A. Iturralde-Vinent of the Cuban Academy of Science, shows that before its disruption in the Tethyan shearing, Cuba was a typical orogen (compare Fig. 62) with miogeosynclinal sediments along the north, overthrust by ophiolites, and overridden in turn by the thick sediments of the eugeosyncline, which are intruded by the axial granites, the whole capped by terminal volcanics. Cuba was then dragged between anastomosing sinistral megashears of the Tethyan torsion, while conjugate dextral shears broke obliquely across it.

Likewise, in Fig. 14, when Europe is moved back westward with respect to Africa to where it has to be for America to wrap around Africa as indicated by the computer fit, the Spanish Peninsula rotates as a block, closing the Bay of Biscay; the Riff orocline unwinds to lie against the Huelva and Algarve coasts (although remaining in contact with the Moroccan coast because the Mediterranean has been greatly widened); the Ligurian sphenochasm closes so that Corsica and Sardinia lie back against the Côte d'Azur, while the southern end of Sardinia joins Minorca (so that the Riff, the Betic, the Balearic, Sardinia, Corsica, and the Ligurian Alps form one continuous belt parallel to the Atlas orogen of North Africa); both the Sicilian and Ligurian oroclines unwind so that the Atlas, Sicily, Calabria, the Apennines, and the Alps also form a single straight belt, combining with the Riff-Ligurian belt to form a single great equatorial orogen. All of these rotations have been confirmed by paleomagnetism since I announced them. Thus the single act of reversing the Tethyan torsion solves all the tectonic complexities of the western Mediterranean. Similar simple tectonic reconstructions apply in Indonesia, New Guinea and the Solomon Islands, and the Caribbean.

Transverse Extension Across Tethys

Plate-tectonic theory has insisted that Africa has been driven north against Europe, crumpling up the Alps, that India has been driven north against Asia, crumpling up the Himalayas, and that South

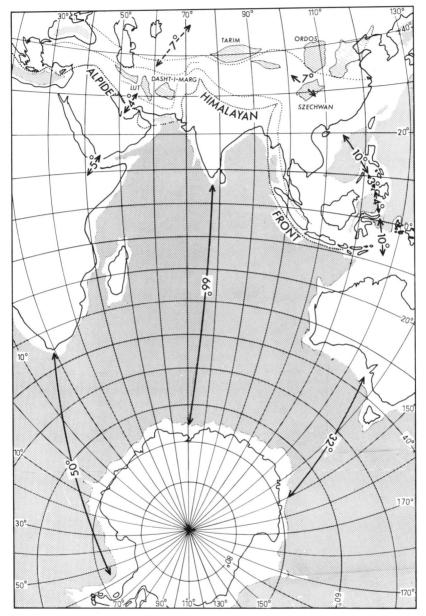

Fig. 84. Distributed extension between Antarctica and China via Australia.

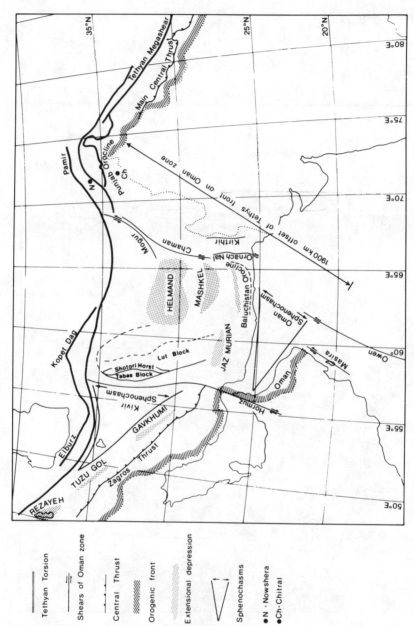

Fig. 85. Tethyan tectonics between Armenia and Tibet.

America has been driven north, crumpling up the Greater Antilles. I have already argued that the Alps and Himalayas developed by diapiric orogenesis in a zone of transverse extension, not compression. Reconstruction of the Mediterranean region indicates that this region has widened transversely by 700 km by the insertion of new oceanic crust (Fig. 14). Professor Tanner of Florida State University has estimated more than 5000 km of meridional elongation between North America and South America when their present position is compared with their positions in Pangaea.

The meridional extension between Australia and China has been 34° (3700 km), as shown in Fig. 84. This is distributed between a series of small extensional seas, including the Ordos Basin in China, which is of the same kind, but has been completely filled with sediments.

The extension between East Africa and central Asia is 22°, distributed between 5° in the Red Sea, 5° in the Persian Gulf, and 12° in the extensional basins (Jaz Murian, Mashkel, and Helmand) which are all now filled with young sediments (see Fig. 85). The 12° between Arabia and Turkmen also shows up in the 12° offset of the Tethyan orogenic front between the Oman Mountains and the Himalayan foothills near Jammu. There was only limited extension during the Himalayan orogenesis, and the whole meridional extension of 66° occurred between India and Antarctica. The extension between Asia and Antarctica is 66° irrespective of whether it is measured via Australia (34° China–Australia and 32° between Australia and Antarctica as in Fig. 84), via India (66° in one opening), or via Africa (22° Africa–Asia as stated above, and 49° between Africa and Antarctica).

The Cause of Tethyan Torsion

The Tethyan torsion seems to be due to the interaction of gravity and rotational inertia. The geoid is the shape of the earth indicated by sea level (as though we cut level canals across the lands to indicate it). The spin of the earth bulges it out at the equator, and many smaller effects, some static and some dynamic, pattern it. Gravity, the strongest force acting on the earth, maintains a close approximation to isostasy everywhere. Just as an iceberg rises to a level where its weight is equal to the weight of the water it displaces, so a continent rises to a level where its weight is equal to the denser underlying sima that otherwise would be there. The center of gravity of a continental slab stands about 2½ km higher with respect to the geoid than an equivalent slab of oceanic crust. Hence each continental slab contributes significantly more to the moment of inertia of the earth than an equivalent oceanic slab.

The interaction of gravity and rotational inertia due to the higher center of gravity of continents, which tends to drive continents west with respect to oceanic crust, shows up on a range of scales. The Pacific Ocean as a whole has tensional basins on the west all the way from pole to pole (Bering Sea, Okhotsk Sea, Sea of Japan, Yellow Sea, East China Sea, South China Sea, Philippine Sea, Sulu Sea, Celebes Sea, Banda Sea, Bismarck Sea, Solomon Sea, Coral Sea, Tasman Sea), but the east Pacific coasts have none. On the west Pacific coasts, the Asian and Australian continents tend to move relatively west, while the oceanic crust tends to move relatively east, causing a greater share of the general expansion to occur there; whereas along the east Pacific coasts, the American continents from pole to pole tend to move relatively west while the oceanic crust tends to move relatively east, hence a smaller share of the general expansion without extensional basins.

On a smaller scale, the Caribbean region, mostly oceanic crust, tends to move relatively east while the two American continents tend to move relatively west, hence the relative eastward drag of the Lesser Antilles arc. Similarly the relative eastward drag of the Scotia oceanic arc between the Antarctic and South American continents, and perhaps also the eastward bow of the Banda arc between Australia and Asia.

During the life of Tethys (essentially the last 200 million years), very much more new oceanic crust was inserted south of Tethys than north of it, and all continents (except Antarctica, which contributes little to the moment of inertia) moved significantly northward. Therefore the proportion of oceanic crust progressively increased in the hemisphere south of Tethys compared to the hemisphere north of Tethys. Hence the moment of inertia of the northern hemisphere progressively increased compared with that of the southern hemisphere. From this difference in inertial moment arose a sinistral torsion that operated along Tethys, with the northern side tending to lag in rotation (that is, creep westward) with respect to the southern side.

A Dextral Conjugate to the Tethyan Torsion

The dextral torsion around the rim of the Pacific was first recognized in 1957 by Hugo Benioff, whose global vision was ahead of his contemporaries in many matters:

On the basis of the evidence here presented, it appears that the principal tectonic movement of the circum-Pacific region is a clockwise rotation of the continents relative to the enclosed oceanic mass. The observed data are not sufficient to determine which of the two structures is moving in an absolute sense relative to co-ordinates fixed with respect to the earth's axis of rotation. The rate of movement has been measured geodetically in one region only—California—and here it amounts to approximately 5 cm per year. If this rate represents a mean constant rate applicable to the whole system, the time required for a complete relative rotation is about 10^9 years.

This circum-Pacific torsion is another great-circle shear zone that cuts the globe into two hemispheres (Figs. 80, 81). The Tethyan torsion is sinistral, whereas the circum-Pacific torsion is dextral, which is the expected relationship of conjugate torsions. They developed simultaneously, with the main movements within the Late Cretaceous and the Paleogene. They are complementary in all respects, mutually at right angles, and crossing in the East Indies and the West Indies, where their interaction produces similar interfering rotations.

Geographers have often commented that the globe has a land hemisphere contrasted with an oceanic hemisphere. The counter-Tethyan torsion separates them. In the land hemisphere side of the torsion belt are all of Asia, Europe, Africa, Greenland, and both North and South America. In the oceanic hemisphere are only Australia and Antarctica, which were one continent until the Early Cretaceous. Also, a dextral splay of the circum-Pacific torsion runs down to the east of Australia through New Zealand to the Ross Sea (the New Zealand Alpine fault), and this was the Pacific edge in the Cretaceous (see Fig. 96). Australia is in the oceanic hemisphere only because of the spreading of the dextral torsion belt from New Zealand to the Ninetyeast Ridge by progressive interaction with the Tethyan torsion.

The fact that the circum-Pacific torsion separates a hemisphere dominated by land from a hemisphere dominated by ocean suggests the cause of the torsion. Granted approximate isostatic equilibrium, the moment of inertia of the land hemisphere must exceed that of the oceanic hemisphere, because the center of gravity of continents stands 2 kilometers higher. Before the Cretaceous rapid increase in expansion, the centers of these maximum and minimum moments of inertia were on the equator (Tethys) about midway between the East and West Indies (that is, the centers of Pangaea and the Pacific). The asymmetric expansion that then set in began tipping the Tethyan equator to eventually reach 40° N in the Mediterranean.

Sectional motions on a rotating gravitating quasi-fluid spheroid that is also expanding are beyond present computing capacity, but it is clear that such a body adjusts until it rotates about its maximum moment of inertia. The present case is further complicated by the two asymmetries already discussed: first, more new oceanic crust is being inserted in the southern hemisphere (increasing the moment of inertia of the northern hemisphere with respect to the south), and second, when isostatic equilibrium prevails, the land hemisphere has greater moment of inertia than the oceanic hemisphere. Although the dynamics are much too complex to quantify, it seems reasonable to suggest that the observed conjugate Tethyan and circum-Pacific torsions were caused by these asymmetries that accelerated during the Cretaceous Period, which disturbed the relationship of isostasy and rotational inertia.

Figure 80 shows both the Tethyan torsion zone and the circum-Pacific torsion zone, each on an oblique Mercator projection, with the primary great circle of each projection along the respective torsion zone. These maps may look unfamiliar, but they contain no more and no less distortion than the common Mercator maps of the world that hang on so many schoolroom and office walls. They differ from the familiar ones in that instead of using the equator as the central great circle of the map, I have in each case used a great circle through the respective torsion zone. These maps and those on a different projection in Fig. 81 demonstrate that the two torsion zones *are* great circles and *are* at right angles to each other, also that one follows the Tethys, and the other separates the land and oceanic hemispheres.

As the Tethyan torsion was equatorial in the Cretaceous, the conjugate torsion passed through the Cretaceous poles, one in far east Siberia, and one opposite this near Bouvet Island in the far south Atlantic. These pole positions have long been indicated by the paleomagnetic data. Owing to the strong dextral shift on the Aleutian trench megashear (as recounted in connection with the Zodiac fan anomaly in Chapter 13) and the similar strong dextral shift on the Bouvet–South Orkney–Scotia megashear, both these Cretaceous poles were migrating during the torsion, so the northern pole began in the angle of the Gulf of Alaska, which was the rotation center for the Alaskan orocline and the axis of the opening of the Arctic and Atlantic oceans (see Figs. 15 to 18).

Another significant fact emerges. This Cretaceous south pole is the center of maximum dispersion of the continents (and hence the orthocenter of the world's new oceans), and the Cretaceous north pole is the center of least continental dispersion. In my 1970 presidential address to the Australian and New Zealand Association for the Advancement of Science, I pointed out that every continental block has increased its distance from every other block. For example, if I stand on Madagascar, I see that Africa has moved further away, and if I turn about, I see that India, Australia, and Antarctica have all moved away from me during the dispersion of Pangaea. If I stand in Venezuela, I see that the distance to North America, to Hawaii, to Africa, Australia, and Antarctica have all increased. It is similarly so for Arabia, Greenland, and every other block, even India when the nature of the Himalayan orogenesis is understood (see Chapter 18). This universal dispersion, of course, can only mean an expanding Earth.

As explained in Chapter 12, I measured the amount of new ocean crust that had been inserted between each pair of continents during the dispersion from Pangaea. Then for each continent I added the separation between it and each of its neighbors, and found its mean separation:

Continent	Mean separation		Continent	Mean separation	
Antarctica	4840		North America	2900	
South America	4820	Mean:	India	2300	Mean:
Australia	4170	4232 km	Europe	2100	2200 km
Africa	3100		East Asia	1500	

These data indicate that the maximum dispersion occurred near the Cretaceous south pole and the minimum dispersion occurred near the Cretaceous north pole, and that the dispersion of the southern continents was nearly double that of the northern continents.

Global Expression of the Conjugate Torsion

Let us now examine some specific examples of the effects of the circum-Pacific torsion. At the 1956 Hobart symposium, I demonstrated that gross dextral shear dominated the west coast of North America, including the Rocky Mountain Trench, the San Andreas shear system, and the Mendocino–Idaho coupled oroclinal drag (which Prof. Donald U. Wise of the University of Massachusetts described in 1963 as an "outrageous hypothesis" although it obviously made sense to him). I also showed that dextral shear dominated the continental margin of East Asia, which had been dragged dextrally to form a series of island arcs and basins separated by megashears, and that this system continued to the Scotia Sea south of Cape Horn.

Figure 86 (from my 1976 book) shows diagrammatically the Mendocino and Idaho oroclines, which are coupled dextrally to form a Z (not an S as in the Tethyan coupled oroclines). A simple shear such as in A could be relieved by a transcurrent fault as in B, which could off-

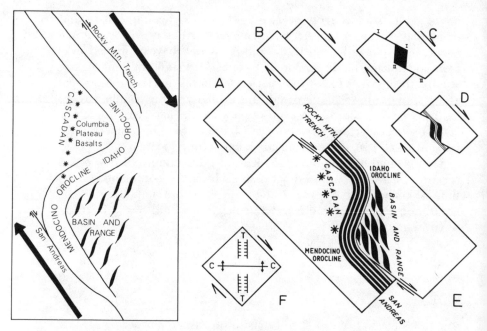

Fig. 86. Genetic relationship of Mendocino and Idaho oroclines, Cascade Range, Columbia Plateau basalts, Basin and Range province, Rocky Mountain Trench, and San Andreas fault system.

set sideways as in C leaving a hole (shown in black), and if part of the block was ductile it could flex across the offset as in D. Diagram E shows this applied to California, where the Rocky Mountain Trench dextral shear sidesteps to the San Andreas dextral shear system. The more ductile orogenic belt flexes to form the dextrally coupled Mendocino and Idaho oroclines. The tensional zone (black) in the concavity of the oroclines forms a series of rifts and ridges, the Basin and Range province, which occurs *only* in that concavity. These rifts trend generally north-south, as they must do as indicated by the tension rifts T–T in diagram F.

Because the flexed orogenic belt was coastal, the tensional zone in the northwestern concavity of the oroclines developed in oceanic material, forming a line of volcanoes (the Cascade Range) trending in the extensional rift direction, and occurring *only* across the oroclinal inflection. The extensional area between the Cascade Range and the oroclinal belt is filled (as should be expected) with basalt. The Cascade belt and the basaltic pile behind it are the marine equivalent of the

Basin and Range province. There are a few anticlinal folds in the basalt plain, and these trend, as expected, in the compressional direction C–C of diagram F.

Paleomagnetic studies have shown that many rocks in the oroclinal belt and in the convex re-entrants associated with them show rotations, always dextral but varying in amount, because of variable degrees of drag within the many fault slices. Older basalts statistically show more rotation than younger ones, because the torsion process has continued through a long stretch of time.

The Rocky Mountain Trench has a large dextral shift farther north (Dr. Gerald Gabrielse of the Geological Survey of Canada estimates more than 900 km of dextral shift, with a further 300 km to the west of it), but the feature peters out entirely at Flathead Lake in Montana. Figure 87 shows how the amount of shift is taken up progressively by a series of extensional rifts, occupied by lakes and major rivers. These rifts trend north-south as in the extensional direction T–T of diagram F of Fig. 86.

The coastal ranges of western Canada have been found during the last decade to consist of a complex "collage" of continental "terranes"

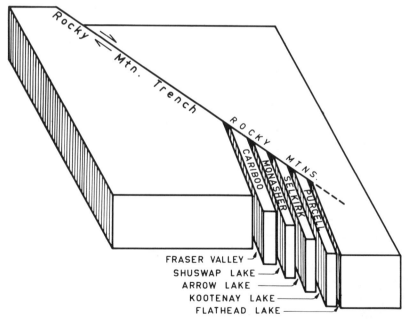

Fig. 87. Southeastern termination of the Rocky Mountain Trench.

that might be described as a "continent breccia," adjacent blocks having discordant fossil faunas and paleomagnetic signatures. In addition to the normal complications of stability of magnetizations, later remagnetization, structural tilts, tectonic rotations, and reversed magnetic polarities (significant here because of uncertainty as to which hemisphere they may belong), gross translations are also involved. Dr. Edward Irving, of the Pacific center of the Canadian Dominion Observatory, has compiled the data in the 1983 number of *Geophysical Surveys*. He concluded:

> There seems, therefore, to be substantial evidence that much of western and central British Columbia (The Coast Plutonic Complex, Vancouver Island, the Stikine block, and also, if the foregoing argument regarding the magnetization of [the Guichon batholith and Copper Mountain intrusion] are correct, including the Quesnel block) moved northwards during the Paleogene perhaps by 1000 km or more. Presumably this motion occurred along the large longitudinal faults (such as those along the Tintina Trench) that are characteristic of this region. These faults are not now active but could have been ancient analogues of the San Andreas fault system. This motion is one of the key kinematic elements of the Laramide orogeny to which the deformation of the Foreland Fold and Thrust Belt (the Rockies) must one day be reconciled.
>
> With one exception, aberrant Tertiary paleopoles indicate clockwise rotations, and in some instances northward displacement. The exception is the result from the Aleutian Islands. A cluster of results, the earliest of which was the data of Cox (1957) from northwest USA, indicate large and variable rotation of the Coast Range of Oregon and Washington [Black Hills, Ohanapecosh Formation, Siletz volcanics, Tyee–Flournoy Formation, Yachats basalt]. Of much interest are the clockwise rotations and northwards displacements of about 1000 km shown by a result from the volcanics of the Transverse Ranges of California; it is mirrored by a result from nearby Late Cretaceous batholithic rocks. Evidently these right-handed and far-sided paleopoles reflect the continuing northward movement of SW California along the San Andreas fault.

All these clockwise rotations and large translations are expressions of the circum-Pacific dextral torsion. More recently, Hildenbrand, Simpson, Godsen, and Kane, of the U.S. Geological Survey, rediscovered the Montana–Florida dextral offset of major gravity features, the Appalachian orogenic belt, and the Florida Shelf (Fig. 88, from Figs. 160 and 162 of my 1976 book), and featured it as a cover picture of the August 1984 issue of the *Transactions of the American Geophysical Union*.

The Aleutian Trench, which is one of the shears connecting the

Global Torsions

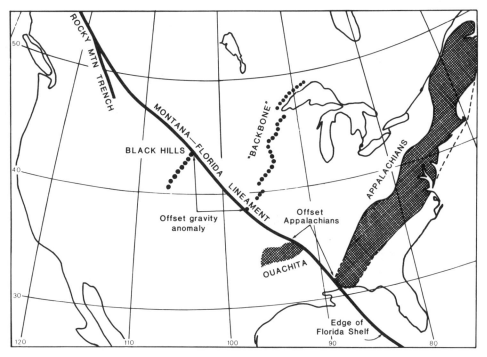

Fig. 88. The Montana–Florida lineament.

Rocky Mountain system to the shear system in east Asia, has a dextral offset of nearly 800 km, as shown by the displacement of the Zodiac fan (Fig. 40).

The whole structure of east Asia changes grossly on a line extending from the Assam orocline axis to the Gulf of Anadyr off the Bering Sea. This line divides coastal China from interior China. Northwest of it, the grain trends east-west. Southeast of it is a great series of northeast-trending dextral faults with extensive horsts and grabens, lowlands and lakelands. A major Tethyan torsion lineament, which trends east-southeast for 2000 km from beyond Ching Hai Lake, through Sian and Nanyang, is truncated south of Nanking by the north-northeast-trending circum-Pacific lineament, which runs for nearly 5000 km from the Gulf of Tonkin, west of Nanking and the Shantung Peninsula, through Shenyang (Mukden) to the Sea of Okhotsk. This is not a single megashear, but a torsional system.

Although north-northeast-trending fractures dominate eastern Asia, there is quite a net pattern of subsidiary faults trending west-northwest and northwest, veering somewhat about these three directions. This net pattern itself implies a torsional system, because the intersections of any of the members are near-vertical traces. In a compressional system, by contrast, the intersections of the conjugate wrench faults with each other, and with their associated tension faults and fold traces, as well as the traces of all of them on the surface, would all be horizontal. The Pacific coast of Asia resembles the Pacific coast of North America in that the gross tectonic pattern is dominated by dextral torsion.

Southeastward from the Assam–Anadyr line, the faults veer increasingly southward, as dextral shift and extensional opening com-

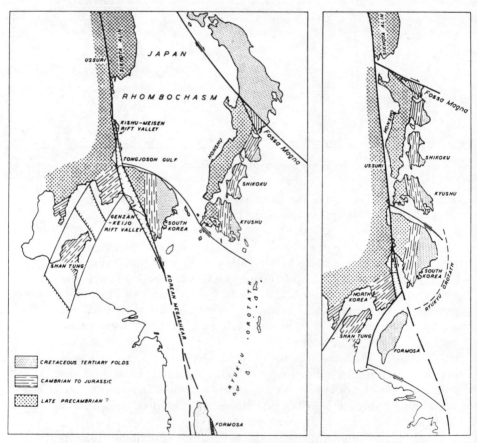

Fig. 89. Tectonics of Japan and Korea. (My 1956 reconstruction.)

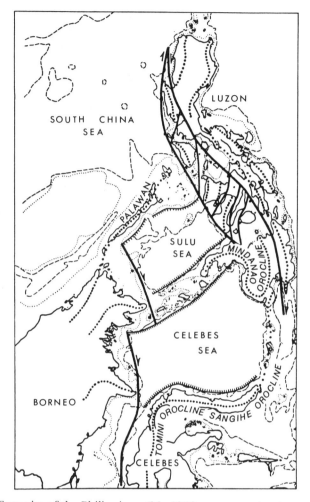

Fig. 90. Tectonics of the Philippines. (My 1956 reconstruction.)

bine to produce rhomboidal seas behind arcs torn from the continental rim—the Sea of Okhotsk, Sea of Japan, Yellow Sea and East China Sea, South China Sea, the Philippine Sea, and the Sulu, Sulawesi, Molucca, and Banda Seas. The pattern of displacement is illustrated by Figs. 89 and 90, which I presented to the 1956 Hobart symposium. At an early stage, before the Pacific margin of Asia was displaced westward by the Tethyan torsion with respect to the Pacific margin of Australasia, these dextral splays continued along this margin via the New Zealand Alpine Fault to the Ross Sea.

The Assam and Sunda dextrally coupled oroclines play a similar role to that of the Mendocino and Idaho coupled oroclines, but they are not a mirror image. The latter drags the coastal orogen northward away from the Tethyan zone, whereas the former deflects the Tethyan zone itself southward. The east-west trend of the Himalayan arc bends sharply in Assam through 90° to trend southerly through Burma, Thailand, and Indochina, then bends again through Sumatra to resume its easterly trend through Java to Flores. Before these bends, Sumatra and Java trended easterly in direct continuation of the Himalayas. The dextral offset is 34°, which is the same as the cumulative extension in the series of tensional basins as shown in Fig. 84, and is the amount Australia has been moved southward with respect to India. When this is combined with the 30° easterly offset of Australia with respect to Asia (Fig. 91), it gives precisely the separation of the Northwest Cape of Australia from the Andhra convexity of India, whence the geology indicates it came (Fig. 31). The oroclinal bends have developed in the young orogens, where the high temperatures

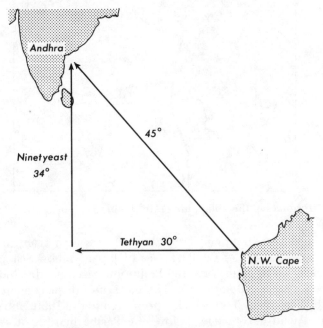

Fig. 91. Combination of Tethyan (sinistral) and Assam (dextral) displacements.

extend deeply into the mantle. In older crust, overlying colder mantle, megashears develop instead of oroclinal bends.

Anthony Hallam, of the University of Birmingham, has suggested, because of the presence of some warm-climate faunas, that some east Asian blocks had moved significantly north since the Permian, which would imply displacement opposite to the circum-Pacific torsion. However, this suggestion is based on a false axiom that the present climatic zonation applied generally in the past. In my 1976 book I discussed the secular change in the obliquity of the earth's axis, which profoundly modifies climatic zonation. For example, cold-blooded reptiles, which could not survive frosts, lived in Tasmania in the Early Triassic within 10° of the south pole. Indeed, George Williams, then of Adelaide University, who has also investigated this question, has shown that with obliquities greater than 45°, glaciation is more likely in low latitudes than at the poles. Williams concluded that the Late Carboniferous and Early Permian were times of large obliquity, and the Jurassic and Early Cretaceous were times of near zero obliquity when uniform climate extended to high latitudes without marked seasons.

The Tethyan and circum-Pacific torsions cross in the East Indies and the West Indies, two unique zones that are not only similarly named, but are similar in morphology and tectonism. In California, Nevada, and Arizona, the two torsions interfere. The Transverse Ranges of California, the Big Pine fault, and the Garlock fault, which are sinistral, deflect the San Andreas dextral system. Recently Gary Calderone and Robert Butler, of the University of Arizona, reported that Miocene lavas in five mountain ranges of southwest Arizona show sinistral rotations averaging 14°, in contrast to the dextral rotations in similar lavas a little farther north. Further east, in Texas, the sinistral shear seems to continue as the Chalk Draw–Carta Valley fault zone, without interference with its dextral conjugate (Fig. 92).

The computer closure of the Atlantic brings southern Mexico against Colombia (AA in Fig. 93), and Florida against Trinidad (BB). This requires more than 1500 km of latitudinal shift (on the Tethyan torsion) and more than 1000 km meridionally. This translation with dextral torsion on the faults west of the Montana–Florida lineament closes the string of extensional basins that form the Caribbean Sea, Cayman Sea, and Gulf of Mexico. The Greater Antilles show strong sinistral paleomagnetic rotations exceeding 40°, and the whole Central American region shows numerous large east-west transcurrent faults (Tethyan), and large northwest transcurrent faults (circum-

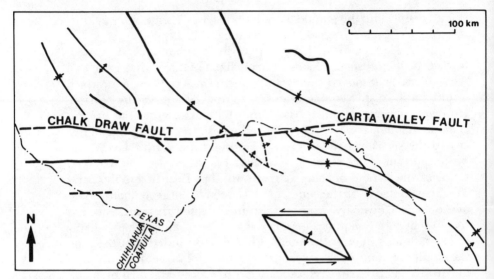

Fig. 92. Chalk Draw–Carta Valley sinistral fault zone in Texas. (After Thomas E. Ewing and W. R. Muehlberger.)

Pacific), with associated west-northwest compressive thrusts and folds and north-northeast tensional rifts. Both the latter continue through the United States into Canada.

In order to unravel the warp and the woof of the two torsions as they interweave in the East Indies, my 1976 reconstruction of the region (Fig. 94) was prepared as follows:

1. Excise the new crust that formed the many small basins in the region (such as the South China, Sulu, Celebes, Flores, and Banda Seas). This step, for example, closes Borneo and Palawan against southern China.

2. Shorten appropriately the orotaths (stretched orogens). For example, the Sunda–Banda orogen becomes progressively more stretched as you proceed along the arc from Sumatra, Java, Bali, Lomboc, Sumbawa, Flores, Lomblen, Pantar, Alor, Wetar, Romang, Damar, Banda, Turtle, Lucipara, thence via a continuous submarine ridge to Salajar, and southwest Sulawesi. Likewise the continuous ridge through Sumba, Savu, Roti, Timor, Moa, Babar, Tanimbar, Kai, Gorang, Seram, Buru, Tukangbesi, Butung, to southeast Sulawesi. Such orogens are elongated partly by fault rift gaps between them (as pointed out by the Netherlands geologist H. H. Brouwer in 1916 and 1929), and partly by flow attenuation. If an orogen standing 1000 m

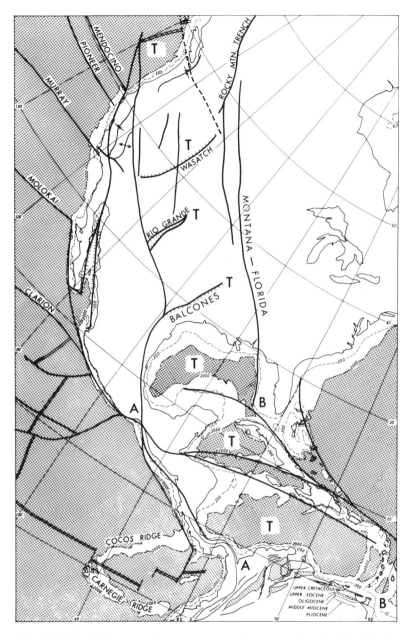

Fig. 93. The Tethyan and conjugate torsions cross in the West Indies. Hatched areas are deeper than 2000 m. Bullard's computer closure of the Atlantic brings A to A and B to B. The three deep basins (Caribbean, Cayman, and Gulf of Mexico), marked by T, and their northwestward continuation as a series of extensional rifts, all close with the AA and BB closure.

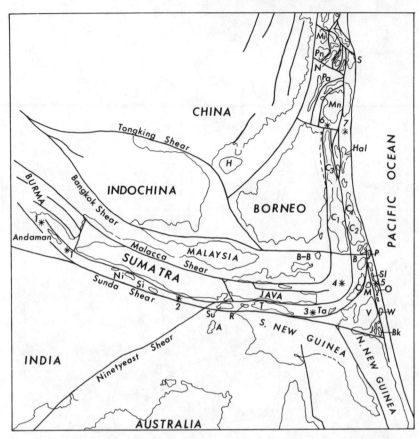

Fig. 94. Cretaceous reconstruction of Southeast Asia. A, Aru; B, Buru; B-B, Bangka-Billiton; Bk, Biak; C, Seram; H, Hainan; Hal, Halmahera; J, Japen; M, Misool; Mi, Mindoro; Mn, Mindanao; N, Negros; Ni, Nias; O, Obi; P, Peleng; Pa, Palawan; Pn, Panay; R, Roti; S, Samar; Si, Siberut; Sl, Sula; Su, Sumba; T, Timor; Ta, Tanimbar; V, Vogelkop; W, Waigeo. Numbered asterisks indicate shortened orotaths: 1, Andaman–Nicobar; 2, Nicobar–Sumba; 3, Timor–Tanimbar; 4, Banda arc; 5, Sula spur; 6, Sangi–Kawaio; 7, Talair.

above sea level were stretched to twice its length, isostatic equilibrium would leave it as a submarine ridge rising to 1500 m below sea level. Such submarine ridges and island chains indicate what points on major islands have to be brought closer, and the degree of necessary longitudinal shortening.

3. Straighten the oroclines, such as the Assam–Sumatra couple, and the loops around the Banda Sea.

4. Check paleomagnetic orientations, rotations, and paleolatitudes. It was found that the preceding steps had automatically corrected these. For example, McElhinny, Haile, and Crawford reported in 1974 that in the Early Cretaceous, the Malay peninsula lay east-west a few degrees north of the equator, which is precisely where the above procedures had left it (Fig. 94). Likewise the 180° rotation of Seram. However, in a region as contused as the East Indies with the crossing of the dextral and sinistral torsions, further data may require adjustments to this 1976 reconstruction, but these will be minor, because the first three steps above determine the gross pattern.

5. Check the large transcurrent faults associated with the Tethyan and circum-Pacific torsions, because quite large translations may have occurred along these. Again, it was found that the foregoing operations had largely corrected for them.

6. Check what had happened to the many faunal and paleogeographic links, such as Wallace's Line, the many faunal connections reported earlier in connection with the gape artifact, and the tight facies and provenance links reported forcefully by the eminent Netherlands geologist J. F. H. Umbgrove in 1938 and more recently by C. J. Pigram and his Indonesian coauthors in 1984, and others. Here again, the foregoing basic operations had already met the requirements, although it is now time to review all these procedures and produce an update of Fig. 94.

Such reconstructions are not like jigsaw puzzles in which a piece may be reassembled where it might seem to fit; severe constraints must be observed to maintain topological homogeneity. No block can leapfrog another block. All blocks must retain the same sequence and mutual relationship though they may have separated by insertion of strips of oceanic crust between them or relatively rotated by insertion of wedges of oceanic crust, or they may have been moved long distances along megashears, and orogens may be bent or very greatly stretched. Once inserted, new oceanic crust becomes permanent. No crust is swallowed or mysteriously eliminated. Continental crust cannot be destroyed or eliminated, nor created except at the widening of active orogens. Reconstruction consists of eliminating oceanic crust, straightening oroclines, shortening stretched orogens, and returning separated blocks along megashears. Figure 14 (the reconstruction of the Mediterranean), Fig. 82 (Venezuela and Colombia), and Fig. 94 (the reconstruction of southeast Asia) are good examples of the operation of these principles.

Integration of Spreading Ridges and Torsions

Both the Tethyan torsion and the circum-Pacific torsion diverge a little from their mean great circles and have to complete the circuit via primary spreading zones. The Tethyan torsion veers to the left and completes its circuit from the Caribbean to Gibraltar via the North Atlantic spreading ridge. The circum-Pacific torsion veers to the right and completes its circuit to Panama via the Southeast Pacific spreading ridge, after splaying into the Udintsov, Tharp, Heezen, and Menard fracture zones, which have a cumulative dextral offset of 1700 km. The global spreading ridges and the two global torsions are one integrated system. The torsional shears also function as transforms where they intersect the spreading ridges, and the latter accommodate differences in displacement along the torsions.

The torsions also interact with each other and with contemporaneous expansions. Hence the amount of offset varies greatly. For example, the offset of the western Pacific at the andesite line is some 5000 km, but much of this is taken up by the spreading of the Tasman Sea, so the offset of Australia against China is only some 3000 km, and most of this is taken up by the Banda S-shaped drag and by stretching of the Sunda arc, so the offset of India against China is minimal although the Indian block has been rotated strongly by the torsion. Likewise, the offset between western Europe and Africa is some 1500 km, but much of this is taken up in the East Siberian sphenochasm, so again the offset of India against China is minimal.

As discussed above, the circum-Pacific torsion zone is relatively narrow in Alaska, where it is farthest from the Tethyan torsion; but proceeding south along East Asia progressively larger pieces are detached with progressively larger extensional seas behind them, as the effect of the Tethyan torsion increases. In the north, there is only one detached piece (Kamchatka), and only one sea (the Sea of Okhotsk) with a width of 600 km. Southward, where the Tethyan torsion crosses, the detached block is very much larger (Borneo plus the Philippines and Sulawesi) and has itself been broken into several pieces, and the extensional seas add up to 1600 km.

The splaying of the circum-Pacific torsion by the Tethyan torsion is really very much wider than that, because, as mentioned above, before the displacement by the Tethyan torsion the western Pacific coast continued directly south via New Guinea, Fiji, and New Zealand (Figs. 95, 96), and there has been substantial dextral shift between New Zealand and Australia; this megashear zone is a splay of the main dextral tor-

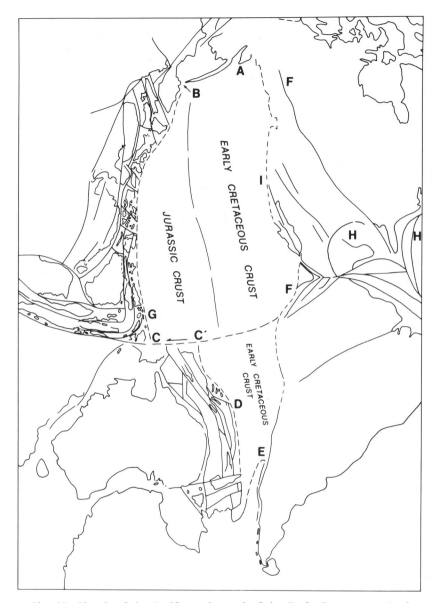

Fig. 95. Sketch of the Pacific at the end of the Early Cretaceous. A, the Gulf of Alaska sphenochasm has already opened substantially. B, the Aleutian nematath has not yet opened, so the Alaska peninsula abuts Kamchatka. C–C′, the Tethyan torsion has already started to offset the west Pacific margin (andesite line—see Fig. 80). D, the southwest Pacific restored as in Fig. 96. E, the Antarctic Peninsula. F, Central America as in Fig. 93. G, the Indonesian region as in Carey's 1976 reconstruction. H, the Gulf of Mexico and the Atlantic have already started to open. I, the Mendocino orocline straightened (see Fig. 86).

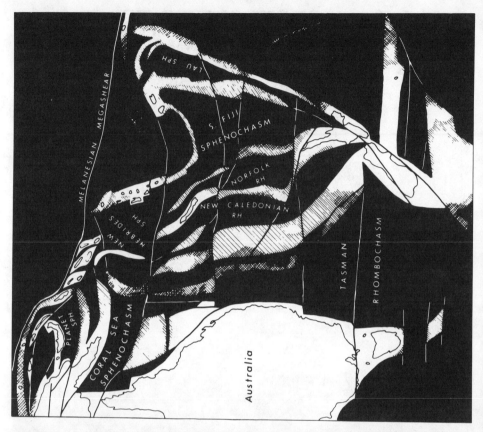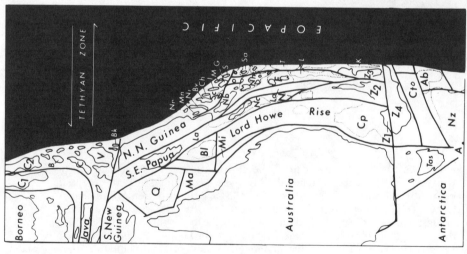

sion and commenced as a direct continuation of the megashear which runs west of Sakhalin, across Honshu via the Fossa Magna, and down the Bonin line to the Marianas. This dextral line is now expressed by the New Zealand Alpine megashear, and by the Ballenny, Tasman, Carey, and George V group of fracture zones, which collectively offset the spreading ridge 1400 km dextrally. Professor Bruce Waterhouse, then of Toronto and now of Queensland University, has long maintained that the Permian fauna of New Zealand indicated warmer waters than the Permian of eastern Australia, which contains tillites and glacial erratics.

The Tethyan torsion has also been offset by its dextral conjugate, which is expressed by the splaying of the large fracture zones across the eastern Pacific, possibly as far north as the Mendocino fracture zone and south to Panama. This was first stated in my 1962 presidential address to the geology section of the Australian and New Zealand Association for the Advancement of Science.

Thus, the relative movements of the crustal blocks is a complex interplay of several motions: (1) the mutual separation as each continent moved out radially as the earth expanded and new oceanic crust developed between them; (2) the Tethyan interhemisphere sinistral torsion; (3) the circum-Pacific dextral torsion; (4) the east-west asymmetry giving greater expansion along the west Pacific margin than the east Pacific margin; (5) the north-south asymmetry of greater expansion of the southern hemisphere than the northern hemisphere.

Fig. 96 (*facing*). Tertiary drift of Australia away from its Early Cretaceous boundary of the Eo-Pacific, leaving behind successive ridges like glacial-retreat moraines, separated by new oceanic crust. C_1, southwest Sulawesi; C_2, southeast Sulawesi; B, Buru; C, Seram; V, Vogelkop; Bk, Biak; Nn, Ninigo group; Mn, Manus; Ni, New Ireland; Bv, Bougainville; Ch, Choiseul; I, Santa Isobel; Sc, New Georgia; M, Malaita; G, Guadalcanal; S, San Cristobal; Sa, Samoa; T, Tonga; L, Lau; K, Kermadec; Nb, New Britain; Nh, New Hebrides; F, Fiji; La, Louisiade Archipelago; Nc, New Caledonia; Lo, Loyalty Group; N, Norfolk Ridge; Z_1, Z_2, Z_3, Z_4, New Zealand; Tas, Tasmania; Q, Queensland Plateau; Cp, Campbell Plateau; Ct, Chatham Rise; Ab, Antipodes–Bounty Islands; Nz, New Zealand Plateau; A, Auckland Island.

CHAPTER 22

Evolution of the Lithosphere

ACCORDING TO the ancient Greeks, Gaea was the goddess of the earth, descended directly from Chaos. Gaea is the patron (or matron) goddess of all geologists, whose perennial task is to draw earthly order out of chaos. Uranus, god of the sea, was Gaea's son by her own father. Gaea bore to her son Uranus many offspring including the twelve Titans, the three Cyclopes (who ruled thunder and lightning), and the three Centimani (each with 100 hands). But Gaea and Uranus also had a daughter, Tethys, who married her brother Oceanus (one of the Titans). Homer portrayed Oceanus as a great river surrounding the ancient world, and Hesiod related that from Oceanus and Tethys sprang all the great rivers. When the great Viennese synthesizer Edouard Suess (1831–1914) extended Neumayr's 1885 concept of a mediterranean sea extending from Mexico via the Alps to the Himalaya, separating a great northern continent from a great southern continent which Suess called Gondwanaland, he recalled the Greek myth and named this equatorial seaway Tethys, daughter of Gaea.

What Is the Tethys?

Suess regarded the present Mediterranean Sea as a remnant of his Tethys, which had dominated global geography from 240 until 25 million years ago. Every paleogeographic reconstruction since has incorporated the Tethys in some form. Suess had taken it for granted that the positions of the continents were fixed and permanent, and that the oceans now separating them had been formed by foundering of intervening lands. Wegener, as we have seen, regarded the Tethys as an equatorial seaway across his Pangaea, from which the present continents had formed by sliding apart with the Atlantic and Indian Oceans filling the gaps. Du Toit replaced Pangaea by two polar continents separated by a wide equatorial ocean, which became narrowed to form the Tethys when Gondwanaland slid toward Laurasia, greatly widening the other side to form the Pacific Ocean. The plate-tectonics theory adopts Tethys as a gape more than 6000 km wide at the Pacific end, tapering to zero in the western Mediterranean, and overlapping by 2000 km in Central America.

In the expanding-earth reconstruction, Tethys existed as intracontinental shallow seas throughout the Paleozoic, but it became equatorial about 380 million years ago, and the distribution of the foraminiferal family Verbeekinidae, particularly the *Neoschwagerina* fauna, proves that Tethys was a throughgoing shallow seaway by 260 million years ago. It was a shallow geosyncline throughout the Mesozoic Era (with a continuous seaway from Spain to New Guinea). The Eocene saw rising Alps and deep rifts in the Himalayas before their rapid rise commenced 15 million years ago, so that later Miocene and Pliocene faunas along the former throughway developed independently.

This Tethys concept has become firmly established, but tectonic geologists have dressed it with widely divergent genesis. Stratigraphers associate "Tethyan" with characteristic facies of sedimentation. Paleontologists define Tethyan faunal realms. Prof. Derek Ager, a leading stratigrapher from Swansea, has said that a persistent problem has been whether Tethys should be defined geographically, tectonically, or in terms of rock suites (lithofacies). He could have added in terms of time, as some would recognize the Tethys throughout the Paleozoic. Indeed, Tethys has become a football for stratigraphers to such an extent that some say the term has become so multiple in meaning that it should be dropped.

In the following pages, I will introduce a genetic meaning. When I examine the Tethys as defined by Suess, the Permian to Eocene throughgoing equatorial geosyncline, I find that such an equatorial orogen, making the tensional and torsional adjustments between the hemispheres, has been repeated at least three times on Earth, and indeed, such a structure occurs on Mars.

Earlier Analogs of the Tethys

The Caledonian–Appalachian–Tasmanide orogenic zone was the early Paleozoic forebear of the Tethys, a globe-girdling equatorial orogen (as indicated by paleomagnetic latitudes as well as its fossils), with its main terminating orogenesis in the Middle Devonian, but with closing activity continuing to the end of the Paleozoic. The sinistral torsion of some 2500 km along this orogenic zone has already been

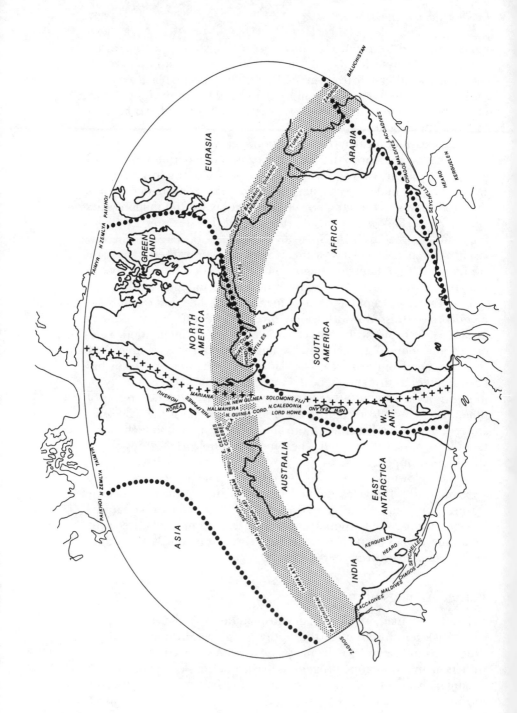

Evolution of the Lithosphere   309

discussed in Chapter 13 (Figs. 38 and 39). The extinction of the Caledonian geosyncline and orogenesis was followed by a whole-crust rotation of 45° and the inception of the Tethys along the new equator.

Still earlier, the Cordilleran orogenic zone of western North America was the Late Proterozoic forebear of the Tethys, functioning as the equatorial orogenic girdle from about 900 million until 600 million years ago, when a 45° counterclockwise rotation transferred this function to the Caledonian–Appalachian belt, which was superseded 300 million years ago when a further 45° counterclockwise rotation transferred this function to the Tethys. A further 45° counterclockwise rotation occurred some 50 million years ago, when the Tethys was tipped from equatorial to 45° north of the equator in the Mediterranean.

Figure 97 shows the general relation of these three generations of the Tethys on a reconstruction of the whole earth when continental crust enclosed it completely. I have extended the continental outlines beyond the bounding oval to show the continuity, but in each case the outline reappears inside the oval in the complementary position on the opposite side of the central line, so the oval does include the whole earth. The right-hand edge is the same line as the left-hand edge, as though I had slit a globe along the back and flattened it out.

The central line of crosses is the Cordilleran generation of the Tethys (equatorial at that time). The line of thick dots is the Caledonian–Appalachian–Tasman generation of the Tethys. Observe that wherever it reaches the edge of the oval it reappears at the complementary point at the twin point on the other side of the oval. The stippled zone is the next generation of the Tethys (the Tethys proper). Observe that where it reaches the edge of the oval, it also reappears at the twin point. You might ask, what happens to the Cordilleran line of crosses, which reach the edges at the top and bottom of the map. The answer is that it runs along both outer edges of the oval and re-enters at the bottom, because there is no place for any line to go outside the oval.

I prepared this projection more than a decade ago but did not publish it because important adjustments still have to be made (such as the relationship of India and Australia, and the position of New Zealand), but each time I move any piece new grids have to be calculated by tedious logarithms for several pieces because the shape of any continent depends on where and in what orientation it lies in the projection. So

Fig. 97 (*facing*). The area within the oval is a preliminary sketch of the whole earth to show the relation of Tethys (stippled) to its predecessors, the Caledonian–Appalachian–Tasman (filled circles) and Cordilleran (crosses). See the text for limitations.

I set it aside until this could be done with computers to soften the fatigue. This is now planned for the immediate future in collaboration with Dr. Ken Perry. Meanwhile this working diagram is useful to clarify the Tethyan genealogy, but it must not be used beyond that or cited as Carey's Pangaean compilation.

Like all useful classifications, this Cordilleran–(rotate 45°)–Caledonian–(rotate 45°)–Tethyan–(rotate 45°) generalization is oversimplified. There were intermediate orogenies, and some stepwise modifications may be necessary. Besides, all polygon boundaries imply diapirism to varying degree up to significant orogenesis, coordinated with the throughgoing equatorial orogenesis. However, it suffices for the present to show the primary pattern.

It is not clear yet whether there was a sinistral torsion associated with the Cordilleran equatorial orogen, as there was with the Caledonian and Tethyan zones, nor is it yet clear whether there were dextral torsions conjugate with the Caledonian and Cordilleran. But it may be significant that all four (Tethys and her conjugate, Caledonian, and Cordilleran) all intersect around Panama. This is not necessarily sur-

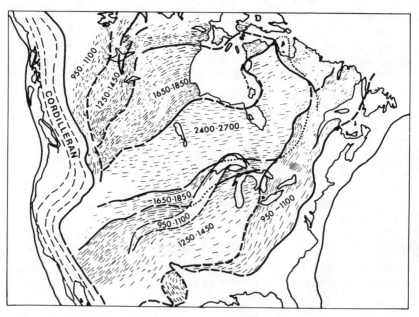

Fig. 98. Truncation of the Precambrian grain by the Cordilleran geosyncline. The numbers indicate age of igneous activity in millions of years. (After R. Gordon Gastil of San Diego State University.)

Evolution of the Lithosphere

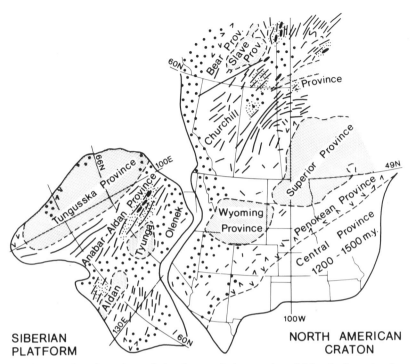

Fig. 99. Continuation of the Precambrian grain of North America into eastern Asia suggested by James W. Sears of the University of Montana and Raymond A. Price of the Geological Survey of Canada.

prising, because once an asymmetry develops, it tends to become locked in by feedback processes, just as when a salt diapir commences in one place, or a thermally driven diapir commences, density feedbacks lock them in position.

Professor Gordon Gastil of San Diego State University has shown that up to about 900 million years ago, the grain of North America was west-southwest. The rifting that initiated the Cordilleran geosyncline (which was the beginning of the Pacific Ocean) cut obliquely across this grain (Fig. 98). The youngest rocks with the older grain are the Beltian series in the United States and the Purcell series in Canada. On the other side of the Cordilleran orogen, we should find the continuation of this truncated grain, where now there is only the Pacific Ocean. Professor James W. Sears of the University of Montana and Dr. Raymond A. Price of the Geological Survey of Canada have looked for it—in Asia!—before the Pacific had opened (Fig. 99).

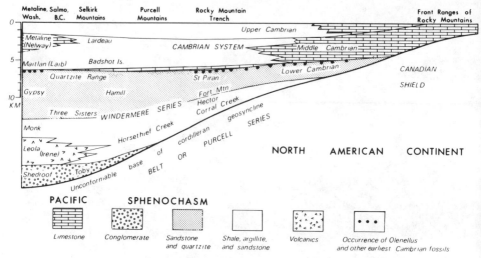

Fig. 100. Facies and thickness of the initial sedimentary cycle of the Cordilleran geosyncline. (After Phillip B. King.)

The first strata to be deposited in the new Cordilleran geosyncline were those of the Windermere Group, which rest unconformably on the truncated Belt and Purcell rocks. Figure 100 shows how Dr. Philip B. King reconstructed them as they were before the folding and disruption that has deformed them since. This is the same as the standard model of a developing orogen (Fig. 62). Where the strata thin out on the Canadian basement, we find limestone and well-worked sandstone and shale, typical of the miogeosyncline, but they thicken rapidly as the continental crust below them thins and are replaced by thick graywacke and volcanics (typical of the eugeosyncline). As the stretching thins the continental basement to zero, these eugeosynclinal sediments rest directly on basalt and other mafic rocks derived from the mantle. This section is just like the deposits on the east coast of North America since Africa rifted away to open the Atlantic Ocean.

Figure 101 summarizes the evolution of the earth. The time intervals between the drawings is not equal, but becomes shorter on a geometrical progression. The expansion of the radius increases in a geometric progression. Four billion years ago, the earth must have been heavily cratered by meteorites and asteroids and looked like the moon looks now, but about twice its present diameter. The next shows the Archean earth with polygonal geosynclines and broad basins as in Fig. 72. Mercury has such a polygonal fracture system, breaking across the pattern of craters.

Evolution of the Lithosphere

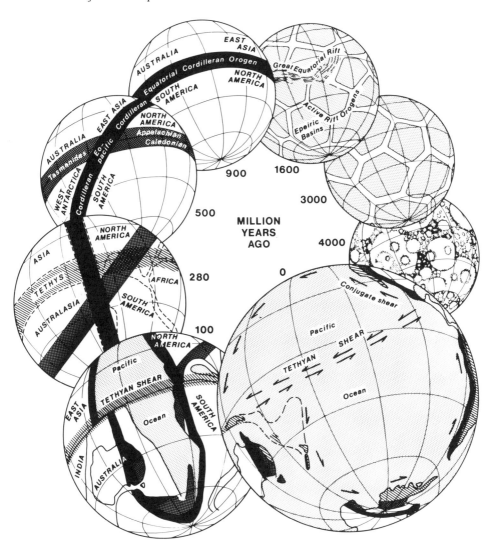

Fig. 101. Stylized diagram of the evolution of the earth.

Birth of the Pacific

The next stage, toward the end of the Early Proterozoic, shows the beginnings of an equatorial extensional fracture system, just like the one we now see on Mars (Fig. 102). With the beginning of the Late Proterozoic about 900 million years ago, the Cordilleran geosyncline

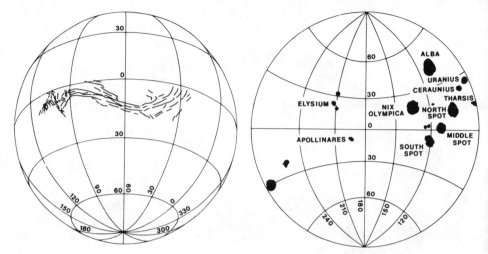

Fig. 102. The asymmetry of Mars. At left, the great equatorial rift system is confined to one-quarter of the circumference, owing to the asymmetry of the expansion. At right, all the volcanoes on Mars are confined to one hemisphere, also indicating asymmetry of expansion.

approximately along the equator forms the first global geosyncline, the beginning of the Pacific Ocean. Less important geosynclines still develop along the polygon rifts, because expansion is accommodated over the whole earth. Broad shallow seas are still found in the basins, as in Fig. 72.

In the next stage, about 500 million years ago, the Cordilleran has rotated 45° counterclockwise, and a new active geosyncline has developed roughly along the equator, subsequently dispersed in four continents as the Caledonian, Appalachian, and Tasman orogenic belts. Orogenic activity still proceeds along the Cordilleran, and along some of the polygon boundaries, especially those branching off the primary orogens. Intracontinental basins still occur.

But there still had been no significant opening of the Pacific Ocean. At 400 million years ago, the Cordilleran orogen was still a fold-mountain range trending across the middle of Pangaea like its Asian continuation today through the Kolimskoye, Stanovoy, Vablonovoy, Tien Shan, and Hindu Kush mountains.

As long ago as 1951, a Japanese geologist, K. Ichikawa, concluded that pebbles in the Triassic basal conglomerate in the Kitikami Mountains came from where the deep Pacific now lies. Several investigators since (most recently Dong Ryong Choi) have confirmed that lands, which the Japanese called Kuroshio and Oyashio, existed on the oceanic side of Japan. Professor Calvin Stevens of San Jose State University found closely related Permian corals in Cordilleran North America and eastern Asia. The affinities of the Permian fusulinids of western North America are with those of the western Pacific, not across the Atlantic, which had not then started to open. Equally anomalous is the occurrence in southeastern North America of the late Paleozoic Cathaysian flora, which is characteristic of east Asia. The simple answer is that the Pacific did not commence to open much until the Jurassic, and that Kuroshio and Oyashio were in fact North America. To explain such anomalies within the plate-tectonic doctrine, traveling "microcontinents" have been hypothesized to ferry small continental masses across the Pacific Ocean, much wider then in their model than the present Pacific.

A similar situation is found along the Pacific coast of South America. Professor Peter Isaacson of the University of Idaho showed that a million cubic kilometers of micaceous sand was deposited during the Devonian Period on Bolivia, Peru, and northern Argentina from a continental source where we now find the deep Pacific Ocean. South American Ordovician trilobites and Devonian brachiopods and conodonts show strong affinities with the countries on the other side of the Pacific—Australia and southeast Asia. Professor Waterhouse has reported that a species of the Permian brachiopod *Attenuatella* found in New Caledonia is closely related to the Mexican species, far across the Pacific. Just like the Japanese above, the French geologist Dr. J. Avias, the leading authority on New Caledonia, concluded that a large continental landmass, which he called Archeofijia, must have existed east of New Caledonia where only the deep Pacific exists today. Later, at the international symposium on the geodynamics of the southwest Pacific in Noumea, New Caledonia, in 1976, Dr. Avias stated that Archeofijia was none other than South America.

A large number of investigators (particularly Dr. B. Dalmayrac) have agreed that the mid-Paleozoic Andean orogen had a mid-continental setting, with continental metamorphic rocks extending far into the area now occupied by the Pacific Ocean. Where is this lost continent? Certainly not foundered. It is Antarctica and Australasia, then a single continent, now separated from South America by the opening of the Pacific Ocean.

By 280 million years ago (Early Permian) a further 45° counterclockwise rotation had occurred, so that the Cordilleran was then meridional and the Caledonian was inclined to the equator, where the

Tethys had taken over as the equatorial geosynclinal zone. At the same time the increasing tempo of expansion initiated the development of the earth's great oceans. Narrow meridional lunes started the north Atlantic and Indian Oceans, and the Cordilleran divided to form the Eo-Pacific, which, by 140 million years ago (Early Cretaceous), had widened to about 3000 km (Fig. 95), separating the east Asian and Tasmanide orogens from the American Cordilleran orogen. The growth of the new oceans was asymmetric, with new growth mainly on the western side of the spreading ridge, so that the oldest part of the Eo-Pacific was on the western side, growing progressively eastward.

V. A. Krassilov, of the Far-Eastern Geological Institute in Vladivostok, concluded that even in the Jurassic the Pacific must have been very much narrower than it is now:

The Cycadeoidea flora is represented on both Asiatic and American margins of the Pacific Ocean. There is a considerable gap in the fossil record of Cycadeoidea between Europe and Central Eastern Asia and the Cycadeoideas from Mongolia and Japan are probably of American origin. Migration across the Bering land bridge is excluded because the northern parts of Asia and America were occupied by the Arctomesozoic [*Phoenicopsis*] flora. There is a striking parallel between the distribution of Cycadeoidea and horned dinosaurs, the latter being represented only in Mongolia and western North America. According to Colbert, the horned dinosaurs crossed the Pacific Ocean but were incapable of surmounting the Lance Sea—a narrow strip of water. Migration along the Bering bridge is also doubtful because of the climatic barrier. These facts enable us to suggest that the migration route lay across the Mesozoic Pacific Ocean at middle latitudes. This means a much narrower, not wider Pacific.

The lateness of this opening of the Eo-Pacific has been supported by Oakley Shields, of the Davis campus of the University of California, who reported at the 1981 Sydney symposium on Earth expansion that

butterfly families are concentrated in the tropics and subtropics of Mexico, Central America, northern South America, west tropical Africa, New Guinea, northern and north-eastern Australia, Indonesia, Philippines, Indochina, northeast India, eastern Tibet, and extreme southern China. This distribution pattern must have occurred prior to continental separations, i.e. before Late Jurassic–Early Cretaceous times.

Shields also suggested that the marsupials, which appear to have originated in North America, entered Australia via Central America before the wide opening of the Pacific, rather than via South America as commonly assumed:

Australia lacked placentals before the Pre-Pleistocene, South America lacks monotremes, and for as long as marsupials have existed in South America, so have placentals. Also, recent comparative morphology and serology work indicates the South American didelphoids and Australian marsupials represent different radiations. North American Late Cretaceous *Alphadon* marsupials have dentitions similar to some of the oldest Australian marsupials and may have given rise to them through a single stock. The diversity of marsupials and the absence of placentals suggests that marsupials entered Australia prior to the Mid-Cretaceous. An Antarctic route poses the problem of adaptation to a polar daytime regime.

Major changes occurred with the transition from the Mesozoic to the Tertiary about 70 million years ago. The Tethyan orogenic zone accelerated its diapiric regurgitation. The Tethyan and counter-Tethyan torsions became active. South of the Tethyan torsion, the spreading ridge along the South American coast, already separated from the Australasian coast by nearly 4000 km of Cretaceous Eo-Pacific crust, changed from asymmetric to symmetrical spreading, and continued in this way between Antarctica and Australia, separating them for the first time. This meridional separation, along with the contemporaneous meridional extension between Australia and east Asia and the bending of the Tethyan orogenic zone in Assam and Indonesia, were effects of the dextral counter-torsion between the Pacific and Asia, which also offset the South Pacific spreading ridge by 2700 km on the Balleny, Tasman, Carey, and George V bundle of shears.

North of the Tethyan shear, spreading continued to be asymmetric from near the Cordilleran coast. The lack of any significant eastward growth, compared with the eastward growth south of the Tethyan torsion, meant that North America moved 1800 km further west than South America (Fig. 80).

As pointed out in connection with the Appalachian clastic fans (Fig. 66), an orogen does not grow uniformly along its length, but tends to form diapiric foci some 600 or 700 km apart, which, if conditions were symmetrical, would form funnel-shaped diapirs rising to ring-shaped orogens at the surface (Wezel's krikogens). Because of the asymmetry along the east Asian coast, oceanic crust grew eastward (as in the eastern Pacific) on the western side of the orogenic spreading zone. The result is a line of basins, with rifted continental crust on the western side, a basin floor of oceanic crust that has grown from west to east, and an orogenic arc on the eastern side.

The Tasman Sea is a more complex version of the same process (Fig. 96). The east coast of Australia corresponds to the east coast of

Asia; the Kermadec trench from Samoa to New Zealand corresponds to the east Asian trenches separating the Pacific proper from the island arcs (the "andesite line" shown by the broken line on Fig. 95). Between, we have a series of extensional basins as Australia retreated westward through progressive growth at the eastern side of the system. Superficially, the Lord Howe, Norfolk, Three Kings, Lau, and Tonga submarine ridges resemble successive moraines left behind by a retreating glacier. The oldest basins are those nearest to Australia, and the youngest are those nearest the Kermadec trench. The relative ages are confirmed by their depths and the heat flow through their floors, because in each case the mantle material that now forms the floors had risen several tens of kilometers, and the thermal relaxation time (the time required for the higher temperature to decline to $1/e$ of the excess) is tens of millions of years. Falling temperature meant subsidence owing to phase changes below. The Tasman Sea system is cut off along the north by the Tethyan torsion, and offset westward 5500 km to the east Asian system (Figs. 80 and 96).

Figure 103 sketches the spreading of the Pacific Ocean. In order to reconstruct the early Pacific, first excise the later ocean floor (stippled in Fig. 103) inserted since the middle Cretaceous, at the same time reversing the Tethyan shear and its conjugate, which yields a condition like Fig. 95. On Fig. 84 it was shown that the southward displacement of Australia relative to China is 34° (made up of the sum of the series of small rift basins); this agrees with the 34° offset of the Sunda and Assam oroclines, and the difference between the northward movement of Australia and India with respect to Antarctica (also 34°). In Fig. 103, the distance between the Hawaiian ridge and the edge of the Melanesian Plateau (between New Guinea and Samoa) is also 34°, so the northward movement of Australia brings the Hawaiian Ridge against the Melanesian Plateau and excises the Central Pacific rhombochasm, which has been inserted by ocean-floor spreading since the Triassic. Thus the two points A in Fig. 103 are brought together, as are the three points B. This movement combines with the dextral global torsion, which closes the Shatsky rhombochasm and brings the points B and the points C together. These C points moved out of the Gulf of Alaska by sliding along the Aleutian Trench (Fig. 41).

According to one of the minor theorems of the Swiss mathematician Leonhard Euler (1707–83), any translation on a globe can be expressed as a rotation about a single pole, so that a 34° extension would decline to zero at this pole. This theorem is not valid for an expanding Earth, on which the 34° extension I have described could apply right

Evolution of the Lithosphere

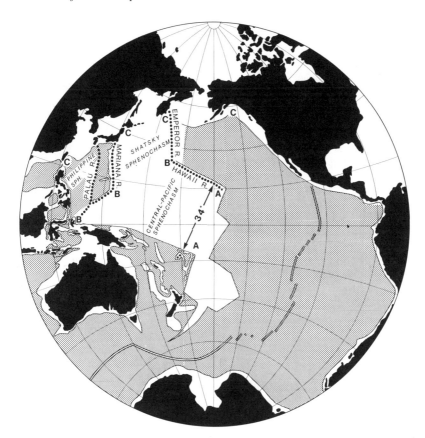

Fig. 103. Spreading of the Pacific. The stippled area has grown slice by slice since the mid-Cretaceous. The broken lines indicate the present active spreading zone. Growth of new crust is much greater south of the Tethyan torsion. The 34° meridional spreading between China and Australia is expressed by (a) the oroclinal offset between the Himalayan front and the Indonesian front, (b) the sum of the small rhombochasms between Australia and China (Fig. 84), (c) the rhombochasm between the Hawaiian ridge and the Melanesian plateau, and (d) the dextral counter-torsion.

around the Tethyan equator. In fact, the Tethyan widening did encompass the globe but not evenly: it has been about 11° across the Mediterranean, about 20° in Central America, and about 34° in the Australian region.

Compare Fig. 103 with Fig. 104, which was generated by Dr. Perry's computer program to model the growth of the earth from 76 percent of its present radius (stereographic projection). Reversal of the Te-

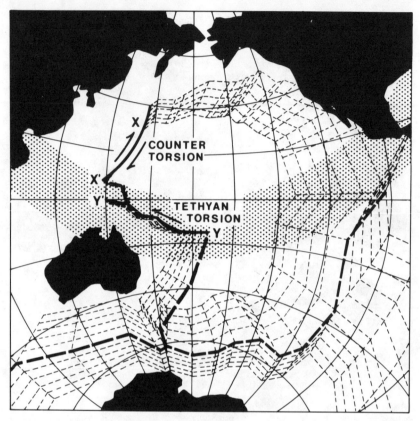

Fig. 104. Evolution of the Pacific as printed by Perry's computer program moving the continents radially outward from Pangaea, which produced the Tethyan torsion YY' and dextral counter-torsion XX', which together displaced X from Y. The program also threw up the termination at the Tethyan torsion of symmetrical spreading from the east Pacific active spreading ridge. The stippling of the Tethyan torsion belt has been added to this printout from Fig. 76.

thyan torsion would bring Y to Y', and reversal of the dextral conjugate torsion would bring X to X', which is also approximately 34°. The torsion shears appear automatically in the computer computation.

In the analysis summarized by Fig. 103, the 34° extension by the counter-Tethyan torsion and the 5500-km shift by the Tethyan torsion are discussed separately from the spreading (shown stippled), as though these were independent sequential events, but this is not so. The spreading and torsional kinds of movements are separated in the diagram only for clarity of analysis. In fact the two kinds of movement were concurrent and complementary. Much more information would be needed for a step-by-step integrated reconstruction to be attempted.

Plate tectonicists place much emphasis on the decrease in the ages of the volcanic rocks along the Emperor and Hawaiian ridges, and interpret them in terms of motion of the Pacific "plate" over a "hotspot" fixed in the mantle. But these ages record the times at which volcanism *terminated*. The lavas at Hawaii are 12 km thick. I suggest that the lava at the base is very much older than those measured at the surface, perhaps as old as any in the Emperor Ridge.

Summary

Paleomagnetic data establish that North America has been rotating counterclockwise since the birth of the Pacific as the Cordilleran rift (Fig. 78). At first the Cordilleran was equatorial and looked toward the present south-southwest to the north pole. By the Cretaceous, when the Tethys was equatorial, the north pole lay to the present northwest, and since then North America has rotated a further 45° counterclockwise. The increasing angles between the 100-million-year intervals in Fig. 78 show that the rate of the rotation has accelerated. More than half the rotation has occurred in the last 150 million years. This corresponds with the fact that more than half the new ocean floor has grown in that time. In fact, every aspect of geotectonics has intensified with time at an accelerating rate.

The great orogens have always been equatorial rifts: the original Cordilleran, the Caledonian–Appalachian–Tasman, and the Tethys. Other orogenic zones have been secondary. When the increasing size and increasing rate of expansion demanded throughgoing meridional rifting 200 million years ago, the most important rift developed along the Cordilleran, which was already meridional, and others developed as new intracontinental rifts to begin the Atlantic, and the Indian Ocean (combined with the East Siberian plain, which has now filled with sediment). The great oceans are a recent phenomenon on the earth's surface; there were none before the mid-Mesozoic. The acceleration of all geotectonic phenomena implies that *now* is not an average index of Earth's mass, volume, physical face, or life. The present is a key to the past only in terms of physical laws, now known— or yet to be discovered.

PART SIX

A PHILOSOPHY OF THE UNIVERSE

CHAPTER 23

The Earth and Cosmology

I MUST, of course, attempt to explain the accelerating expansion I have described. However, if the explanation I offer should turn out to be invalid, that explanation should be rejected, not the reality of expansion. Let us remember that Wegener presented his empiricism of continental displacement together with his physical explanation. When the latter was found inadequate, the whole idea of continental displacement was rejected for several decades, even though in the long run it was found to be essentially correct.

Let us first review five explanations of expansion that have been offered by others.

1. Cyclic pulsation of the earth, whereby the expansion phase opens or widens ocean basins and the contraction phase causes orogenesis, has been favored by several Russians and some others (for example, Shneiderov, Steiner, Khain, Kropotkin, and Milanovski). This is an alternative to simple expansion rather than an explanation of cause. It does not satisfy exponentially waxing expansion; furthermore, it is based on the entrenched false axiom that orogenesis implies contraction of the earth's crust. Indeed, other Russians, led by Beloussov, have been in the forefront of the rejection of crustal contraction, asserting that orogenesis is primarily a vertical process driven by gravity, a view I wholly support. Nevertheless, I do see compelling evidence that orogenesis is cyclic and pulsed, probably on a hierarchy of time scales, whereby expansion waxes to a crescendo, then wanes perhaps to zero before the next wave of expansion. I see no evidence demanding crustal contraction. All the great overthrusting seen in the Alps and Himalayas is an inevitable consequence of diapiric orogenesis.

2. Accretion of meteorites and asteroids has been proposed during recent decades (for example by F. Dachille, S. V. M. Clube and W. M. Napier, L. S. Myers, and others). Certainly Earth has received addi-

tions to her mass in this way throughout her history, and substantial arguments have been advanced that a large asteroid impacted Earth at the end of the Cretaceous Period, some 60 million years ago. Indeed, each day sees some addition of micrometeorites, and each year well-known swarms arrive on their predicted dates, which at least in some cases are the dates that Earth passes through the orbits of comets. The present rate of accretion is many orders of magnitude too small to make any significant contribution to Earth's volume and radius. Evidence from other planets and satellites indicates that a few billions of years ago the infall was substantial. However, this cannot be the primary cause of Earth expansion, because the meteoritic flux rate has diminished exponentially with time whereas expansion has waxed exponentially with time.

3. The third and most popular theory postulates that originally Earth had a core of ultradense matter, which has changed slowly to "normal" material, causing progressive expansion. Various models of this kind have been proposed by several authors in the United States, Canada, Australia, Hungary, Britain, Germany, and Russia. It has been pointed out in earlier chapters that solids recrystallize to progressively denser materials as confining pressure is increased: graphite becomes diamond with 50 percent increase in density, quartz changes to coesite, thence to stishovite, with a similar increase, basalt (which consists mainly of feldspar and augite) changes to denser eclogite consisting of garnet and jadeite. The pressure at depths of only a few tens of kilometers is enough to cause these changes, and it is suggested that at the pressures reached at the core the density could be many times greater. Certainly, fantastically high densities are believed to exist in white dwarfs and neutron stars, but critics argue that the pressure within the earth has never been great enough to produce the ultradense core postulated.

This criticism is sidestepped by claiming that the ultradense state was inherited from an earlier stage (before Earth separated from its stellar progenitor) and remained in a metastable state. For example, ordinary glass is metastable, and over the centuries it devitrifies to a crystalline state. Likewise eclogite, stishovite, and diamond are metastable at the earth's surface, although the relaxation time (the number of years to see recrystallization about one-third complete) is quite long. So, according to this hypothesis, the metastable ultradense core has been changing to less dense materials, causing large expansion. There remains, however, a crucial obstacle for all such theories, which postulate a primitive Earth with the same mass as now but about half the diameter—the force of gravity at the surface would have been about four times its present value, and this would have shown up in many geologic processes. Professor Stewart of Reading University has rejected Earth expansion on this very ground. But essential to his rebuttal is the assumption that the mass of the earth has not significantly changed.

4. According to Newton's law of gravitation, the attraction between two bodies is proportional to the product of their masses divided by the square of the distance between them, all multiplied by the "gravitational constant" G. Over the last 50 years, several influential astronomers, including Dirac in Britain, Jordan in Hamburg, Dicke in Princeton, and D. D. Ivanenko and R. M. Sagitov in Russia, have suggested that this gravitational "constant" has not been really constant but has been diminishing. This would have caused the earth to expand. The interior of the earth everywhere is compressed elastically by the weight of the overlying rock. The progressive reduction of G claimed by these astronomers would have reduced the weight everywhere, reducing the load and allowing all rocks to expand elastically.

In addition, the diamond-graphite, stishovite-quartz, and eclogite-basalt phase changes described above all depend on pressure, and there are a very large number of such phase transitions at all depths in the earth. As G diminished, the depth of each of these transitions would move down to the relevant pressure, so that still more expansion would occur. Many now believe that such change of G is valid and therefore that the implied expansion of the earth is inevitable. Two difficulties remain. Quantitatively, it is difficult to get by this mechanism the amount of expansion needed. Also the crucial obstacle faced by those who advocate constant Earth mass not only remains, but is intensified, because if G were greater in the past, the force of gravity at the surface would be even greater than it would have been from radius change alone. Hence secular decline in G cannot be the primary cause of the expansion.

5. A cosmological cause of expansion linked with secular increase in mass was first proposed in Russia and has been consistently developed there (Yarkovski in 1889, followed by Kirillov, Neiman, Blinov, and B. I. Vaselov). In 1933 in Berlin, Hilgenberg independently adopted secular mass increase, and in 1976 I also came to this conclusion, because of the crucial obstacle of unacceptably high surface gravity when the diameter of the earth was smaller. Whereas the other explanations discussed above are soundly based and must have in some degree contributed to expansion, the limits of surface gravity in the

past left me no alternative but to join the Russians and Hilgenberg in their conclusion that not only had the volume of the earth increased at an increasing rate, *but so had its mass.*

It was then that I realized that Earth was not alone with this enigma. The spatial expansion of the universe was recognized half a century ago, but the implications of Hubble's law (discussed later in this chapter) forced me to the conclusion that everything in the universe has suffered the same accelerating increase in mass. Therefore to understand the expansion of the earth, we must seek to understand the expansion of the universe. Should I then, as a geologist, simply bow out and leave the problem of Earth expansion to cosmologists? Unfortunately, if I did so, that is where the matter would languish and wither.

Academia partitions knowledge into physics, chemistry, geology, geophysics, astronomy, cosmology, and so on. Scientists, evolving toward ever narrower specialization, with diminishing knowledge of other fields, scorn the capacity of interlopers and resent their intrusion. Doctrines become inbred into creeds, and are stated as fact to others. Fundamental problems tend to be swept under the mat at the edge of the field. But Nature herself knows no such barriers. All science is simply common sense, and stripped of jargon and notations should be understood by any thinking person. Scientists have a duty to ensure consistency, not only within their own specialty, but also across the whole of nature. The most fundamental problems require input from disparate sources. The data of geology contain clues just as significant for the formulation of new physical principles as do the laboratory experiments of physics; moreover, the scales of size, mass, and time are far beyond any accessible to laboratory experiment. Physics is the poorer when it neglects the potential input from geology, as when Newton scorned Hooke, Kelvin ignored constraints on the age of the earth, Jeffreys rejected continental drift, and contemporary paleomagneticians dogmatically plot paleopoles without allowance for gross change in the earth's radius. In 1970, I concluded my presidential address to the Australian and New Zealand Association for the Advancement of Science with these words:

Our ancestors believed for millennia the obvious truth that the Earth was flat. Later we believed the obvious that as the sun, moon, and stars rise in the east and set in the west, the Universe revolves round the Earth. We have believed for millennia the obvious that the Earth's diameter has been essentially the same since primordial consolidation except for cooling shrinkage, but now find that the Earth has been expanding and is expanding at an increasing rate. The sooner physicists receive the lesson of these models the sooner they will seek new principles necessary to accommodate these facts. Herein is the clue for vast new discovery.

Thus, in order to pursue the meaning of the increase I find in both diameter and mass of the earth, I, a geologist, must launch into the ethereal space-time of cosmology, for if I leave it to the specialists, it won't be done, at least not in my lifetime. Besides, the lesson of the past is that jumps from accepted dogma commonly have to await an interloper from another field.

Hubble's Law and the "Big Bang"

Early in this century, astronomers debated whether the numerous nebulae were inside our Milky Way galaxy or far beyond it. Edwin P. Hubble settled this in 1924 when he identified Cepheid variable stars in the Andromeda nebula, which he proved to be a separate galaxy rather like our own Milky Way galaxy and a million light-years outside it. He went on to identify hundreds of other such galaxies even farther away, and in 1929 announced that the more distant a galaxy is from us, the faster it is receding. This was formulated as Hubble's law, that the velocity of recession of galaxies increases by about 30 kilometers per second for every million light-years it is distant from us. Hence the universe is expanding. Estimates of the Hubble constant have varied somewhat according to the method of measuring it, but this does not affect the principle. Dirac then pointed out the obvious: this means that if we reverse the record and look back in time, the further back we go, the nearer all the galaxies become. To quote Dirac (1937), "the universe had a beginning about two billion years ago when all the spiral nebulae were shot out from a small region of space, or perhaps from a point!"

What a massive point that would have been! All the matter of the whole universe concentrated at a point! One hundred billion galaxies, each containing some 10 billion billion stars, condensed to a point! I found this ultraheavy embryo of the universe inconceivable. Black holes, many of which are said to exist within our own galaxy, each so dense that even radiation cannot escape, would be thistledown compared with that initial egg. Lemaître called it the primeval atom. Gamow called it *ylem*. But surely nothing, not matter, not even radiation, could escape such a mass concentration. Modern cosmologists might agree, then go on to assert that nothing *has* ever escaped, and that neither time nor space existed outside that embryo or exists outside the present universe so begotten. Certainly in the initial stages the whole of the cosmos would have been within the Schwarzschild radius and hence would have remained the blackest of black holes ever conceived.

What then does Hubble's law mean? It means that what I found to apply to the earth also applies to the whole universe—volume expansion and mass increase go hand-in-hand.

Regression of Hubble's expanding universe must involve diminishing mass, so that the initial embryo, far from the unthinkable mass contemplated by Dirac, Friedmann, Lemaître, Arthur Eddington, and Gamow, had negligible mass, or *no mass at all*! The "Big Bang" (as this concept is now nicknamed) is a fiction, a fantasy. Like most models of the universe, it assumes as an axiom that all the matter in the universe existed since initial creation. On what evidence? This is a gratuitous assumption.

The "Big Bang" myth must be rejected on even more fundamental grounds. According to this theory, all the matter of the whole universe, the raw material for 100 billion galaxies, appeared instantaneously from the void, from nothing. This transgresses the first axiom of physics, the law of conservation. Bondi has pointed out that "the beginning" is regarded as a singular point on the border of the realm of physical science. Any question that refers to antecedents of the beginning, or the nature of the beginning, can no longer be answered by physics and is not a proper question for it. The laws of physics come into being with creation, but these laws do not consider creation. Physics begs the question of creation by sweeping it under the mat. As McRea put it, creation happened once and for all, and therefore is not subject to ordinary scientific discussion but determines the initial conditions for the rest of the discussion.

Such physics is rootless. I find this unacceptable. For me, the laws of nature, including the conservation laws, must be universally true. This is only possible if the universe is a state of zero, as proposed in 1973 by Edward Tryon on cosmological grounds, and quite independently by me in 1978 through projection from the expansion of the earth. The universe always was, is, and always will be a state of zero. Creation of matter and energy resembles the creation (from a state of cash zero) of a bank loan, which brings into being an asset, permitting all sorts of activity, but with it an equal and opposite debt. This concept of a null universe will be developed further below.

Philosophically, the Big Bang always involved Anaximander's dilemma, later debated by Aristotle, of the unique initial term, or the uncaused cause. As there was no logical escape from this enigma (except the null solution), divine intervention, not subject to natural laws or to logical sequence, had to be invoked. Logically, this did nothing to solve the dilemma of the uncaused cause, for it only substituted the problem of initial creation with the problem of the origin of the creator, leading on to a chicken-and-egg infinite succession of creators creating creators. However, since the dawn of intelligence, what could not be explained by observation and reason has always been assigned to gods, who could do anything they chose. As most philosophers to this day believe in an ultimate divinity, a rootless physics was acceptable to them; certainly to Einstein, who repeatedly affirmed his belief in God and who could not accept the uncertainty principle because "God doesn't play dice."

The uncertainty principle has been seized upon by the "new cosmology" to escape the dilemma of the origin of the Big Bang. Not only did all the matter and energy of the whole universe come into being from nothing at that instant, but so did time and space. Neither time nor space nor God existed before that instant. The vacuum itself fluctuates randomly about zero, resulting in a probability of the creation of matter at any instant; in this way the whole universe appeared as ultracondensed plasma at that instant and has been expanding ever since. Good experimental evidence exists that random quantum fluctuations do indeed occur in the subatomic domain in accordance with the Schrödinger wave function, with important practical applications for semiconductors (such as tunnel diodes).

However, two fatal flaws demolish this model. First, whereas the probability is significant at the subnuclear scale, it diminishes exponentially with scale, and to invoke a vacuum quantum fluctuation to create the whole universe according to a subatomic wave function is surely an absurdity. Second, the Big Bang model assumes with Einstein that mass and energy are interconvertible aspects of the same essence; before that beginning the mass-energy total was zero, but an instant later the mass-energy total was inconceivably large, indeed, the total mass-energy of the present universe. But in the quantum fluctuation model, the mass-energy total is unchanged before and after the event, the role of the fluctuation being no more than to jump over (or tunnel through) an energy barrier so that a particle appears on the other side of the barrier at the initial energy level. The ghosts of Anaximander and Aristotle have not been exorcised.

This absurdity of the "new cosmology" has arisen only because the Big Bang concept has latterly been adopted as creed, and hence, as John Archibald Wheeler has said, since the Big Bang happened it has to be explained, and quantum fluctuation seemed to be the only candidate. Although Werner Heisenberg conceived the uncertainty principle solely for subatomic phenomena, where the process of physical measurement introduces seemingly inescapable uncertainty for the simultaneous determination of position and velocity of a subnuclear particle (or is it a wave?), the quantum theorists have taken greater and greater liberties in the macroscopic world until current models of the convolutions of space and gravity have become psychedelic (see for example Bryce De Witt's review of quantum gravity in the 1984 *Scientific American*). Mathematicians revel in harmless sophisticated fantasies, and new-cosmologists buy them as real estate.

Over recent decades, there has been debate over the relative merits of the Big-Bang theory (according to which all matter and even time and space began less than 20 billion years ago) and steady-state theories (which see neither beginning nor end). Latterly the Big Bang has been favored because there may be more radio galaxies and quasars per unit volume with increasing distance (as judged by signal strength, and by red-shift in a few cases where an optical image coincides with the radio source). Also, there seems to be about four times as much hydrogen in the universe as helium, which fits the theory of the genesis of the elements based on the Big Bang. But a marked shift toward acceptance followed the accidental discovery in 1965 by two Bell Laboratories physicists, R. W. Wilson and A. A. Penzias, of a universal background radiation resembling that emitted by a "black body" near absolute-zero temperature. Such residual radiation had been predicted by George Gamow as the attenuated residue of the flash of the Big Bang. The validity of this universal radio noise has been confirmed, but its attribution to an initial Big Bang remains merely a speculation. Instead, I suggest that the background radiation is the inevitable expression of what was called Olbers' paradox.

Olbers' Paradox

Heinrich Matthäus Wilhelm Olbers (1758–1840), a Bremen physician and amateur astronomer famous for his discovery of several asteroids and comets, drew attention in 1826 to a paradox (which had been reported 82 years earlier by a Swiss astronomer, Philippe de Chéseaux, and indeed by several others dating back to Thomas Diggers in 1596—Olbers did not mention that!): if stars were uniformly distributed throughout infinite space, rays in all directions would eventually meet stars, and the sky would be flooded with lethal light like an infinite searing Sun. Intervening absorbing matter would not

The Earth and Cosmology

protect us, because eventually it would radiate as much as it received. Lord Kelvin's answer to the paradox was that as the light output of stars came from their gravitational contraction (as he believed), which limited their lives to 100 million years, the transit time of their light was much longer than the duration of their luminosity, so that even if they extended to infinity, we would receive the light from only a small fraction of them at any one time. It was Hubble's discovery that all galaxies are receding from us, with a velocity proportional to their distance from us, that finally solved Olbers' paradox, because of an effect explained in 1842 by an Austrian mathematician and physicist, Christian Johann Doppler (1803–55).

When a train approaches at high speed, the pitch of its whistle is raised to a higher tone, but the pitch drops as the train passes and speeds away. The same happens to the color of the train's headlight—the approaching velocity shifts the color a little along the spectrum toward higher frequencies—but because the approach velocity of the train is so small compared to the velocity of light, it would be very difficult indeed to detect the approach-blueing and recession-reddening of the train's light.

But the velocity of recession of distant galaxies is not so small a fraction of the velocity of light, and the Doppler color shift toward the red end of the spectrum is quite significant and measurable. Indeed, for the most distant observable galaxies, their light has already been shifted to longer and longer wavelengths beyond the red end of the visible spectrum into the infrared. But at this distance they have first become faint beyond the limit of detectability of optical telescopes, and then beyond the limit of detectability of radio-telescopes, and so the universe fades out.

This is only a limit of our technology. The gaps between rays that would meet a galaxy (if we could see them fainter and fainter and at longer and longer wavelength) get less and less, until the receding cloud of galaxies begins to appear continuous to the coarse screen of microwaves. Then Olbers' continuous light dawns, not with blinding brightness and searing heat as he imagined, but as a universal black-body radiation at wavelengths near the limit of radio detection and at temperature less than 3° above absolute zero. As this radiation does not originate from a single emitter but from a vast cloud of galaxies too close together in direction to be differentiated by the antenna, but at vastly different distances and red-shift, the combined radiation would be uniform across the spectrum, and hence appear as black-body radiation. I call this peephole on the remotest realms of the universe "Olbers' Window."

Then even this fleeting farewell flash of the receding galaxies finally fades forever as the radio wavelength stretches beyond our technology. The absolute limit of the universe we could ever detect by any future means would be reached when the velocity of recession approaches the velocity of light. At that limit the wavelength of the light reaching us would approach infinity, the time taken for it to reach us would also approach infinity, and its intensity would approach zero. Whatever may happen beyond is forever outside our physical ken. In other words, we would see nothing whatever, even though astronomers in those galaxies beyond our knowable universe would see a universe receding from them just like what we see. Light that left our galaxy some 10 billion years ago is now reaching the galaxies on the edge of our knowable universe as microwave black-body radiation. To an observer there, we would be on the limit of his knowable universe. Just as on the high seas a mariner on our horizon would see us on his horizon, but the ocean would extend beyond him outside the limits of our vision, so, beyond our knowable universe, the cosmos goes on and on forever. The boundary of our universe is our own artifact, no more a real boundary than a mariner's horizon.

Newton Attraction and Hubble Repulsion

If Newton's law of universal gravitation were truly universal, all galaxies would attract one another, however far apart they might be, which is contrary to Hubble's empirical law of galactic recession. This really should not surprise us, because Newton's law was also purely empirical. There was no *a priori* reason why it should be so. It was based on the detailed observations of Tycho Brahe, which enabled Johannes Kepler to announce in 1619, after six years of agonizing working and reworking of Brahe's data, that the orbit of Mars is an ellipse with the sun at one focus. Newton showed that this would happen if each body attracted every other body with a force that is directly proportional to the product of their masses and inversely proportional to the square of the distance between them, and further that this assumption explains the weights and motions of bodies on the earth's surface. Hence Newton's law. But, as I emphasized in discussing the effects of scale on physical phenomena, such an empirical law should not be assumed necessarily to remain valid when extrapolated beyond the range of the observations on which it was based, in this case the solar system.

I can best illustrate this by comparing the Newton and Hubble laws with the empirical laws of deformation. Hooke found that in elastic deformation, stress is proportional to strain, which was expressed as

The Earth and Cosmology

$$s = \frac{p}{\mu} \qquad (1)$$

where μ is the elastic modulus, called the rigidity, s is the deformation, called the strain, and p is the deforming force, called the stress. At about the same time Newton found that flow deformation is proportional to the viscosity and the duration of the loading, which was expressed as

$$s = \frac{pt}{\eta} \qquad (2)$$

where s and p are the same as before, t is the duration of the loading, and η is the empirical viscosity modulus. Where both elastic and viscous deformation occurred together, the Newton and Hooke laws were combined by Maxwell, just as I combine the Newton and Hubble laws where both are involved. So we have the Maxwell deformation equation:

$$s = \frac{p}{\mu} + \frac{pt}{\eta} \qquad (3)$$

Further experiments with the deformation of steel, ice, and marble showed that whereas elastic deformation is due to straining the relative positions of atoms from their equilibrium positions and recovers when the stress is removed, and whereas viscous deformation is due to diffusion of atoms, which happens statistically to atoms with higher thermal vibration energy, still other deformation modes occur such as slippage on glide planes in crystalline grains. So the empirical equation was further extended to the form:

$$s = \frac{p}{\mu} + \frac{pt}{\eta} + \beta t^{\frac{1}{3}} \qquad (4)$$

where β is another empirical constant. Physics is full of examples where such empirical laws fail when they are extrapolated beyond the range of the experiments that gave birth to them.

What has all this to do with the apparent conflict of Newton's law of universal gravitation and Hubble's law of galactic recession? It simply means that although both laws are valid, the Hubble term was too small to have been detected in Brahe's observations of the orbit of Mars on which Kepler founded his law, on which Newton in turn formulated his law of universal gravitation. In my Johnston Memorial Address in 1976, I combined the Newton and Hubble laws in a single equation, just as Maxwell had combined the elastic and viscous laws:

$$F = Gm_1 m_2 \left(\frac{1}{d^2} - \frac{ad^2 H^4}{c^4} \right) \qquad (5)$$

G is the gravitational constant, m_1 and m_2 are the masses of the two bodies, H is the velocity of galactic recession found by Hubble, c is the velocity of light, and a is a pure number to be determined empirically. H and c have to appear in the fourth power to maintain the consistency of dimensions. The term "dimensions" is used in physics and applied mathematics in a special technical sense to indicate the powers at which mass M, length L, and time T are involved in a physical property. It has nothing to do with its "magnitude." Thus velocity is distance divided by time, that is $L^1 T^{-1}$ (irrespective of whether it is a large or very small velocity), energy is mass multiplied by the square of velocity, that is $M^1 L^2 T^{-2}$, and so on. In any valid equation such dimensions must balance, which requires that in the combined Newton–Hubble law, H/c must be in the fourth power.

What does equation (5) mean? If we omit the last term, the equation simply reverts to Newton's law. If we omit the $1/d^2$ term, we are left only with Hubble recession. Relatively, H is an extremely small number and c is a very large number, so the fraction H/c is an extremely small fraction indeed; when raised to the fourth power it is a fantastically small quantity, so that when the distance d is only of the scale of the solar system, that term is so near zero that it cannot be detected in the planetary orbits, and what is left is Newton's law as he found it. However, the term $1/d^2$ gets smaller and smaller as the distance d gets larger, while the only part of the second term that changes is the d^2, so that term, although it starts so extremely small, gets larger and larger, until a distance must be reached where the two terms cancel to zero, and the attraction becomes zero.

This brings me to the numerical constant a. When I applied the above equation to the actual numbers in the universe with the numerical constant a at unity, d turned out to be the radius of the knowable universe. In order to apply it to the observed recession of the galaxies, a has to be 10^{20} so that Newton attraction and Hubble recession cancel each other at 10^5 light-years. I call this distance the Newton–Hubble null and illustrate it qualitatively in the graph of Fig. 105.

So my equation (5) combines Newton's empirical law and Hubble's empirical law satisfactorily. The distance at which the attraction becomes zero is the initial distance between galaxies, because at shorter distances matter tends to gravitate together to form a galaxy, whereas at greater distances cosmic repulsion occurs, which explains the otherwise curious fact that the trillions of galaxies appear to be roughly the same size—or more precisely, a Gaussian distribution about a mean. At still greater distances bodies start to repel each other, and the re-

The Earth and Cosmology

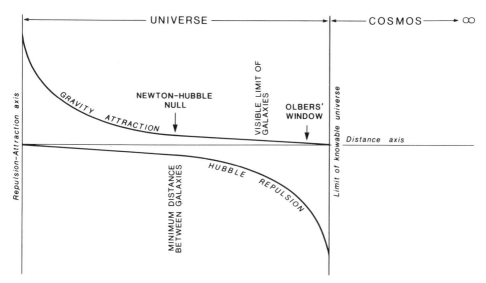

Fig. 105. Modified law of gravitation.

pulsion gets greater as the distance increases. The nearest galaxy to our Milky Way galaxy, the Greater Magellanic Cloud, is a little further than the Newton–Hubble minimum, and our nearest spiral neighbor is the great nebula M31 in the Andromeda constellation, which is about 2 million light-years away.

The velocity of light c and Hubble's constant H are precisely the constants we should expect to appear in such a generalization of Newton's law of gravitation; they are the two most fundamental constants relevant to cosmic distances, because c/H is the radius of the knowable universe; its second power is the cosmic constant, which Einstein reluctantly found it necessary to introduce into his general relativity field equations to unify the electromagnetic and gravitational fields; and its fourth power is essentially the volume of the universe in four-dimensional space-time (called the "hyper-volume"). "Knowable" refers not only to mind, but knowable by any physical criterion—energy, mass, radiation. Hence, $(H/c)^4$ in the above equation means that the rate of decline of gravitational attraction with distance, the minimum distance between galaxies, and the hyper-volume of the knowable universe are intrinsically related.

Just as more refined data made it necessary to add an additional term to Maxwell's equation, and just as the formula for the figure of the earth became progressively refined (as recounted in Chapter 2), so further observations could require refinement of the Newton–Hubble law as expressed in equation (5). The curvature of the graph of Fig. 105 does not fit closely Hubble's law as currently stated, that the velocity of galaxies is H km/s/megaparsec. But the value of H is still only roughly determined, and varies greatly from different methods of assessment.

Further study could find gross distribution patterns for the galaxies, perhaps calling for the addition of still another term. For example, three Harvard astrophysicists (John Huchra, Valérie de Lapparent, and Margaret Geller) have reported what looks like a pattern of voids some 100 million light-years across, which seem to be devoid of galaxies. "If we're right," they say, "these bubbles fill the universe just like suds filling the kitchen sink." They certainly think big to visualize soap bubbles larger than a thousand galaxies, but their "kitchen sink" could hold a billion billion of such "bubbles"!

The Null Universe

Probably as early as 5000 years ago, the philosophers of northern India denied specific creation of matter. I quote J. Arunachalam's translation from the Sanskrit: "asdva idam agra aseet; tha do vai sata jayatah: . . . tasmat swayama kurutha uchyata iti" (the universe was originally formless, later it differentiated into hundreds of things; . . . the universe is self-created—came by itself). In the second century B.C., before the Christian straitjacket on thought, Titus Lucretius Carus wrote in his *De rerum natura* (On the nature of things) that nothing can be created by divine power out of nothing. This resurfaced in Immanuel Kant's dictum (1787), *Igni de nihilo nihil, in nihilum posse reverti* (Nothing comes from nothing, nothing can revert to nothing). This I would amend to: *Omnia de nihilo gemina nasci, in nihilum gemina posse reverti* (All things are created from zero as mirror pairs, which as pairs can revert to zero).

Just like the credit and debit of a new bank loan, the null universe requires that everything in the universe cancels—matter, energy, charge, momentum, magnetic fields, spin—everything. It is natural to conceive that all charge in the universe cancels, and so also magnetic poles, and all momentum. A single particle in a universal void has neither velocity nor momentum. Once there are two particles, each has velocity and momentum relative to the other. The null universe also requires that matter and energy be opposites that mutually cancel. Matter and energy are like the two sides of a coin; one cannot exist without the other.

Dr. C. Møller, at the Brussels Solvay Conference in 1958, formulated a consistent expression for the total energy density of the universe, consisting of a matter part and a gravitational part. When this expression for the energy density is applied to the case of the metric for a homogeneous and isotropic universe, "the energy density is zero at all times. This means that the positive matter energy is constantly counterbalanced by a corresponding amount of negative gravitational energy."

Einstein commented that the most surprising fact of nature is the equivalence of inertial mass and gravitational mass. Indeed, this is the basis of his relativity. Edward Tryon, a cosmologist of the City University of New York, made the same point: "It is one of the most striking features of our Universe that, within observational uncertainties, the independently measured values of G [gravitational constant], M [mass of the observable Universe], and R [Hubble radius] satisfy the relation $GM/R = c^2$," where c is the velocity of light. Tryon commented that there was no *a priori* reason to expect this relation to be so, because G, M, R, and c are independent physical constants. But this fact, "surprising" to Einstein and "most striking" to Tryon, is inevitable in the null universe, where mass and potential energy are inseparable *gemini* that mutually cancel at all times. The above relation simply means that the inertial mass-energy (mc^2) of any body in the universe is always equal to the potential energy of the whole universe in its field (mGM/R). In other words, whenever a new mass is added to the universe, the potential energy of the universe is increased by the energy equivalent of that mass. Mass and energy are mutually cancelling opposites. Beginning from void-zero, mass and energy are added *pari passu*; their sum remains zero at all times.

This can be stated in another way. If a particle of mass m is added to the universe, it has, according to Einstein, an internal energy:

$$E_{int} = mc^2$$

At the same time the addition of the particle creates new gravitational potential energy due to its interaction with all the other particles in the universe:

$$E_{pot} = -\Sigma \frac{Gmm_1}{r_1}$$

where m_1 is the mass of the particle at a distance r_1 from the new particle m. But

$$\Sigma \frac{Gm_1}{r_1} = \frac{GM}{R}$$

which as Tryon pointed out (above) has been found empirically to equal c^2, so

$$E_{pot} = -mc^2.$$

Hence,
$$E_{int} + E_{pot} = 0$$

That is, the mass and energy added to the universe are equal and opposite; they came from zero and cancel to zero.

Mach and later Einstein agreed that the potential energy of the entire universe is the *direct cause* of inertial mass. According to Mach's principle, a single particle in a void has zero inertial mass. Infinitesimal force accelerates it. Energy is zero because inertial mass is zero, so that potential energy mgh and kinetic energy $\frac{1}{2}mv^2$ are each zero. Inertial mass is the effect of all matter in the universe. Einstein also emphasized that a single particle in a void could not have inertia, for there can be no inertia of matter against space, only inertia against other matter. Energy is a direct first-power function of inertial mass. The primitive form is mgh, which may convert to any of the other energy modes. Universal potential energy (and from this the derivative energy modes) is directly proportional to total inertial mass of the universe.

Hermann Bondi, in his 1960 book *Cosmology*, reached a similar conclusion by way of Dirac's dimensionless numbers (pure numbers, independent of length, time, and mass). In 1937, Dirac wrote: "Any two very large dimensionless numbers occurring in Nature are connected by a simple mathematical relation, in which the coefficients are of the order of magnitude unity." From what was said before about dimensions, it is clear that dimensionless numbers come only from comparing similar things, for example, a mass to a mass, an energy to an energy, a force to a force, a potential to a potential. In the following table from Bondi, e is the charge of an electron, m_e is the mass of an electron, m_p is the mass of a proton, γ is the gravitation constant, c is the velocity of light, ρ_0 is the mean density of matter in the universe, and T is the reciprocal of Hubble's constant (which has the dimensions of time). Expression (1) below is the ratio of the electrical attraction to the gravitational attraction between an electron and a proton. Expression (2) is the ratio of the radius of the "knowable universe" to the "effective" radius of an electron. (3) is the ratio of the mass of the "knowable universe" to the mass of a proton (which would be the number of atoms in the "knowable universe" if they were all hydrogen). Expression (4) divides (3) by the product of (1) and (2):

$$\frac{\text{force}}{\text{force}} = \frac{e^2}{\gamma m_p m_e} = 0.23 \times 10^{40} \tag{1}$$

$$\frac{\text{length}}{\text{length}} = \frac{cT}{e^2/m_e c^2} = 4 \times 10^{40} \tag{2}$$

$$\frac{\text{mass}}{\text{mass}} = \frac{\rho_0 c^3 T^3}{m_p} = 10^{80} \qquad (3)$$

$$\frac{\text{energy}}{\text{energy}} = \gamma \rho_0 T^2 = 1 \qquad (4)$$

The last expression shows that the potential energy of the "knowable universe" in the field of mass m equals its inertial mass:

$$\frac{m\gamma \times \text{mass of Universe}}{\text{radius of Universe}} = \frac{m\gamma \rho_0 c^3 T^3}{cT} = mc^2 \gamma \rho_0 T^2 = mc^2 \text{(from (4) above)}$$

which is the inertial mass of m. Thus, the mass and potential energy of a body are equal, the same conclusion we have reached from other routes. Mass and energy are two sides of the same coin, which come into being together, increase equally, and cancel together.

Perhaps we should not be surprised that my new formulation of the Newton–Hubble law adds another to the Dirac "large numbers," because the ratio of the radius of the knowable universe to the radius of a galactic realm is 10^{20}, and the number of stars in the universe is about 10^{20}.

Clearly, 10^0, 10^{20}, 10^{40}, and 10^{80}, are colossally separated numbers. Bondi commented: "The likelihood of coincidences between numbers of the order of 10^{40} arising for no reason is so small that it is difficult to resist the conclusion that they represent the expression of a relation between the cosmos and microphysics, a relation which is not understood." But this relation *is* now understood, because it follows automatically from the null universe.

Willem de Sitter's uniform steady-state solution of Einstein's equations was the first to predict the recession of the galaxies with speeds proportional to their distances, but it was only viable if the mass-energy density was zero; in other words, the de Sitter model was an unreal empty universe. This apparent defect vanishes if mass-energy was, is, and always will be zero.

Newton's law and Hubble's law are both empirical laws. No *a priori* reason is known why either of them should be so. It is just that the universe is found to behave that way. Newton himself was puzzled by this empirical truth, for he wrote in one of his letters to Richard Bentley:

That gravity should be innate, inherent, and essential to matter, so that one body may act upon another at a distance through a *vacuum*, without the mediation of anything else, by and through which their action and force may be conveyed from one to another, is to me so great an absurdity that I believe no man who has in philosophical matters a competent faculty of thinking, can ever fall into it.

Philosophically, we should not be surprised about the reciprocal Newton and Hubble laws, for in a null universe everything pairs and cancels; so having discovered gravitational attraction we should expect to find a complementary repulsion, governing behavior on the other side of their null point as Fig. 105 shows them to be. Newton's gravity field spreads outward from each mass; Hubble's field spreads inward from the whole universe. Newton's and Hubble's laws are *gemini*, essential to each other. Gravity acceleration is independent of the mass of the body accelerated. Similarly, the Hubble rate of recession increase is independent of the mass of the body accelerated. Applying Parkinson's rationalization of fundamental dimensions discussed below, the Newton acceleration and the Hubble recession have the same dimensions: T^{-1}.

The Cosmological Principle

The steady-state concept goes back to Heraclitus of Ephesus (535–475 B.C.), whose living universe had neither beginning nor end, with constant turmoil, generation, and destruction, but without overall progressive change. A century later, Plato reverted to a definite beginning when God created the world and established the laws of nature. Then Aristotle rejected a beginning of time and argued that land and sea and the heavens above had existed forever in a steady state of flux without progression.

In 1948, the steady-state concept was resurrected by Hermann Bondi and Tom Gold, and independently in the same year by Fred Hoyle, stating *ex cathedra* the Cosmological Principle, that an observer anywhere in the cosmos sees the same general picture in all directions as an observer anywhere else in the cosmos. This was extended further to require that an observer anywhere sees the same general picture as any other observer, not only now, but anywhere at any past or future time. This is the Perfect Cosmological Principle, a postulate that survives so long as observations do not deny it.

We need new definitions here. "Universe" and "cosmos" are commonly used as synonyms. When Hubble proved that the nebulae were distant galaxies far beyond our Milky Way, it was fashionable to call them "island universes" as some still do. So I propose that galaxies should only be called galaxies, that universe should mean our physically knowable universe, and that the cosmos should be the limitless whole, limitless in space and time according to the Perfect Cosmological Principle. However, if the Hubble recession can never reach the velocity of light, the universe and cosmos may be identical and infinite.

The Bondi–Gold–Hoyle steady-state concept still bore the lethal gene that matter has to be continuously created, which was contrary to the conservation laws. This dilemma only vanishes with the conclusion a decade ago by Tryon and by me that matter and energy are cancelling opposites. A steady-state theory must also maintain a balance between decaying radioactive elements and their accumulating stable derivatives.

The Big-Bang disciples, assuming that the velocity of recession is what is left of the velocity each galaxy got in the initial explosion, debate whether the universe will go on expanding forever or will slow down and eventually collapse under Newton's universal gravitation, perhaps to a Big Crunch whence another Big Bang would incubate still another phoenix universe, destined in turn to crunch and bang again and again like a cosmic diesel. In their view, the issue turns on the total mass of the universe: if it is high enough, no matter will escape, and the collective mass will eventually slow each receding galaxy to a halt to begin its accelerating return to the center. Current estimates of the total mass of the universe leave them without resolution whether the universe is a hyperspheroid, which must eventually collapse, or a hyper-hyperboloid, which will go on expanding forever.

The "new" cosmologists make much of their finding that the mass of the universe is so finely balanced, just enough mass to block infinite expansion and just enough to prevent eventual collapse, so finely tuned, they say, that some governing purpose is implied.

The null universe leaves no doubt on all this. There never was a Big Bang. The universe, as defined above, is constant in mass and radius. New matter is forever appearing at the Newton–Hubble null (and at other singularities as explained below), but is forever passing beyond our knowable universe. The general dogma that the universe is expanding is not really true. Certainly, every galaxy is receding, but this only expresses the perpetual accession of new galaxies, their perpetual recession, and the perpetual loss of galaxies beyond the knowable limit. This steady state is like the steady state of a stretch of river, which remains perennially unchanged although new water never ceases to enter in balance with old water going on its perennial way.

Imagine a balloon covered with spots: as the balloon expands, every spot sees every other spot receding from it, and the farther away it is the faster it is receding. This is a model of the Hubble recession, except that the balloon is a two-dimensional surface deployed in three dimensions, whereas the universe is a three-dimensional system deployed in four dimensions of space-time. If you now imagine that whenever the distance between adjacent spots gets large, a new spot appears between them, you have a steady-state universe, which is always statistically the same even though it is continuously expanding. But the mass, size, and number of galaxies in the physically knowable universe remain forever about the same, because as new galaxies constantly develop in the voids, other galaxies continuously vanish beyond the knowable horizon.

In 1966 Dr. Richard Stothers, of the Goddard Space Center, "postulated that matter is created *where* it is lacking (between galaxy clusters) *because* it is lacking (due to the universal expansion)."

The Newton–Hubble null is a singular point where the gravity acceleration is zero, but the potential energy of a particle to fall to a galactic center is a maximum. Hence this region is swept clear of matter to maintain the rarest vacuum in the universe. Matter may enter here by random quantum fluctuations, but it does not remain there. None of the fatal flaws of the Big-Bang "new cosmology" remain. Because there is no force at the null and hence no acceleration, newly created matter should accumulate there as tenuous gas from which stars are born, seeds of a nascent galaxy, destined to grow until its domain reaches its own Newton–Hubble null. So galaxies have a statistical size throughout the universe, a fact hitherto unexplained.

The locus of maximum potential energy is a surface normal to the line joining the two nearest galaxies; perhaps this is why galaxies tend to be planar. Because adjacent galaxies beget new matter between them, there should be a tendency to form galaxy clusters, as indeed is found to be so.

In contrast with the Newton–Hubble null, any center of mass, be it of a planet, a star, or a galaxy, has a minimum in potential energy, where vacuum fluctuations face zero or at least a minimal energy barrier. At the center of the earth the gravity acceleration is zero. Hence matter should be entering there by random quantum fluctuations and at all other such places. In 1928, Sir James Jeans wrote in his *Essays on Cosmogeny*: "The type of conjecture which presents itself somewhat consistently, is that the centres of nebulae are of the nature of singular points at which matter is poured into our universe from some other and entirely extraneous spatial dimension, so that to a denizen of our universe they appear as points at which matter is being continuously created."

This view has been supported by Professor William McCrea of Sussex University. The energy barrier just outside the singularity rises ever more steeply as the mass concentration increases, so the rate of creation would not rise linearly with mass concentration. Herein could perhaps lie the explanation for my empiricism that the expansion of the earth proceeded very slowly before rapid acceleration 100 million years ago; this rate cannot be extrapolated linearly to the sun.

Thus the apparently contradictory opinions of Jeans and Stothers on where new matter enters the universe may both be right. But there is a difference: matter entering at the Newton–Hubble null has a large potential energy to fall to the nearest galaxy, so this region always tends toward absolute vacuum; in contrast, near any center of mass, matter enters at a potential minimum, so this region always tends toward waxing mass concentration.

Of course, matter plus energy may leave the system at such singularities as well as enter it, because quantum fluctuations should be reversible. Here again the process should not be expected to be linear. As mass concentration increases, this very fact may increase the rate of the reverse process.

In what form does new matter appear in the universe? Dirac has debated the relative merits of "multiplicative" creation, where existing atoms beget new atoms, and "additive" creation, where matter appears first in much simpler subatomic form. He "presumed" that the nascent matter of multiplicative creation "consists of the same kinds of atoms as those already existing." Chao-Wen Chin and Stothers then asked: "Why are well-preserved Precambrian and Early Cambrian fossils [buried more than 500 million years ago] in essentially perfect shape if their masses have increased by a significant percentage?" According to the null model, the probability of quantum fluctuations begetting new matter is vanishingly small except at intergalactic nulls or near the centers of mass concentration where the gravity acceleration approaches zero. Hence expansion occurs at the center of the earth, and the Cambrian fossils and other rocks of the lithosphere are not affected, except by the circumference increase to accommodate the growing core.

Moreover, Dirac's presumption that silicon atoms would beget silicon and that atoms generally would beget their kind is surely improbable in the highest degree. The likely form of matter arising from quantum fluctuations would be in the simplest state conceivable—"infra-quarks" or whatever, quarks being the simplest form of subatomic particles so far conceived. We might then expect that the first atoms would be atomic hydrogen, the simplest atom, and the most abundant atom in the universe. However, even these self-evident facts (on which axioms are based) could be false clues.

All atoms are made up of nucleons (protons and neutrons) and electrons, only one nucleon in hydrogen, 56 in iron, and 238 in uranium. Of all elements, iron has the least energy per nucleon. Less configurational energy is needed per neutron and proton to build iron atoms than for any other atom. So could iron rather than hydrogen be the first element to appear? Certainly incandescent iron observed in the universe is less than 1 percent, and current dogma holds that iron is only produced via slow nuclear cooking or in supernova explosions. But it is a fact that, so far as we know the universe, *small cold* bodies, such as meteorites, asteroids, and planets, do have a much greater proportion of iron, and even among our own planets, the proportion of iron diminishes and the proportion of hydrogen increases with their size. Much recent work suggests that there is at least as much invisible mass in the universe as the visible hydrogen-dominant stars. Could this be small dark iron-rich bodies?

Could it be that as terrestrial planets grow, the energy available from increasing pressure at the core leads to progressively lighter elements absorbing more energy per nucleon, until the Jupiter stage is reached, when hydrogen is already dominant, and the planet begins to radiate like a star? All bodies we can see in the universe appear to be mainly hydrogen, but did they have terrestrial precursors dominated by iron?

The minimum-energy status of iron is also relevant to the release of nuclear energy. Uranium and transuranium elements, which have the highest configurational energy per nucleon, fission to form a pair of much smaller elements, nearer to iron in the periodic table, that have less energy per nucleon than uranium, and this is the energy that is released. Likewise, atoms of small elements like helium and hydrogen, which also have much more energy per nucleon than iron, potentially may fuse to form elements nearer to iron, with less energy per nucleon than they had, and this energy difference is released. In both fission of heavy elements and fusion of light elements, it is only configurational energy of the nucleus that is accessible, never the inertial energy of the constituent nucleons. Neither fission nor fusion energy could ever be released from iron.

In this respect, nuclear energy is analogous to various forms of chemical energy. We charge a battery by adding electrical energy that raises the oxidation state of lead oxide. We get this energy back as the lead oxide reverts to the lower oxidation state. We burn gasoline (which had stored solar energy from long ago in the electron shells of its constituent atoms), and we exhaust carbon dioxide and water, which have less energy in the electron shells of the atoms than had the gasoline, and the energy difference accelerates our car. We can store elastic energy in a bow, and release it to shoot an arrow. In nuclear energy, the energy is stored in the configuration of the nucleons and mesons within the atomic nucleus; in chemical energy, the energy is stored in the configuration of the electron shells outside the atomic nucleus; in elastic energy, the energy is stored in the configuration of whole molecules with respect to each other.

Elton's Planetary Evolution Process

Sam Elton, a geophysicist of Manhattan Beach, California, I. V. Kirillov and V. B. Neiman of Moscow, V. F. Blinov of Kiev, and Branislav Ćirić of Belgrade, and, more recently, Jacob Ehrensperger of Winterthur, Switzerland, have each independently proposed a startling supplementary proposition, whereby not only are all stars, planets, and satellites in a state of increasing mass, but the solar system was at an earlier stage like Jupiter with his moons is now, and also that the present solar system is the embryo of a future galaxy.

In Elton's model each galaxy has grown from a solar system, which had grown from an embryonic star, which had evolved from a primitive gas cloud, which had accumulated from the spontaneous quantum fluctuations at the Newton–Hubble nulls between galaxies. Stated in reverse order, this concept is precisely what is implied by my statement at the beginning of this chapter that "regression of Hubble's expanding universe must involve diminishing mass, so that the initial embryo, far from the unthinkable mass contemplated by Dirac, Friedmann, Lemaître, Eddington, and Gamow, had negligible mass, or *no mass at all!*" Such a progression has been going on forever in the past, and will continue forever into the future. Elton has emphasized that we must abandon static concepts of a fixed amount of matter impulsively created in a single act in the "beginning" and instead recognize the universe as a continuing steady-state *process*.

In such a process, not only stars but also galaxies would exist in all stages of evolution, each ruling its region out to its Newton–Hubble null, some young and full of gas and nascent stars, others mature spirals, and finally old-star ellipticals with little residual gas. Although each galaxy has its full life cycle, the overall pattern of the universe would be the same if we could see it a thousand million years ago or a thousand million years hence—just as a Paleolithic family or one today would each have its babes, its parents, and its moribund senescents. As most galaxies would commence between a pair of mutually retreating galaxies, matter accretion would be dominated by them in two opposite streams, which would become spiral as they gravitated toward the slowly rotating center; but more complex patterns could occur.

Reverting to Elton's model, certainly Jupiter is more like a dim star than a planet. Dr. T. R. McDonough, of Cornell University, wrote in 1974: "Jupiter resembles the Sun more than the Earth. In composition, energy production, differential rotation, field eccentricity, and the interaction of its plasma with its magnetic field, Jupiter displays truly stellar properties."

The subsequent NASA flyby confirmed Jupiter's embryonic star status. The systematic spacing of Jupiter's satellites mimics that of the sun. In 1772, the Prussian astronomer Johann Daniel Titius stated his famous empirical law of planetary distances from the sun, now better known as "Bode's law" after the editor of *Astronomisches Jahrbuch*, who published Titius's discovery (and apparently took the credit).

	Mercury	*Venus*	*Earth*	*Mars*	*Asteroids*	*Jupiter*	*Saturn*	*Uranus*	*Neptune*	*Pluto*
A	4	4	4	4	4	4	4	4	4	4
B	0	3	6	12	24	48	96	192	–	384
C	4	7	10	16	28	52	100	196	–	388
D	3.9	7.2	10	15.2	28	52	95.4	192	307	395

In this tabulation, each planet is given an initial 4 (line A), to which a geometric series starting with 3 and multiplying by 2 (line B) is added to give line C. For comparison, line D gives the actual distances of the planets from the sun, taking the earth's distance as 10. The match is good enough to have predicted the existence of Uranus and the asteroids before their discovery, but is not perfect, because to be consistent in line B Mercury should not be 0 but 1½, and Neptune and Pluto have to be taken together, which may not be unreasonable because they are in mutual resonance-captive orbits, as are Earth and Moon. But as I pointed out a decade ago, this Titius–Bode law hinges on the earth's distance, whereas any but an egoistic earthling would have recognized the special status of Jupiter rather than of Earth. When I took Jupiter as the fulcrum, the initial line of 4s was not needed, Mercury became regular, and Neptune and Pluto were less aberrant:

	Mercury	Venus	Earth	Mars	Asteroids	Jupiter	Saturn	Uranus	Neptune	Pluto	?
A	$\frac{1}{48}$	$\frac{1}{24}$	$\frac{1}{12}$	$\frac{1}{6}$	$\frac{1}{3}$	1	3	6	12	24	48
C	$\frac{1}{49.2}$	$\frac{1}{19.3}$	$\frac{1}{11.9}$	$\frac{1}{6.3}$	$\frac{1}{3}$	1	2.5	7.1	13.9	20.9	?

This tabulation can be expressed in the form $T_n = 3J \cdot 2^{n-1}$, in which J is Jupiter's sidereal period (that is, his period as seen from a star, as distinct from his synodic period, which is his apparent period as seen from Earth) and T_n is the period (as seen from Jupiter) of the nth planet beyond Jupiter; the reciprocal of this expression applies to the planets inside Jupiter. For symmetry, another planet is predicted beyond Pluto. I called this the Jove–Titius law. The interesting point in connection with Elton's model is that a law similar to this law for the solar system can be written for Jupiter and his satellites, and also for Saturn and for Uranus. Each of these satellite systems presents the same enigma as the solar system, namely that if they condensed from a cloud mass as postulated by the nebular hypothesis, the primary body should carry most of the angular momentum instead of the small fraction actually found. This enigma vanishes if the mass of the central body has mainly grown from spontaneous creation of new matter there rather than by condensation from a rotating gas cloud.

If we accept the secular increase in mass for all these bodies, then Jupiter, Saturn, and Uranus are destined to become "solar systems." Indeed, four physicists from Queen's University at Kingston, Ontario, have reported in *Nature*: "Using the at present accepted value of Hubble's constant, $H = 100$ km/s/megaparsec, which is 1.65×10^{-4} mm per year per mile, and substituting the value of the Earth's radius in the Hubble equation, $v = RH$, we obtain a radial expansion for the Earth of 0.66 mm per year." A Swedish physicist, H. B. Klepp, followed this up, also in *Nature*, with the observation that applying the same Hubble law to the moon's orbit would imply expansion of the orbit at 3.9 cm per year.

General expansion of the planetary orbits, together with implied progressive change in the luminosity of the sun, should have shown up in the solar heat received at the surface of the earth in the geological past. We do know that seas existed and rain fell and rivers flowed as far back as thousands of millions of years ago, along with intermittent glaciation; no systematic climatic trend has been observed in the geologic record. Two NASA cosmologists, Chao-Wen Chin and Richard Stothers, investigated this question, assuming that mass grows in proportion to its own concentration, and reported in *Nature* "the rather surprising result that [these solar models] are nearly the same as those based on standard theory! This occurs in spite of the widely disparate masses, occasioned by a full range of choices for t_0. The reason for this similarity is that the effect of a larger G in the past is to increase the luminosity, whereas lower stellar mass decreases it." They went on to investigate the effects on the orbits and concluded that the deduced temperatures at the surface of the earth do not conflict with the paleogeographic evidence.

In summary, the unified Newton–Hubble law governs all motion in the universe, determining the volume of a galactic domain and the hypervolume of the universe. The universal null reinstates the steady-state universe as an everlasting aspect of zero without the crucial enigma of previous steady-state models of matter continuously created from nothing. Infinitely in the past, matter and energy appeared from random vacuum fluctuations at subnuclear level at quasi-zero energy barriers as canceling opposites, continued to grow and continued to condense to stars and growing galaxies, which have mutually dispersed. The mass, energy, and radius of the knowable universe have always been the same, limited by the velocity of recession reaching the velocity of light, and of the propagation of the gravity potential energy field. All laws of nature and the Perfect Cosmological Principle remain universally true in an infinite cosmos with no unique time, no unique place, no beginning, and no end—the ultimate steady state.

CHAPTER 24

Philosophical Speculations

SEEKING COSMIC harmony for my conclusion that both mass and volume of the earth had increased through time led me into strange new fields along the meandering path to a coherent philosophy. Along the way many exotic propositions, from hard-nosed logic to Disneyland fantasy, lured me to digress and contemplate them. My first intention had been to end this discourse at the point just reached, but, as these speculations were a by-product of the quest, I decided to close with a medley of them for love or scorn.

Antimatter and Black Holes

Conventionally "antimatter" is regarded as the opposite of matter, as indeed this name would indicate. This has fogged realization that it is energy that is the opposite of matter. But the name "antimatter" was ill-chosen and has channeled thought up a blind alley. The positron is the "antiparticle" of the electron, but it is "anti" only in charge. The gram measures its mass, not the *marg*, and the erg measures its energy, not the *gre*. "Negative matter" would have been a little better name. When Tryon first proposed that the universe was a quantum fluctuation of zero, he predicted (because of this false lead) that it should contain equal quantities of matter and antimatter, which is not supported by observation. What he should have predicted is that the universe should contain equal amounts of matter and energy, which of course is clearly so.

Another very badly chosen name is "black hole." A schoolboy definition of a hole is "nothing with something all round it," which is accurate in its conception and rather neat. A "black hole" is as far from this as it could possibly be, being a prodigious concentration of matter, certainly not a "hole." Indeed, it would be more appropriate if it were given the name of its discoverer, Schwarzschild, which happens to mean "black shield."

Gravity Waves

The conclusion that the inertial mass of a body is caused by and is equal to the potential energy of the whole universe in its field implies that when new matter appears, its gravitational field is established throughout the universe. How does this new gravity field propagate? When is the existence of the new field felt at a distant galaxy? Instantaneously? Or does it reach there traveling at the velocity of light? This would be "instantaneous" on relativistic "world lines." Is the propagation of the field in the form of gravity waves? Such have been sought but, so far as I am aware, not yet unambiguously detected.

Gravity waves should be like the propagation of light—"waves" traveling through vacuum space without attenuation, at the velocity of light. But gravity waves could not be wholly like light waves, which, because they involve the interaction of two parameters (electric and magnetic), oscillate in the plane of two axes normal to the ray. That spectrum is wholly occupied, from the lowest frequency limit in the very long wavelength radio waves, through radar, heat, light, and x-rays, to the highest frequency limit. But the longitudinal wave spectrum (the vacuum equivalent of sound and seismic compressional waves) is vacant and unknown. Like the missing elements in Mendeleev's periodic table, the very vacancy itself proclaims the need for new discovery. Surely, such waves should exist, and surely that is where gravity-field propagation, which involves only one parameter on one axis, could be found. The experiments seeking to detect gravity waves have involved two masses in the plane normal to the waves sought. If my induction is correct, this would not detect them even if present, because the real action would be in the direction of the ray. Also, gravity waves should be transients, propagating from any new mass to establish a gravity potential field.

Complementary electric, magnetic, and gravity fields, mutually at right angles, and propagated by transverse electromagnetic energy waves and longitudinal gravity potential waves would fill the theoretically possible states and would complete the elegance of the matter-energy *gemini*—contrasted, inseparable, and mutually canceling to zero.

Unfortunately, this induction is contrary to the general theory of relativity, which plots the gravity field with two axes normal to the gravitational force vector, like the electromagnetic field. Huxley lamented the great tragedy of science, the slaying of a beautiful hypothesis by an ugly fact—or in this case, by a beautiful theory.

One of the unanswered questions of the universe is why antimatter is generally absent. In the null universe paradigm adopted here, most matter enters the universe by quantum fluctuations near mass concentrations where there is a consistent potential energy gradient toward the center of mass. Therefore the tensor defining the site of creation of a particle would be asymmetrical, in that gravity waves in the direction normal to the electrical and magnetic parameters would always be propagating away from the center of mass, never toward it. It is conceivable that this asymmetry would also bias the electrical polarity and yield matter rather than antimatter.

Dimensional Equivalence of Mass and Energy

The apposition of mass and energy as mutually canceling opposites involved the apparent anomaly that mass and energy are not dimensionally equivalent, as mass is M and energy is ML^2T^{-2}. In this I had taken it for granted that the dimensions L, M, and T were fundamental and universally valid. However, Dr. W. D. Parkinson removed my blindfold by pointing out not only that such dimensions are indeed arbitrary, crystallized from our human concepts of reality, but that the classical adherence to them involves anomalies in electromagnetism, which can be avoided by making T and L codimensional, leaving only two dimensions, L and M.

I repeat Parkinson's argument verbatim, but this paragraph may be skipped by any reader unfamiliar with the physical theory and mathematics involved.

One of the reasons for introducing S.I. units [Standard International Units] was the incompatibility of e.m.u. and e.s.u. [electromagnetic units and electrostatic units]. Electromagnetic units are based on the magnetic pole (m) such that the force of interaction is

$$F = \frac{m_1 m_2}{r^2} \quad (1)$$

in vacuo, giving dimensions

$$(m) = M^{\frac{1}{2}} L^{\frac{3}{2}} T^{-1}$$

and for magnetic field

$$(H) = M^{\frac{1}{2}} L^{-\frac{1}{2}} T^{-1}$$

Electrostatic units take electric charge (q) as fundamental, such that the force of interaction *in vacuo* is

$$F = \frac{q_1 q_2}{r^2} \qquad (2)$$

giving dimensions

$$(q) = M^{\frac{1}{2}} L^{\frac{3}{2}} T^{-1}$$

and for the electric field

$$(E) = M^{\frac{1}{2}} L^{-\frac{1}{2}} T^{-1}$$

But electromagnetic induction indicates the following relations:

$$(H) = M^{\frac{1}{2}} L^{\frac{1}{2}} T^{-2} = (E) L T^{-1}$$
$$(E) = M^{\frac{1}{2}} L^{\frac{1}{2}} T^{-2} = (H) L T^{-1}$$

This incompatibility can be removed by inserting a dimensional constant into either (1) or (2) with the dimensions $L^{-2} T^2$. The S.I. overcomes this incompatibility by assigning the constant μ to (1) and $1/\varepsilon$ to (2), such that $(\mu)(\varepsilon) = L^{-2} T^2$ and takes electric current as a fourth fundamental dimension. This is necessary because the above does not assign dimensions to either μ or ε but only to their product.

There is another way out of this impasse. If length and time are assigned the same dimensions, say L, then the discrepancy disappears. All physical quantities can then be expressed by the two dimensions M and L only. We must then redefine the unit of time. If this is taken as the "light-meter," i.e. the time taken for light to travel one meter *in vacuo*, then velocity is non-dimensional, being a fraction of c (the quantity usually expressed by β in relativity theory). A suitable name for the unit of velocity would be *stein*. Light travels at a velocity (in German) of *Ein-stein*.

The dimensional equivalence of distance and time is implied by the null universe, because every increment of mass implies an equal increment of energy, that is, mass M equals energy $ML^2 T^{-2}$. Canceling M leaves $L^2 = T^2$.

An important consequence of Parkinson's simplification also is that mass, momentum, and potential energy have the same dimensions, and Hubble's constant has the same dimensions as gravity acceleration. The significance of this revision goes far beyond the elegant removal of the incompatibility of the electromagnetic and electrostatic units, which hitherto had been adjusted by an arbitrary improbable fiddle. Indeed, the dimensional equivalence of length and time is implied by the four-dimensional space-time of relativity, where time and distance are interchangeable; whereas the coordinates were written in the form x, y, z, and ict (i is the imaginary square root of minus one, which appears in the four-dimensional equations); t now has the same dimensions as the other three, c is a pure number, and i is attachable equally to any one of the four.

The reduction of fundamental dimensions from four to two is a step toward Eddington's philosophical hope of linking universal constants to reduce their number, hopefully to a mutually canceling pair, and toward Milne's ideal cosmology, which has no constants with dimensions, and also toward Einstein's faith that all universal constants should have logical inevitability and that no dimensionless constants should be arbitrary.

By Parkinson's simplification, the velocity of light c becomes a pure number like π and e, whose values are absolutely determined, π geometrically and e arithmetically. Two pure numbers, the velocity of light and absolute zero, are the physical limits of existence.

Appropriately, the recent International Conference on Weights and Measures has redefined the meter, the basic unit of length, in terms of time, as the distance traveled by light in a vacuum during 1/299,722,458 of a second. This defines the meter 10 times more accurately than the earlier definition in terms of the krypton-86 orange line.

Cosmic Rotation

The dimensions of Hubble's constant are intriguing in relation to cosmic rotation. One of the most striking facts of the universe is that almost everything in it rotates. Even the smallest subnuclear particles have spin. The nucleus of every hydrogen atom spins, and this is used in the proton-precession magnetometer to measure the strength of a magnetic field. Planets rotate on their axes and revolve in their orbits. Galaxies and spiral nebulae rotate. Quasars rotate in a day or so, and pulsars rotate in a second or so.

Why should all this be so? What is implied? Laplace assumed that rotary motion was an inherent property of matter. Kant even more vaguely assumed random rotary movements in his primitive nebula, which had gradually become uniform; but if such rotations had been random on a cosmic scale, consolidation of them should have canceled to zero. According to the universal null, all the angular momenta of the universe cancel to zero. Four decades ago, P. M. S. Blackett hypothesized that magnetism was a fundamental property of rotating bodies, and proceeded experimentally to prove that his induction was wrong.

Physics still does not know why rotation is so universal, so some new principle may remain to be discovered. Perhaps Hubble's constant may be involved. It is an interesting fact that the "dimensions" of Hubble's constant, $M^0 L^0 T^{-1}$ or simply T^{-1}, is the same as the dimensions of angular velocity, which is measured by distance traveled along the circumference divided by the radius (hence L^0) divided by the time taken (hence T^{-1}). Is this a mere coincidence, or does it have a physical meaning? Although angular velocity and Hubble's constant have the same dimensions, there is a vectorial difference: as we move out along the radius of a rotating body, linear velocity at right angles to the radius increases directly in proportion to the radius; the same is true with the Hubble recession, except that the waxing velocity is *along* the radius instead of at right angles to it. Are angular rotation and cosmic recession fundamental properties of the universe, related to each other in a way similar to magnetic and electrical vectors?

Mass, Substance, Energy, and Mind

To most philosophers, right up to this day, mass and matter implied "substance," whereas energy was a mental concept, real but not "substance." But my apposition of mass and energy as mutually canceling opposites implies that ponderable and tangible "substance" (matter) is equated to imponderable intangible energy. This instinctive barrier of the human mind had already been removed by Einstein, but not without intuitive distress to Max von Laue and some other contemporary physicists. David Bohm, biologist-philosopher of the University of London, has said that physics deals with "real properties" such as mass, length, time, charge, etc., which are supposed to exist "out there" independently of human beings, while qualities like harmony and conflict, beauty and ugliness, are supposed to exist only in the eye of the beholder. But those supposedly "real properties" have been created in the mind of man. A few thousand years ago, nobody felt that these qualities are what is "out there."

The solidity of substance is but a deception of our senses. We *feel* a quartz crystal and *know* its solidity. But light and x-ray diffraction show that quartz is mostly empty space, with distances between the atoms very much greater than their size. But at least the atoms are solid—until it is shown that an atom is mostly empty space, with the distance between its constituent electrons and nucleus very much greater than their own size. But at least the atom's electrons and nucleus are matter with mass and substance, until the nucleus is found to be mostly empty space like a planetary system of protons, neutrons, and many kinds of mesons. By now it is quite apparent that the proportion of empty space to the total volume of the fundamental particles of the "solid" quartz crystal is at least as much as the proportion of empty space between the stars of a galaxy! But at least the protons, neutrons, electrons, and mesons are solid. Are they? Contemporary particle physicists say that they in turn are composed from 15 kinds of quarks.

We can do repeatable physical experiments to detect and quantify properties we call mass, charge, and spin (as well as other charming strange properties that dumbly have been called "charm" and "strangeness"), but we have no idea what any of these properties really are. The more fundamentally we examine these properties the more intangible they become. Charge and spin involve energy at the macroscopic level, and kinetic energy adds relativistic mass, and according to Mach's principle, the inertial mass of a single particle in a universal void would be zero, and its mass only comes into being by virtue of its potential energy to move toward other particles. An electron and a positron, each of which has mass, can wed and give up their ponderability to become massless imponderable photons, which are pure energy! Clearly our intuitive axiom of substance has deceived us.

By contrast, the Irish metaphysical philosopher, Bishop George Berkeley (1685–1743), in revulsion from emergent French materialism and prevailing free thinking, asserted in his 1710 book, *Principles of Human Knowledge*, that matter has no existence except in the mind of man. A thing that is not perceived cannot be known, and not being known, cannot exist. Berkeley's sophistry flows from his axiomatic creed of the divinity of mind. Younger than Hooke and Steno, but older than Cuvier, Bishop Berkeley had no inkling of fossil proteinoid microspheres in the Isua Quartzites, the most advanced life nearly 4 billion years ago, long before any human mind could contemplate and materialize anything.

Fogs of Notation

Our habitual channels of thinking can conceal simple truths even though the thought channel itself be quite valid. To multiply 1444 by 3888 is easy for us, but to the schoolboys of Rome to multiply MCDXLIV by MMMDCCCLXXXVIII was a problem quite beyond them, and their masters. The Augean task facing the Roman student was not intrinsic to the integers, but stems entirely from the manmade notation for identifying them. The Roman notation is nevertheless perfectly valid, and for some purposes could be the ideal notation. To write the motions of the planets as elliptic orbits about the solar focus, then transfer the origin of coordinates to the earth, would describe their motions in a way very similar to Ptolemy's, but the algebraic expressions would be awkward, although perfectly valid. Unfortunately our ancestors started deeply grooved in such a channel of thought, which obscured the simpler truth from them for millennia.

There is something in common with Achilles' race with the tortoise. There never was such a race, but philosophers loved to argue about the outcome. As the story goes, Achilles has to give the tortoise a substantial start, and a definite time must elapse before Achilles reaches the tortoise's starting point, and in that time the tortoise must surely have made some progress. So some more time must elapse before Achilles reaches this next point, but in this time the tortoise must have made some more progress, so some more time must elapse before Achilles . . . I need a broken phonograph record to keep repeating this forever! And so the philosophers argued. Achilles can never catch the tortoise because whenever Achilles gets to where the tortoise was last, that wily reptile has gone on a bit further, and there would never be an end to it. All quite true, but we have stated the problem in the form of an infinite series that has as many terms as we have time and patience to state, and still there remains an infinity of terms ahead of us.

Now all this is very simple and naïve. But so often in science we have done just that sort of thing. We look at a problem in a way that is perfectly correct, but we can never reach an answer because we have sidetracked ourselves into an infinity or a blind alley. With Roman arithmetic, the geocentric universe, and Achilles' race, the complexity is not intrinsic to our problem but is introduced by our manner of thinking about it. In the first, it was simply a choice of notation unsuitable for that problem, in the second an inappropriate choice of origin of coordinates, and the third found an infinite number of terms introduced only by our method of looking at it.

Similarly, our intuitive conviction of a fundamental difference in kind between matter and energy has misled us. Do you remember the wrangle between the wind and the sun as to who was the more influential? The wind (matter) blew on his skin and the man felt cold; the sun (energy) shone on his skin and he felt hot. Similar, but opposite! Do not try to rebut this by pointing out that wind on sails blows the ship along while the sun can't, because Crookes's radiometer has "sails" of polished metal foil blackened on one side, swiveling in a vacuum, driven by the absorption of massless photons (energy) on the black side and their reflection from the polished side.

Philosophical Speculations

I cannot resist the temptation, with my tongue only partly in my cheek, to snipe at modern nuclear physics. Could it be that the cycles of complexity of atomic, nuclear, and subnuclear particles (or are they waves?) demanded by observations reflects the constraints of our way of thinking about them?

When I was a student, we knew of 92 atoms as unsplittable immutable fundamental particles, which fitted into a periodic table so elegantly that members as yet undiscovered proclaimed their absence by their obviously vacant places. Then the unsplittable was split into protons and electrons, and then neutrons. With these new subatomic particles, the atom took on an elaborate shelled structure with scores of particles in complex groups of orbits—solar systems in miniature, each with a central nucleus which in some unknown way blended protons and neutrons into a single central sun. Then a new mysterious entity popped out of this nucleus—*the* meson. Just after World War II, a Japanese physicist was scorned when he claimed he had found *two different* mesons. But the denunciation of him was short-lived as mesons multiplied like rabbits until now we have nearly a hundred such subatomic particles, which form a shelled orbital structure within the nucleus, and which fit into a table of properties so reminiscent of the periodic table of atoms that Mathews could not resist chanting the old rhyme about big fleas having little fleas to bite 'em, while little fleas have lesser fleas and on *ad infinitum*.

Do we now begin another cycle of this circus? Will someone soon show that the "fundamental" particles (that's what we were told atoms were) are themselves still smaller solar systems? The accuracy of the laboratory observations is not in question, but does the cyclic complexity of properties demanded by the observations arise from some false axiom from way back that ossifies our way of looking at things? Some time ago, theoretical physicists discussed the state of matter and the synthesis of elements one minute after that cataclysmic Big Bang. Then after one second. Then after the lapse of one millisecond. Then after a millionth of a second—then a billionth, a trillionth, and now after only a quadrillionth of a second—quite seriously, without any thought of laughing at themselves! Does Achilles' tortoise run again? In the soundest of mathematics, Achilles will never catch that tortoise spiraling down the asymptote of an infinite series of ever-diminishing intervals. The theoreticians, now only a quadrillionth of a second or less away from that stupendous instant when time and space began and God created God, will never reach their mirage.

Can you hear with Pythagoras the harmony of the spheres? Through a thousand years, meticulous and accurate observations by thinkers as acute as any now demanded ever more crystal spheres to bear the heavenly bodies (79 of them indeed), and, as Tycho Brahe pontificated, the alternative proposed by Copernicus implied phases on Mercury and Venus, and parallax for the stars, which were positively denied by the consistent observations of the most competent astronomers in the best-equipped observatories.

David Bohm, mentioned previously, who embraced physics via biology (and hence escaped indoctrination), claims that there are no fundamental particles; that the image was all a mistake stemming from Democritus in the fifth century B.C., and thence from the interpretation of Newton in terms of a billiard-ball universe.

Are We Alone?

One of the puzzles of the universe is why we have not yet recognized signals from other thinking beings. We are flooded with diverse radiation from all over the galaxy and beyond, but have found no sign of wilful modulation. Surely, with so many quadrillions of suns comparable to ours, it is highly improbable that we are the only cognitive beings, or the most advanced. The more so if the age of the universe is infinite! Why have we not heard from them?

One answer is that they would not have seen any evidence of us—yet. A Christian fundamentalist would answer that not only the earth but the whole universe was created solely for man, so there could be no other inhabited planets. Or is eventual self-destruction inherent in any competitive system soon after it achieves the technical competence to do so, so that there never has been, nor ever will be, a race much more advanced than ours?

Then again, perhaps we are not searching the right medium. A century ago, we were quite unaware of radio waves. Perhaps when we master the modulation and demodulation of coherent light, we may find all the messages we seek. After all, light can carry vastly more information than any radio channel and may well be the preferred medium for those who know how to use it. Shall we suddenly break into a vast communication system of the galaxy? Perhaps we have to learn to communicate by some other propagating field, for example, gravity radiation, or even some medium yet to be conceived. Do minds intercommunicate, as telepathists maintain? I do not know that they do, but certainly do not know that they could not.

Contemplation of alien cognitive life is severely befogged by the assumption that life anywhere in the universe would, like ours, be based on amino acids, forgetting that our model results from 4 billion years of random chemical evolution and natural selection by our planet's particular temperature-pressure hydrosphere and atmosphere history. In other conditions, silicon might replace carbon as the central element, yielding polymers based on SiO and organofluorsilicones with silane (SiH_4) and tetrafluor silane (SiF_4) replacing methane. At moderate temperatures and pressures, silicon displays a complex array of hydration and fluoridation states, and given billions of years of random reactions in appropriate environments, who would predict the outcome? Current organo-metal research is yielding a bewildering range of stable synthetic complexes that had not emerged from natural evolution, but would have done so in different environments, and their potential horizon is far beyond current contemplation. Silicon metal, which is more electropositive than tin, is an obvious candidate for such new chemistry.

Our computers, completely devoid of organic compounds, already point to other pathways of complex reasoning, with increasing levels of redundancy, randomness, self-repair, learning by experience, and even "compassion." True, they still lack self-replication and self-generation of their energy, but surely we cannot assert that other routes are impossible.

However bold our speculations, have no doubt that our proudest achievements in science are trivial against the vast unknown beyond us.

Epilogue

WE HAVE TRACED the philosophy of Earth and the universe from primitive man to Einstein and beyond, and witnessed the step-by-step elimination of axioms that seemed obviously true and hence were not even questioned.

Common sense through the millennia knew the earth to be flat, until Pythagoras recognized its shadow on the moon; but his spherical earth was rejected for another century. Through another 16 centuries savants still knew that the earth was the stationary hub of the universe until Copernicus subordinated it to a mere satellite of the sun, soon seen to be insignificant among billions of billions of other suns; but his insight was rejected by contemporary astronomers and theologians. Living species were unique, immutable creations until Darwin substituted an evolutionary process; but evolution is still passionately ridiculed by faithful believers in the literal Bible. Organic chemicals could only be made via the "vital force" of living organisms until Friedrich Wöhler synthesized urea; but it was decades later before the false creed was abandoned. Man, the living image of God, for whom the whole universe was created is only now being rationalized as the transient apex of an evolutionary process, destined to go on to still inconceivable lineages through the countless eons ahead. Right up to this century the fixity of the relative positions of continents was axiomatic, until Wegener astonished the world by demonstrating that assembly of the continents produced surprising coherence, only to be scorned and ridiculed because physicists insisted on the impossibility of such continental plates sliding on their substratum. When American oceanographers established beyond any doubt that each of the oceans was spreading at surprisingly rapid rates, the physicists' objections were swept away before the flood of acceptance of plate tectonics. But in hindsight the physicists were correct. The continents do not move over their underlying mantle. The truth had been occluded from both geologists and physicists by another false axiom, the gratuitous assumption that the radius of the earth was constant. Ocean growth expands the earth without moving the continents with respect to the mantle below them.

Have we now finally cleared away all our spurious axioms? Of course not. We must now have the courage to emulate Pythagoras, Aristotle, Leonardo da Vinci, Newton, Darwin, Wegener, and Einstein to emancipate our minds from ever more false axioms inherited inviolate from our primitive past—things we think we know and take for granted, without really asking whether they are valid.

First, orthodoxy has always assumed that the universe was created with its complete inheritance of matter, which thereafter has remained constant. Likewise, that all the matter in the present solar system was present in the initial gaseous nebula that spawned the sun and its satellites. Similarly it is taken for granted that all the matter of the earth has been inherited from the time of its initial accretion. Each of these cognate assumptions is false; matter is created continuously and spontaneously at all levels.

Second, current cosmology clings to the myth that a few billion years ago time and space came into being and the matter of 20 billion stars appeared from nothing in a Big Bang, which has been blowing apart ever since.

Third, orthodox dogma teaches that the total mass energy of the universe is stupendously large. In contrast, Tryon and I independently realized that the cosmos always was, is, and always will be a state of zero. Philosophically, there is no other way the cosmos could come into being.

Fourth, Einstein expressed surprise that inertial mass and gravitational mass should turn out to be precisely equal without any fundamental reason why this should be so. He also deduced that mass equals energy divided by the square of the velocity of light (the latter being a pure number, following Parkinson's insight). Mass and energy are thus of the same kind, indeed complementary gemini like the two sides of a coin, born together and canceling together. If a new mass is added to the universe, potential energy precisely equal to its inertial mass is also added, and vice versa.

Fifth, Newton's and Hubble's laws are complementary, one fitted to the observed behavior of the solar system, the other to the observed behavior of galaxies. When combined into a single equation they govern the dynamics of the universe, and automatically throw up Einstein's cosmic constant and determine the mean size of galaxies.

Each of these five propositions deletes a false dogma we have taken on faith from the beginning, and thus ranks with the spherical earth of Pythagoras, the heliocentricity of Copernicus, the evolutionary principle of Darwin, and the continental dispersion of Wegener.

The fugue of this narrative is the recurrent obstruction of progress by creed, be it religious doctrine, the renaissance straitjacket of Aristotle, or the veto of the contemporary establishment. The greatest thinkers have been blinkered by their beliefs. Creed is the narcosis of vision.

The more radical the advance from current orthodoxy, the more certain will it be scorned and rejected. Prestige is the canker of the great, because it has been the innovators like Werner, Newton, Kelvin, Jeffreys, Bailey Willis, Gaylord Simpson, and Tuzo Wilson who have led lesser lights into withering rejection of new wisdom.

The emergent generation has not arrived too late. The glory of science, so marvelous to us, will be eclipsed, and eclipsed again, at an accelerating rate, in the least expected places, and each new success will light up new horizons beyond, *ad infinitum.*

But do not expect to be hailed as a hero when you make your great discovery. More likely you will be a ratbag—maybe failed by your examiners. Your statistics, or your observations, or your literature study, or your something else will be patently deficient. Do not doubt that in our enlightened age the really important advances are and will be rejected more often than acclaimed. Nor should we doubt that in our own professional lifetime we too will repudiate with like pontifical finality the most significant insight ever to reach our desk.

Should we then give credence to every heretic and iconoclast with the naïveté or the zeal or persistence to challenge the established order? Of course not! Most heresy is doubtless false—yet latent there are the gems of the age. To discriminate unerringly within doctrine and within heresy needs a keener mind than any yet—but this must be our ever-unattainable goal.

Glossary

Glossary

Italics indicate that an ordinary English word is used in a special technical sense. Boldface indicates a cross-reference to a Glossary entry of particular interest.

actualism: synonym for **uniformitarianism.**
alluvium: sediment deposited by a river when the rate of flow is no longer sufficient to transport it.
amphibole: an important family of blackish minerals (usually green in thin slices), which are silicates of magnesium, iron, and aluminum, with also hydroxyl. Many other elements may be present in small amounts. Chemically they are very similar to the **pyroxenes** but are formed in a more hydrous environment.
amphibolite: a metamorphic rock consisting mainly of amphibole and plagioclase feldspar, sometimes with a little quartz.
andalusite: a mineral of composition Al_2SiO_5, which develops in low-grade thermal metamorphism of clayey rocks. Sillimanite and kyanite have the same composition but crystallize in different environments.
andesite: a category of volcanic lava (so named by von Buch in 1826 from its common occurrence in the Andes mountains) found to be characteristic of volcanoes along orogenic belts. Andesite commonly contains feldspar (appropriately named andesine), intermediate between the soda and lime ends of the plagioclase series, and mafic minerals such as pyroxene, amphibole, or mica.
anhydrite: a mineral with the composition of anhydrous calcium sulfate.
anticline: literally, a fold whose limbs slope away from each other, and hence a fold convex upward; but more accurately, a fold in which the core strata are the oldest stratigraphically.
aragonite: a mineral with the composition calcium carbonate (which is the same as the commoner mineral calcite).
archaeocyathid: a group of lime-secreting animals superficially resembling sponges and corals, abundant in Early Cambrian time.

Archean: the oldest era in the history of the earth, extending from about 4000 million years ago until the beginning of the Proterozoic 2500 million years ago.

asthenolith: a body of melted or partially melted mantle material which, because of its lower density, rises diapirically (q.v.) from the **asthenosphere** through the **lithosphere**.

asthenosphere: a name coined in 1914 by Joseph Barrell for a weak yielding zone (ἀσθενής, weak) 100 or more kilometers below the earth's surface at which the weights of all sections of the earth's crust balanced in hydrostatic equilibrium; this is the principle of **isostasy**. The asthenosphere separates the lithosphere above from the centrosphere or *mantle* below. Later, seismologists found a zone at appropriate depth which transmitted earthquake waves more slowly, hence referred to as the *low-velocity zone*. With the adoption of "plate tectonics" in the 1960's, the asthenosphere was accepted as the zone of decoupling at which continental and oceanic "plates" moved independently of the underlying mantle. Some refer to the asthenosphere as "the upper mantle." The asthenosphere is an elastic solid for impulsive stresses, but flows when stresses are maintained (compare the flow of a glacier).

astrobleme: a scar on the earth's surface resulting from the ancient impact of an extraterrestrial body.

augite: a common black or dark-green mineral of the pyroxene group.

basalt: the most abundant extrusion from volcanoes, a dark fine-grained rock which microscope study shows to consist mainly of plagioclase feldspar and pyroxene, but olivine may also be present. Its common quarry name is "blue metal."

batholith: an intrusive mass of igneous rock (usually granitic) some tens of square kilometers in area, and without known base.

Benioff zone: a surface that extends down from Pacific ocean trenches at a steep angle to depths between 300 and 700 km, along which earthquakes are frequent. Named after the distinguished seismologist Hugo Benioff.

Big Bang: the hypothetical cosmic explosion that initiated time and space and all matter in the universe, from which the universe has been expanding ever since.

black body: a body which, when raised to incandescence, emits a continuous spectrum (carbon and tungsten approach closest to this ideal). Black-body radiation is what would be emitted by an ideal black body with an absorbtive power of unity and zero reflective power.

blastoid: an extinct invertebrate group related to the crinoids with fivefold symmetry, which were abundant in the Paleozoic.

blueschist: a loose name for a group of metamorphic rocks with a bluish color due to the presence of sodic amphibole (glaucophane or crossite), and commonly bluish-gray lawsonite; aragonite and quartz are often present.

boudinage: a structure common in **similar folding** that involves considerable

Glossary 371

increase in the surface area of strata, so that stronger beds tend to break up into pancake or sausage shapes called boudins.

brachiopod: a member of an invertebrate phylum superficially resembling mussels, but more primitive and growing its shells on the right and left instead of front and back.

Caledonides: Ancient orogenic belts are indicated by the suffix *-ides* (for example, the Alpides and Tasmanides). The Caledonides refer to an early to middle Paleozoic orogenic belt found in Scandinavia, the British Isles (hence Caledonia), Spitzbergen, eastern North America, and northwestern Africa, now scattered by the Mesozoic opening of the Atlantic Ocean.

Cambrian: the time interval between 570 and 505 million years ago (see table, p. 79). Named by Adam Sedgwick for rocks of this age in North Wales.

Carboniferous: the time period between 360 and 286 million years ago (see table, p. 79). So named because of the abundance of coal of this age in Europe.

catastrophism: the concept that the physical and biological evolution of the earth has been dominated by violent epochs of gross change on a scale outside our experience, in contrast to the **uniformitarianism** concept that slow processes such as those currently acting have been sufficient.

Cathaysian flora: a late Paleozoic flora of East Asia (Cathay is an old name for China).

Cenozoic: literally (time of) recent organisms; the time interval between 66 million years ago and the present (see table, p. 79).

center of gravity: the point in a body at which its weight may be assumed to act, and at which the body may be supported in neutral equilibrium.

cephalopod: a molluscan group which includes squids, octupuses, cuttlefish, and pearly nautilus, and the extinct belemnites, ammonites, and many nautiloids.

chert: a hard, very compact, siliceous rock, too fine in grain for individual mineral grains to be seen with a hand lens.

cleavage: In geology, cleavage is a grain developed in rocks during crystalline flow, so that they split or cleave readily in a plane normal to the excess pressure that induced the flow. The capacity to cleave in this plane is due to the crystallization during flow of minerals (particularly micas), which grow with specific crystal axes facing the maximum stress. In mineralogy, cleavage refers to the crystallographic planes along which particular minerals cleave readily because of the tesselated arrangement of their atoms.

coesite: a mineral with the same composition as quartz, which needs a pressure of 20 kilobars to form at surface temperatures, hence it is only known at the surface at **astroblemes**.

colloid: literally, a glue-like substance. A suspension of particles of submicron size which results in surface-charge density dominating the physical behavior of the substance.

columnar jointing: systematic prismatic *joints* formed in relief of tensional stress that exceeds the tensile strength during the cooling contraction of a lava. Where the conditions are very uniform, the columns are perfectly hexagonal with systematic joints normal to the columns (see Fig. 6).

concentric folding: folding in which the orthogonal thickness of beds remains constant. (See Chapter 16.)

conjugate shears: Any state of non-hydrostatic stress within a body may be represented by three "principal stresses" mutually at right angles, which in turn may be replaced by two conjugate (literally "married together") planes, mutually at right angles (which intersect along the intermediate stress axis where the shear stress is a maximum), one dextral and one sinistral. If failure ensues, the angle between the conjugate planes of shear *failure* is reduced from a right angle by the angle of friction, although the angle between the planes of maximum shear *stress* remains a right angle.

conodonts: tiny marine fossils, toothlike in form but not in function, abundant in Paleozoic and Triassic strata, and very useful for correlation.

convection: Strictly, convection refers to the transfer of heat borne by a moving fluid. As most fluids expand on heating and become less dense, colder columns sink and hot columns rise, then lose heat at the surface, and convective circulation ensues. On the time scale of geology, crystalline ice and crystalline *mantle* are fluids and flow as in a glacier, or as convective circulation of the mantle.

Cordilleran: Cordillera is a general term for a prominent mountain range system. Here Cordilleran refers specifically to the mountain range system of North America facing the Pacific Ocean.

Coriolis effect: As the earth rotates, points on the surface nearer the equator move eastward faster than points nearer the poles. Therefore any body moving toward the equator moves eastward more slowly than the surface toward which it travels. Hence the path of such a body is deflected westward. This effect causes currents in the atmosphere or oceans or in the fluid interior of the earth to circulate clockwise in the northern hemisphere and counterclockwise in the southern hemisphere.

cosmic rays: very high energy radiation consisting of atomic nuclei, which bathe the earth from beyond outer space. Their origin is not yet understood.

cosmological principle: The cosmos as observed from any point in the cosmos is statistically the same as observed from any other point. The *perfect cosmological principle* holds this to be true for any past or future time.

cosmos: generally a synonym of the universe, but herein universe is defined as the physically knowable universe (limited by the Hubble recession reaching the velocity of light) whereas cosmos is infinite in time and space according to the perfect cosmological principle.

couple: a pair of parallel but opposite forces or stresses.

Cretaceous: the time period between 144 and 66 million years ago (see table,

Glossary

p. 79). So named because of the common occurrence of chalk during this period (Latin *creta*, chalk).

crinoid: an invertebrate group popularly called sea lilies, with mobile appendages rising from a body contained in a calyx made of calcareous plates with a fivefold symmetry carried on a stem or column.

critical temperature: the temperature at which the surface tension drops to zero, so that the liquid becomes a gas, irrespective of how great the pressure may be.

Curie point: The strength of ferromagnetism diminishes with rising temperature, and the temperature at which it becomes zero is called the Curie point, which varies up to 700°C for different magnetic materials.

declination: See **magnetic declination**.

décollement: a surface of detachment between more folded strata and underlying less folded strata; like a gently dipping fault, except that the upper beds are more folded than the lower.

dextral: right-handed or clockwise, in contrast with **sinistral.**

diabase: a rock name that has had a variety of meanings in the older literature, but now a synonym of **dolerite.**

diapir: an intrusive rock body that has pierced upward through the overlying rock.

dike (American), dyke (English): a wall-like igneous intrusion formed by magma entering and widening a planar crack transverse to the strata.

diluvium: an obsolete term for glacial drift when these gravel and boulder deposits were thought to have been deposited by floodwater, and particularly by Noah's flood reported in *Genesis*. Hence the "Diluvial Period."

dimensions: an index of capital letters showing the powers at which mass, length, and time are involved in a physical parameter. For example, a velocity is a distance divided by time, hence LT^{-1}; an acceleration is a velocity divided by time, hence LT^{-2}; a force is a mass multiplied by an acceleration, hence $ML^{-1}T^2$. The dimensions of pure numbers (such as e or π, and digits) are one, and are usually omitted, but if included at all they would be $M^0L^0T^0$. Dimensions do not indicate in any way the size of the parameter.

dolerite: a dark hard rock consisting mainly of plagioclase feldspar and pyroxene and sometimes with olivine, which solidified from basaltic magma intruded into sedimentary strata. *Diabase* is the preferred term in America.

drag: the bending of rocks when blocks move relative to each other; hence "drag folds."

e: the base of hyperbolic or natural logarithms, defined as the limiting value of $(1 + 1/m)^m$ as m approaches infinity; e has the value 2.718282.... The pure numbers e and π are fundamentally important in many fields of mathematics and physics, and are joined herein by c, the velocity of light.

earthquake epicenter: the point on the earth's surface vertically above the focus of the earthquake.

earthquake focus: the location within the earth where the rupture occurred that generated an earthquake.

Earth's crust: conceptually equivalent to the crust of a loaf of bread, but used ambiguously in geologic writings: (a) the outer shell of the earth, more than 100 km thick above the **asthenosphere**, which tends to behave like a rigid plate (synonym of **lithosphere**); (b) the shell above the **Moho**, 10 km or less thick below the oceans, but 30 km or more below the continents, determined by a jump in the velocity of compressional seismic waves from 6–7 km/s to 8 km/s; (c) the shell above an isopotential surface some 50 km below sea level above which the weight of overburden per unit area is the same everywhere, also called "the depth of compensation"; (d) in the usage of earlier writers, such as Lyell, that part of the earth's exterior accessible to human observation in surface exposures, in mines and bores, and physical instrumentation, and from direct inference from such observations.

eclogite: a crystalline rock consisting of garnet and soda-rich pyroxene and minor amounts of rutile, kyanite, and quartz. The overall composition might be the same as that of basalt or diabase, and eclogite may develop from these at the high pressure in the mantle, and vice versa.

entropy: a thermodynamic quantity that measures the gross disorder of a physical system. If a substance undergoing a reversible change takes in a quantity of heat dQ at absolute temperature T its entropy increases by dQ/T. For example, when crystalline ice melts to liquid water its molecular disorder (entropy) greatly increases, continues to increase during further heating, and increases greatly when liquid water boils to steam.

Eocene: literally "dawn of recent (organisms)." The time epoch between 58 and 37 million years ago (see table, p. 79). So named in 1833 by Charles Lyell.

epoch: used in stratigraphy in a more restricted sense than in ordinary English, as a time unit intermediate between an *age* and a *period*. Geological periods last a few tens of millions of years, whereas geological epochs last a few million years.

erratic: a pebble, boulder, or larger mass transported by a glacier or floating ice and dropped in an exotic place when the ice melted (Latin *erratus*, wandered).

ether: the hypothetical medium, filling all space, with the property of transmitting electromagnetic waves and gravitational attraction, but without any other properties of matter.

eugeosyncline: See **geosyncline**.

eukaryote: In contrast with **prokaryotes**, eukaryotes are slightly more advanced organisms having a vesicular nucleus, various membrane-bounded organelles, and more advanced protoplasmic organization.

Euler pole: One of the many theorems enunciated by the Swiss mathematician Leonhard Euler was that any movement of part of a spherical sur-

Glossary 375

face over itself is the same as a rotation of that part about a pole on the surface. Hence, in continental drift, any displacement of a continent is the same as a rotation of it about the relevant Euler pole.

eustasy: process of sea-level change caused by change in the balance of the total volume of seawater and the total volume of the ocean basins, hence *eustatic*. Synonym: *eustatism*.

evaporite: sedimentary rock formed by precipitation during evaporation, particularly where seawater can continue to enter an evaporating basin. Minerals in evaporites include gypsum, anhydrite, rock salt, dolomite, and more rarely, various nitrates and borates.

exponential: In algebra, an exponent is the power to which a number is raised (for example, in 2^3, 3 is the exponent). In an exponential relation, the number sought (let us say y) varies as some power ($y = m^x$, as x varies from 0, 1, 2, 3, . . .). Very many natural phenomena vary exponentially: gravity force diminishes according to the *square* of the distance and the weight of a sphere increases with the *cube* of its radius, to take two examples.

fabric: a specific usage defined in 1930 by the Geman petrologist Bruno Sander for description of rocks under the microscope, to include all aspects of the pattern of the mineral constituents, including the statistical shapes, distribution, and orientation of the crystal axes of minerals. "Fabric" is used in other fields in specialized senses.

facies: a particular category, commonly environmental, applied to a rock formation, best illustrated by example. For the sediments deposited at a specific time we might refer to the intertidal facies, the nearshore facies, the outer shelf facies, the continental slope facies, the abyssal facies.

fan: a cone-shaped deposit of sediment accumulated where a sediment source emerges to a broad area with flatter gradient.

fault: a fracture in rock at which displacement has occurred along the fault surface.

feldspar: a very abundant group of minerals that are silicates of aluminum and another cation. The commonest feldspars are orthoclase (a silicate of aluminum and potassium) and plagioclase, in which the cation is sodium (albite) or calcium (anorthite), or any mixture of sodium and calcium.

feldspathoids: literally, "feldspar-like." A group of minerals that look like feldspars and have quite similar chemical composition, except that their silica content is systematically lower. They do not occur in any rock that contains quartz (crystalline silica).

figure of the earth: the precise shape of the earth.

flexure folding: term commonly used in America for **concentric folding**; should be discontinued on grounds both of priority and ambiguity, as well as an involved misconception concerning the nature of folding (see Chapter 16).

flysch: a characteristic facies deposited in active geosynclines prior to a main orogenic paroxysm, in contrast with the "molasse" facies, which follows the paroxysm. The flysch facies consists of poorly fossiliferous sandy and calcareous shale and mudstone, repetitively grading from sand to mud.

fold, folding: In geology, folding refers to the bending of the planar bedding surfaces, or less commonly the deformation of other surfaces.

foraminifera: single-celled, mostly marine protozoan animals with a shell (usually of calcite) of one or several chambers, usually about the size of a pin's head, but a few grow to as much as a couple of centimeters across.

fossil: evidence preserved in rock of a former living organism.

fusulinid: a group of single-celled **foraminifera**, shaped like a grain of wheat, abundant in the Late Proterozoic.

gabbro: a coarsely crystalline dark-colored igneous rock in which the dominant minerals are calcic feldspar and pyroxene. Olivine may also be present and hornblende may substitute for pyroxene. Gabbro is the plutonic equivalent of basalt lava.

galaxy: the largest system of stars and interstellar matter, which forms a self-gravitating unit and repels other such units. An average galaxy contains 10^{12} stars; it is estimated that there are 10^{10} galaxies to the limit where they become too faint to detect.

garnet: a family of cubic silicate minerals with the composition $A_3B_2(SiO_4)_3$ in which A may be Ca, Mg, or Fe and B may be Al, Fe, Mn, V, or Cr.

gemini: inseparable twins.

geoid: literally, "earth-like." The actual sea level surface, including its extension through land areas. The geoid is perpendicular everywhere to a plumb-line.

geometrical progression: a series in which each member is related to the previous member by a constant multiplier (for instance, the series 1, 3, 9, 27 . . .).

geosyncline: a long narrow depression in the earth's crust, which continues to sink for millions of years while it fills with sediments to a thickness of several kilometers. The more active belt within it (the eugeosyncline) suffers volcanism, seismicity, faulting, generation of granites, deformation, and uplift, that is, "orogenesis." A parallel belt (the miogeosyncline) has steady subsidence to depths of several kilometers without the volcanism, seismicity, or orogenesis.

glaucophane: a bluish fibrous or prismatic **amphibole** rich in sodium, usually derived from the flow metamorphism of soda-rich igneous rocks.

Glossopteris flora: an assemblage of plants including the gymnosperm *Glossopteris* characteristic of **Gondwanaland** during the Permian.

gneiss: a coarsely crystalline rock with distinct folia or bands, usually containing quartz, feldspar, and mica, but not restricted to these minerals, formed by the recrystallization of either sedimentary or igneous rocks during flow in the solid state.

gneiss dome: a **diapir** in which a core of crystalline gneiss has intruded upward by solid-state flow, draping the pierced rocks into a dome configuration.

Glossary

Gondwanaland: late Paleozoic supercontinent composed of South America, Africa, Madagascar, India, Australia, and Antarctica. See **Laurasia**.
gore: a lune-shaped gape between great circles on a sphere.
gouge: soft, finely ground material in a fault zone.
graben: a long depressed zone on the earth's surface, commonly bounded by faults. Antithesis: **horst**. Geomorphologic synonym: rift valley.
granite: As a general term, granite is the coarsely crystalline rock that forms the bulk of the large **plutons** intruding the cores of fold mountains, and which contains quartz, feldspar, and usually some dark minerals such as mica or hornblende. More strictly, only some such rocks are granite, those with quartz, potassium feldspar, and commonly mica and perhaps hornblende. Other granites in the broad sense are called adamellite, granodiorite, quartz monzonite, and so forth, according to the kind of feldspars present and the amount of quartz.
granulite: a coarse-grained metamorphic rock formed at high hydrostatic pressure without the mineral alignment of **gneiss**. A variety of minerals may be present according to the composition of the original rock: quartz, feldspars, garnets, pyroxenes, and amphiboles.
graphite: the stable form of carbon at the earth's surface. A pressure of 50 kilobars is needed to form diamonds from graphite.
gravitational constant: When Newton found that the gravity attraction between two masses was *proportional* to the product ($m'm''$) of their masses and *inversely proportional* to the square (d^2) of the distance between them, this implied that a constant could be found so that the attraction was *equal* to $m'm''/d^2$ multiplied by this "gravitational constant." Its value is 6.670 ± 0.005 × 10^{-11} newton m^2/kg^2.
gravitational mass: the attractive force exerted by a body on a standard body at a specified distance.
gravity anomaly: the difference between the gravity acceleration observed at a point and the value expected at that latitude, longitude, and altitude calculated from an adopted formula for the shape of the earth and rock-density distribution.
great circle: any circle on the surface of a sphere that divides the surface into hemispheres. The equator and all meridians are great circles, but any plane that passes through the center of the sphere cuts the surface along a great circle.
greenstone belt: elongated belts of dark-green metamorphosed mafic igneous rocks common in older Precambrian terrains. The green color is due to minerals like chlorite amphiboles, epidote, and serpentine.
Grenville: a major orogenic belt of eastern North America which terminated about 1000 million years ago.
guyot: a seamount rising from the deep ocean to a flat top within 1000 m of the ocean surface.
gypsum: a mineral with the composition hydrated calcium sulfate.
heat flux: the rate of flow of heat out of the earth, measured in microcalories

per square centimeter per second or milliwatts per square meter. The mean heat flux through continents and oceans is 60 and 95 mW/m² respectively, although variable from place to place.

hornfels, hornstone: a tough dense fine-grained rock, commonly formed by the alteration of clayey rock by heat and permeating fluids adjacent to an igneous intrusion. In the past hornstone was used more widely for any fine-grained tough rock including **chert**.

horst: a long uplifted zone bounded by faults. Antithesis: **graben**, rift valley.

Hubble's law: Galaxies are receding from each other at H km/s per megaparsec, where H is the Hubble constant. The value of H varies in different methods of measuring it; 75 is a commonly cited value, but other methods give values between 5 and 8.

hypersphere: a "sphere" in four dimensions. Thus, where $x^2 + y^2 = c$ is a circle (in two dimensions) and $x^2 + y^2 + z^2 = c$ is a sphere (in three dimensions), the equation $x^2 + y^2 + z^2 + t^2 = c$ describes a hypersphere.

Ice Age: a loosely used term, referring in some contexts only to the last glaciation stage, which lasted from 30,000 to 10,000 years ago, but in other contexts to the glacial epoch which commenced more than a million years ago and has not yet ended. The present time is in a rather cool interglacial stage with another glacial stage likely to follow not many thousands of years ahead.

igneous: rocks formed by solidifying from a molten state (Latin *ignis*, fire). Some rocks, which have been generally regarded as igneous, are considered by some to have formed from sediments without passing through a molten state, but the term igneous may be extended to include rocks so formed, so that "igneous" becomes a descriptive, not genetic, term.

inertia: resistance of a body to being moved or moved faster or rotated.

inertial mass: a measure of inertia, that is, the resistance of a body to being accelerated by a force.

isobath: literally, equal depth: a contour on the ocean floor through points of equal depth below sea level.

isostasy: conceptually, the application of Archimedes' principle to the earth's crust, a term proposed in 1889 by the Astronomer Royal, C. E. Dutton, to explain why the warping of the level by the attraction of the Himalayas was found by the Surveyor-general of India, Sir George Everest, to be only a third of what would be expected from the mass above sea level of those mountains, that is, why the Himalayas were not pulling their weight. The principle of isostasy (Greek ἰσοστάσιος, equal balance) states that above a "level of compensation" equal areas of the earth's crust have equal weight, so that low areas contain denser rocks than areas of higher altitude. If an additional load (such as sediments or a lava flow or an ice sheet) is added, that area will subside to maintain isostatic balance, just as loading a ship causes it to float lower in the water. Because the viscosity or stiffness of the **asthenosphere** gets less as the temperature rises, the rate of adjustment to isostatic equilibrium varies from region to region. For

Glossary 379

normal (nonvolcanic) regions, the departure from balance reduces to about half in about 5000 years, and to a quarter in about 10,000 years, an eighth in some 15,000 years, and so on.

isotherm: surface of equal temperature.

isotope: a specific variety of an element which has chemical properties identical with other varieties of the same element, but differs from them in having a different number of neutrons in its nucleus, and hence a different atomic weight. Isotopes of an element cannot be separated chemically, but can be separated by a physical process which is sensitive to the difference in atomic weight.

jadeite: a green mineral of the pyroxene group that only forms at high pressure, composition $Na(Al, Fe)Si_2O_6$.

jasper: a **chert** colored pink, red, or purple by hematite (the sesquioxide of iron).

joints: systematic sets of usually planar fractures present in all rocks, formed in relief of tensional or shear stress that exceeded the strength of the material. Displacement is limited to the relaxation of the elastic strain present before fracture. The name is derived from superficial similarity to the joints in a masonry wall, but the analogy is false, because in masonry the blocks have been joined whereas in this jointing, rock formerly entire has separated.

Jurassic: the time period between 208 and 144 million years ago (see table, p. 79). So named after the Jura Mountains between France and Switzerland, where rocks of this age were first studied in detail by von Humboldt in 1795.

kinetic energy: energy possessed by a body by virtue of its motion.

klippe: an isolated erosional remnant of a **nappe**.

krikogen: a term introduced by F. Wezel for the ring-shaped surface expression of an orogenic diapir.

Laurasia: a contraction of Laurentia + Asia, for the late Paleozoic supercontinent consisting of North America, Greenland, Europe, and Asia (excluding India). Laurasia and its twin south of the Tethys (Gondwanaland) together formed Wegener's **Pangaea**.

law of superposition: Any stratum that has been deposited on another stratum records a later event than the underlying stratum; any fault or fracture or deformation or intrusion or other geological feature affecting a stratum or feature records a later event than the violated stratum or feature. This law, which is the fundamental law of stratigraphy, was first stated unambiguously by Steno in 1669.

lawsonite: hydrous calcium aluminum silicate, a grayish-blue metamorphic mineral.

lineation: used in a general sense for any kind of repetitive linear structure in rock, including slickenside striae, microfold axes, intersection of cleavage with bedding, elongation of pebbles or ooids, wrinkles, streaks, and so on. In a more restricted genetic sense, it refers to structures developed in the direction of crystalline flow during solid-state deformation, including elongated crystals which crystallize with that orientation to the stresses, and other cognate structures.

lithosphere: etymologically, the "rock sphere," applied to the crust of the earth, 100 km or more thick, which behaves rigidly in contrast to the underlying **asthenosphere**, which, although still solid, yields under sustained loads, much like the flow of a glacier. In early writings, lithosphere meant the whole of the solid earth in contrast with the hydrosphere and atmosphere, but this usage has been abandoned.

Mach's principle: The motion of a particle is only meaningful when referred to the rest of the matter in the universe, and inertial properties are meaningless for empty space. Hence the inertial mass of a particle in empty space is zero, and increases *pari passu* with the total mass of the universe.

mafic: adjective applied to minerals and igneous rocks with a high content of **ma**gnesium and iron (**f**errum).

magma: generally, the molten material that solidifies as igneous rock. But the Greek μάγμα means a fluid paste rather than a melt, and Elliston (Chapter 4) argues that granitic magmas are really colloidal pastes rather than melts, and still uses "magma" in this sense.

magnetic declination: the horizontal angle between magnetic and geographic north at a point (called by some the *variation*, although strictly variation is the annual change in the declination).

mantle: In geophysics, this term always refers to the solid earth between the fluid core about 3000 km below the surface, and the **Moho** only 10 to 100 km below the surface. In other contexts, mantle is used as a synonym for *regolith*, the weathered material at the surface; in petrology, mantle is used in the phrase *mantled gneiss dome* (following Eskola in 1948) for the hood of metamorphosed strata layered and foliated parallel to the surface and foliation of a diapiric gneiss dome; in paleontology, the folded lobe of the tissues draped over the main body of a mollusk or brachiopod, which secretes the outer shell, is called the mantle.

massif: a block resistant to tectonic deformation.

megashear: a transcurrent fault along which the horizontal displacement of part of the earth's crust exceeds significantly the thickness of the crust.

mélange: a chaotic medley of large blocks, ranging in size up to as much as a kilometer, forming a mappable unit. There are at least two categories of mélange: sedimentary and tectonic. The former results from major slumps. Some tectonic mélanges occupy megashear zones where large transcurrent displacement occurs with transverse tensional conditions.

meson: a generic name for any hadronic particle with baryon number zero, that is, a family of subnuclear particles much smaller than protons and neutrons, and which strongly interact. They include π-, K-, and D-mesons and their antiparticles.

Mesozoic: literally, "(time of) middle organisms." The time interval between 245 and 66 million years ago (see table, p 79).

Glossary

metamorphism: etymologically, "change of form," the process whereby rock recrystallizes to a different form as a result of heat, pressure, shear, or permeation by fluids, or some combination of these. Commonly different minerals result (but not when a pure limestone is metamorphosed to marble, or a clean quartz sandstone is metamorphosed to quartzite). The chemical composition may change by the introduction of new elements or the excretion of others.

metastable: a state temporarily stable because of the slowness of adjustment to changed conditions. Devitrification and recrystallization and solid-state flow are diffusion phenomena, and depend on the relevant **relaxation time**.

metonic cycle: the 18.6-year period of nutation (from Latin *nutans*, nodding) of the earth's axis caused by a cyclic variation in the combined attraction of the sun and moon on the earth's equatorial bulge as the line of intersection of the orbital planes of earth and moon regresses with the **precession** of the equinoxes. Called after its Athenian discoverer, Meton, in 432 B.C.

migmatite: literally, "mixed rock." A streaky rock with alternating veinlets of granitic and metamorphic minerals, found in the axial regions of orogens where granite plutons infiltrate into the intruded metamorphic rocks in thin layers.

Miocene: literally, "a minority of recent (organisms)." The time epoch between 24 and 5.3 million years ago (see table, p. 79). So named in 1833 by Charles Lyell.

miogeosyncline: See **geosyncline**.

Mohorovičić discontinuity (= Moho): a discontinuity in the lithosphere where the velocity of seismic compressional waves jumps from about 6 km/s to about 8 km/s.

molasse: originally a soft green nonmarine sandstone associated with marl and conglomerate of Miocene age in the Alpine foothills, subsequently applied generally to similar nonmarine or deltaic facies derived from the erosion of a newly-risen orogenic zone. (From a French word meaning soft, flabby, or lacking in body.)

moment of inertia: a measure of the resistance of a body to angular acceleration. It is the sum of the products of the mass of each particle and the square of its distance from the rotation axis (Σmr^2).

moraine: rock material deposited by the melting of a glacier that had transported it to that site (compare ***till***).

mountain building: here, the processes involved in the development of fold ranges. Apart from higher altitudes, such mountain ranges have specific characteristics: millions of years of rapid sedimentation to a thickness of many kilometers, epochs of strong folding, volcanism and seismicity, and anomalies in the gravity field. Synonym: orogenesis (which merely means the same in Greek).

mylonite: literally, milled rock. Finely ground rock at the shearing surface of a fault, where the pressure and friction temperature have been sufficient to cause recrystallization to a compact cherty rock without cleavage.

nappe: literally, a sheet. A sheet of rock that has been pushed over other rock on a flat or gently dipping surface a distance on a scale of kilometers.

Neogene: the time period between 24 and 1.6 million years ago, embracing the Miocene and Pliocene Epochs.

Neptunism: Werner's eighteenth-century doctrine that all rocks were precipitated from a universal ocean. (See Chapter 4.)

neutron star: a star containing about 1½ solar masses compressed as neutrons into 10 km diameter, rotating in seconds or even less than a second; such stars are pulsed radio emitters (**pulsars**).

nucleon: a collective name for the proton and the neutron.

nuée ardente: literally, glowing cloud. An incandescent cloud of steam and dust, erupted in a violent volcanic explosion, which flows at great speed down the volcano slope, the rock fragments being borne by the violent turbulence.

obliquity: the inclination of Earth's axis to the ecliptic (the plane of Earth's orbit).

Oligocene: literally, "few recent (organisms)." The time epoch between 37 and 24 million years ago (see table, p. 79). So named by Charles Lyell in 1833.

olivine: a greenish mineral, $(Mg, Fe)_2SiO_4$, which is the most abundant constituent of the upper mantle, and derivatives from it. (Formerly called peridot.)

ooze: in marine geology, a deep-sea sediment containing 30 percent or more of calcareous or silicious remains of pelagic organisms with fine clay.

ophiolite: a group of mafic and ultramafic igneous rocks and their greenstone derivatives rich in chlorite, epidote, amphiboles, and serpentine, commonly found in the core zones of orogens.

Ordovician: the period between 505 and 438 million years ago. Named in 1879 by Lapworth for rocks of this age in North Wales, formerly inhabited by the Celtic Ordovices tribe (see table, p. 79).

orocline: literally, "mountain bend." An orogen deformed in plan.

orogenesis: synonym of **mountain building**.

orotath: an orogen that has been greatly stretched so that it has become a string of islands on a submarine ridge.

overthrust: a large-scale gently dipping thrust fault.

Paleocene: the time epoch between 66 and 58 million years ago (see table, p. 79).

Paleogene: the time period between 66 and 24 million years ago, embracing the Paleocene, Eocene, and Oligocene Epochs.

paleopole: a pole of the earth at some former time.

Paleozoic: literally, "(time of) old organisms." The time interval between 570 and 245 million years ago (see table, p. 79).

Pangaea: literally, "the whole earth." The name given by Alfred Wegener in

Glossary

1912 to the single continent which existed at the end of the Paleozoic, which subsequently broke up to form the present dispersed continents.

parallax: the apparent displacement of a heavenly body on the celestial sphere owing to change of position of the observer. Parallax may be *diurnal*, if observed during the rotation of the earth, *annual* if observed during Earth's orbital motion, and *secular* when longer periods are involved.

parameter: a number which measures some characteristic property for a case under consideration, but which differs for other similar cases, for example, the elastic parameters of a particular rock.

peneplain: literally, "almost a plain." Defined by W. M. Davis in 1889 to mean a surface of low relief resulting from prolonged erosion.

peridotite: literally, "olivine rock." A coarse-grained plutonic rock in which the dominant mineral is olivine, together with pyroxene (or sometimes amphibole), chromite, and magnetite.

period: defined in stratigraphy as specific time in the hierarchy *age, epoch, period, era*. A geological period lasts a few tens of millions of years.

Permian: the time period between 286 and 245 million years ago (see table, p. 79). So named after fossiliferous rocks of this age exposed in the Perm Basin of Russia.

petrology: literally, the study of rocks, but with a bias toward their genesis, chemistry, and structure, compared with the descriptive bias of *petrography*, studying thin sections with a microscope.

phase: Phase is used in several specific senses in different fields of geology. Herein it refers to the different sets of minerals assumed by a given rock composition under different conditions of pressure, temperature, and water availability; for example, the phases of carbon as graphite and diamond, or of silica as chalcedony, quartz, coesite, and stishovite, or of the rock as basalt, gabbro, and eclogite.

piedmont: an area or other feature extending from the base of a mountain range.

plate: in plate-tectonics theory, a large thin segment of the lithosphere, assumed to move horizontally and join other plates along tectonically active zones, which may be collisional, extensional, or transcurrent.

platform: a continental area covered by flat-lying or gently deformed sediments or volcanic rocks resting on a basement which had previously been consolidated and peneplaned.

Pleistocene: literally, "most recent (organisms)." The time epoch between 1.6 million and 10,000 years ago (see table, p. 79). So named in 1839 by Charles Lyell.

Pliocene: literally, "a majority of recent (organisms)." The time epoch between 5.3 and 1.6 million years ago (see table, p. 79). So named in 1833 by Charles Lyell.

pluton: a general term for an igneous intrusive body, other than specific simple forms such as a **dike** or ***sill***.

plutonism: eighteenth-century doctrine that attributed a major role to the earth's internal heat in the evolution of surface rocks.

poise: the unit for measuring absolute viscosity, equal to one dyne-second per square centimeter indicating that a force of 1g will maintain unit rate of shear of a film of unit thickness between surfaces of unit area. Named after the physicist Poiseuille.

polar wander path: the path of the pole with respect to a single continent as indicated by paleomagnetic latitude data from rocks of that continent over a substantial interval of time.

porphyroid: literally, porphyry-like. Foliated rocks containing "eyes" of feldspar in a sheared groundmass of minerals such as chlorite, mica, amphibole, and epidote. Regarded by some as sheared porphyry lavas or tuffs, but by others as having formed during shearing of pasty clayey sediment with the crystallization of feldspars.

porphyry: an igneous rock containing a generation of larger crystals of quartz and/or feldspar in a more finely grained groundmass. The adjective porphyritic is used irrespective of the kind of mineral forming the larger crystals.

potential energy: energy possessed by a body by virtue of its position relative to other bodies.

precession: the gyration of the axis of a spinning top, so called because similar gyration of the earth's axis causes the equinox to come a little earlier each year, that is, it "precedes" the expected time (by 50.37 seconds in longitude or 3.75 seconds in time).

primordial lead: lead with isotope ratio similar to that in meteorites, which is assumed to be the same as in lead at the time of the formation of the solar system. The proportion in lead ore of dilution with lead isotopes derived from the radioactive disintegration of uranium and thorium gives a measure of the age of the lead ore, and of the age of the earth.

prokaryote: primitive single-celled organism without a vesicular nucleus or organelles within membranes.

Proterozoic: literally, "(time of) early organisms." The time between 2500 and 570 million years ago (see table, p. 79).

pulsar: a celestial radio source emitting intense very regular short radio pulses ranging from one-third of a second to 4 seconds; they are believed to be neutron stars originating from the supernova collapse of a star a little heavier than the sun to about 10 km diameter.

pulsation: a geotectonic theory which claims that the earth suffers periods of expansion lasting millions of years followed by periods of shrinking that cause orogenesis.

pyroxene: a common family of blackish minerals (green in thin slices), which are silicates of iron or magnesium or both, but an important subfamily of them contain aluminum also. Their composition is very similar to the amphibole family, except that the latter contain hydroxyl and form in a more hydrous environment.

Glossary

quantum fluctuation: a phenomenon arising from the wave-particle duality at the subnuclear scale and the Schrödinger wave function whereby the probability is real that an energy barrier, impassable according to classical physics, may be bypassed. Examples are the ejection of alpha particles from a radioactive nucleus and the "tunnel diode" in semiconductors. The "new cosmology" relies on such phenomena on a mega-scale to initiate time and space and create instantaneously the total mass and energy of the cosmos, and so to trigger the **Big Bang**.

quark: a theoretical basic constituent of which fundamental particles are made, according to the evidence of some experiments. Six quark "flavors" differ in mass, charge, baryon number, spin, strangeness, and charm.

Quaternary: the youngest division of geological time in the old Primary–Secondary–Tertiary–Quaternary nomenclature. The Quaternary Period covers the last 1.6 million years, and includes the Pleistocene and Holocene Epochs.

red-shift: the reduction in the frequency of light from a star retreating from the observer so that spectral colors seen are redder than the color emitted by the star. The amount of red-shift is a measure of the velocity of retreat.

relativity: a universal law, Einstein's special relativity, which states that the laws of mechanics remain valid notwithstanding uniform rectilinear motion of the coordinates to which they are referred, and that the velocity of light is constant and independent of the motion of the observer.

relaxation time: When a load is applied to a body so that the principal stresses in the body become unequal, the difference between the maximum and minimum stresses is dissipated at an exponential rate by internal diffusion. The time required for the stress difference to decay to $1/e$ from its amount at any point in time is the relaxation time. In fluids like water with low viscosity the relaxation time is measured in picoseconds, in glycerine in nanoseconds, in bouncing putty in seconds, in ice in hours, in rock salt in centuries, in the upper mantle in thousands of years. Fluid behavior prevails in each of these materials if loads are sustained for times significantly longer than their relaxation time. Similar relaxation-time concepts apply to many behavior fields, such as decay of electric charge or magnetization, with consequent behavior thresholds on appropriate time scales.

revolution: As applied to a planet, revolution refers to its orbital motion, while rotation refers to its spinning on its axis.

Rhaetic: an epoch of approximately 200 million years ago, regarded by some as the end of the Triassic Period, and by others as the beginning of the Jurassic Period.

rheid: There are three states in which matter deforms according to fluid laws: gas, liquid, and solid where the duration of the phenomenon is significantly longer than the **relaxation time**. Examples of rheid flow of crystalline solids are the flow of a glacier ice, the rise of rock salt in a salt dome or salt glacier, and the diapiric rise of a gneiss dome. See Chapter 17.

rheopexy: the sudden gelation of a colloidal sol as the flow declines to a critical rate and flocs congeal to each other.

rhombochasm: a rhomb-shaped extensional opening in the continental crust occupied by oceanic crust.

rigidity: In ordinary English, rigidity refers to the resistance of a substance to elastic or viscous or plastic deformation, whereas in physics rigidity has been defined as the modulus of resistance to elastic deformation *only*. Hence ice (viscosity about 10^{13} poise and shear modulus about 10^9) has lower rigidity than water, which has a very low viscosity (about 10^{-2} poise) but high elastic shear modulus (about 10^{10} dyne/cm^2). It is therefore essential to be sure whether rigidity is used in the English or technical sense.

Rocky Mountain Trench: a long corridor stretching from Montana to Alaska (when the cognate Tintina Trench is included) drained in succession by the headwaters of several river systems. It is the locus of strong dextral transcurrent faulting.

root: the basal part of a **nappe** where it emerges from the rising diapir and begins to flatten out.

rotation: As applied to a planet, rotation refers to its spinning on its axis, as compared with revolution, which refers to its orbital motion.

salt dome: a **diapir** with a roughly circular piercing core of rock salt a couple of kilometers in diameter, which has risen several kilometers through the overlying sediments.

"salt glacier": a glacier-like flow of rock salt that develops if the piercing core of salt extrudes at the surface more rapidly than it can be removed by solution; hence they are rare except in arid regions.

schist: foliated metamorphic rock that has been recrystallized during solid-state flow, which causes the crystallization of "schistophile" lamellar minerals, particularly micas and amphiboles, and acicular (needle-like) minerals pointing in the direction of flow; hence the adjective *schistose*.

Schwarzschild radius: the distance from a star within which the general theory of relativity implies that, because of the curvature of the gravity field, radiant energy cannot escape.

sediment: clay, silt, gravel, or other matter deposited by any geological transporting agency such as river, currents, glacier, wind, volcanic eruption, or by chemical precipitation, or by organisms.

Senonian: the youngest stage of the Cretaceous Period in Europe, ranging from 91 to 67 million years ago.

sensitivity: the ratio of the unconfined shear strength of a clay or silt material before and after remolding at constant water content. Compare this with the relative stiffness of cream before and after whipping, or the relative strength of the gel and sol stages of a **colloid**.

serpentine: a group of green minerals, mainly hydrous silicates of magne-

Glossary 387

sium, commonly derived from the alteration of **mafic** and **ultramafic** rocks. Rock consisting predominantly of serpentine may also be called serpentine, but *serpentinite* is more correct and unambiguous.

shear folding: a term commonly used in America for similar folding, but which should be dropped on the grounds of priority and a misconception involved concerning the nature of folding (see Chapter 16).

sial: an acronym from **si**lica and **al**umina, used as a general term for the granitic continental crust, which is richer in alumina and silica than oceanic crust and subcrust below continents, where the silicate minerals are richer in **ma**gnesia and are hence called **sima**.

sigmoidal: S-shaped.

sill: an igneous body intruded concordantly with bedding or foliation.

sillimanite: a needle-like metamorphic mineral with the composition Al_2SiO_5, which is the same as andalusite and kyanite. Sillimanite forms at much higher temperature than andalusite, and in more hydrous conditions than kyanite.

Silurian: the time period between 438 and 408 million years ago (see table, p. 79). Named by Sir Roderick Murchison for rocks of this age in North Wales, formerly inhabited by the Celtic Silures tribe.

sima: an acronym from **si**lica and **ma**gnesia, in contrast with **sial**.

similar folding: deformation mode in which the thickness of beds remains constant in the direction of flow.

sinistral: left-handed or counterclockwise, in contrast with **dextral** or clockwise.

slate: a fine-grained rock which cleaves readily into thin plates independently of the original bedding; formed during the deformation flow of mudstone and shale, which develop the *slaty cleavage* normal to the maximum stress causing the flow.

solstice: literally, "sun stands." Either of the two points of the earth's orbit where the sun and the earth's axis are coplanar. At the solstices the sun is vertically above either the Tropic of Cancer or of Capricorn.

sphenochasm: a wedge-shaped sector of oceanic crust between blocks of continental crust, formed by the rotation of one block relative to the other.

stereographic projection: projection of one hemisphere onto a tangent plane from the opposite end of the diameter from the tangent point (see Fig. 32).

stishovite: a mineral with the same composition (SiO_2) as quartz and coesite, but only formed at pressures above 100 kilobars; hence, it has only been found at the earth's surface at **astroblemes**. Named after the Russian geochemist S. M. Stishov, who synthesized it in 1961 shortly before it was found naturally at the Arizona Meteor Crater.

stratum: a layer of sedimentary rock.

stress difference: The state of stress at a point can be represented by three "principal stresses" mutually at right angles. If these are not equal (hydrostatic), the difference between the maximum and minimum principal stresses is called the stress difference.

subduction: the postulated process whereby one lithospheric "plate" descends below another into the mantle. In a less specialized sense, the concept goes back to Ampferer and E. Kraus's *verschluckung*.

superposition: See **law of superposition**.

suture: In plate-tectonics theory, a narrow zone between two lithosphere plates after intervening oceanic crust has been eliminated by subduction.

syncline: literally, a fold in which the limbs slope toward each other, hence a fold convex downward; but more accurately a fold in which the youngest strata stratigraphically are in the core.

synclinorium: a major synclinal structure composed of smaller folds.

syneresis: literally, weeping. While flowing in the sol state, colloids tend to clot as like charges on the submicron particles link and subsequently continue to draw the particles more closely into an orderly formation, while intervening water weeps out from the clots as they crystallize.

syzygy: for astronomic bodies, the state of conjunction or opposition. (Greek συζύγος, paired.)

Tertiary: formerly a division of geological time in the old Primary–Secondary–Tertiary–Quaternary nomenclature. In current terms, the Tertiary Period extends from 66 million years to 1.6 million years ago, and includes the Paleocene, Eocene, Oligocene, Miocene, and Pliocene Epochs.

Tethys: Mesozoic equatorial seaway separating **Laurasia** and **Gondwanaland**, so named by Suess after the daughter of Gaea the goddess of the earth and Uranus the god of the sea. The Tethys developed in the Early Permian and was terminated in the Oligocene by the Alpine–Himalayan orogeny.

thermodynamics: the mathematical balance of heat and mechanical energy, expressed in two laws: (1) a quantitative balance exists between heat transferred and joules gained or lost; (2) heat can't pass spontaneously from a colder to a hotter body.

thixotropy: the property of colloids to change from the gel to the sol state when sheared and to regain the gel state after cessation of the disturbance. Although the total positive and negative charges on the submicron particles balance, positive parts migrate to link with negative parts of neighboring particles to give the substance overall strength (gel), which is destroyed by shearing so that positive repel positive and negative repel negative and the substance becomes a mobile fluid (sol) until the particles slowly regroup to return to the gel state.

tholeiite: a category of **basalt** rather high in silica and high in iron relative to magnesia, regarded as an indicator of tectonic environment.

thrust: an overriding of a rock body over another rock body, usually on a rather large scale.

tidal friction: The attraction of the moon draws the ocean waters of the earth to a bulge both on the side facing the moon and on the side furthest from the moon. As the earth rotates, these bulges are dragged eastward over the surface, causing a friction drag that acts like a brake on the rotation.

Glossary

There are also running elastic bulges in the solid earth that add to this tidal friction.

till: sediment deposited under a glacier due to melting of the ice between the sediment particles, which are not further redistributed by subglacial meltwater. A glacier melts from its base upward yielding till, and also from the surface downward yielding moraine. When these two surfaces meet, till grades upward into moraine.

tillite: rock formed from the consolidation of *till*.

tonalite: a granitic rock in which plagioclase feldspar is more abundant than orthoclose feldspar.

torsion: state of strain resulting from twisting one part of a body against another part.

transcurrent fault: a fault at which one side has been moved a substantial distance horizontally with respect to the other side.

transform fault: a specific category of transcurrent fault introduced by Tuzo Wilson. See Chapter 10 and Fig. 25.

trench: a long narrow trough in the ocean floor a couple of kilometers deeper than normal ocean floor. The **Benioff zone** of earthquakes reaches the seafloor in a trench. *[wrong]*

Triassic: the time period between 245 and 208 million years ago (see table, p. 79). Named in 1834 by Alberti from its threefold division in Germany: *Bunter, Muschelkalk,* and *Keuper*.

trilobites: an extinct group of Paleozoic arthropods with a three-lobed exoskeleton.

tuff: volcanic dust showered during an eruption.

turbidite: a sediment deposited from a succession of turbidity current flows, that is, of water carrying much suspended matter, commonly set off by a storm or seismic jolt that converts a bottom sediment colloidal gel into the sol state.

ultramafic: adjective applied to minerals and rocks with a very high content of magnesium and iron.

unconformity: a surface in a rock sequence where a stratum rests on underlying rock that had been formed, then deformed and eroded, during the interval before the deposition of the overlying stratum.

underthrust: a thrust fault where one rock mass is thrust under another rock mass.

uniformitarianism: the concept that processes currently acting have been responsible for the physical and biological evolution of the earth.

universe: generally used as a synonym of cosmos, but herein defined as the physically knowable universe, limited by the distance where the Hubble recession equals the velocity of light, in contrast with cosmos, defined as infinite space-time in accordance with the perfect **cosmological principle**.

vacuum fluctuation: See **quantum fluctuation**.

variation: See **magnetic declination**.

Wallace's Line: a line that separates the quite distinct faunas and floras of Asia and Australia. Alfred Russel Wallace, who first defined this line (between Bali and Lomboc thence through the Straits of Macassar between Borneo and Sulawesi, then east of the Philippines), was influenced most by the distribution of birds. Max Weber, who was more concerned with mammals, drew the line further east. A somewhat different line can be drawn for each animal pair competing for an ecological niche.

whinstone: a vernacular term in Britain for dolerite, diabase, or basalt.

white dwarf: a star of very low luminosity and very high density. For example, Sirius, the brightest star in the sky, has a white dwarf companion whose mass is about the same as the sun but whose luminosity is only $1/360$, radius about 1 percent and density of about 30,000 times that of the sun. Because of their low luminosity, white dwarfs further than 15 light-years away are too faint to be detectable.

Wilson cycle: a tectonic cycle postulated by Tuzo Wilson, involving rifting of a continental plate, the separation of the parts by ocean-floor spreading between them, consumption and final excision of this ocean floor by subduction, and formation of an orogen at the suture. New rifting may start the cycle again.

wobble: Precession and nutation of the earth's axis are caused by external torques, and the latitude of an observatory remains constant; in contrast, a wobble of the earth's axis is caused by redistribution of mass (such as continental drift or diapirs) and does result in change of latitude.

Index

In this index an "f" after a number indicates a separate reference on the next page, and an "ff" indicates separate references on the next two pages. A continuous discussion over two or more pages is indicated by a span of page numbers, e.g., "pp. 57–58." *Passim* is used for a cluster of references in close but not consecutive sequence. A page number in boldface type (**126**) indicates a major discussion. A page number in italics (*126*) indicates a figure.

Aar massif (Switzerland), 248
Absolute zero, 355
Académie des Sciences, 36
Académie Française, 36
Academy dei Lincei, 36
Achilles' race with tortoise, 358f
Actualism, 45, 58, 369
Adam, 138
Adelaide University (Australia), 297
Aegean Sea, 236
Afghanistan, 159f, 162, 250
Africa: fit against Americas, 96, 156, 158, 166, 278, 282; rift valleys, 113f, 175; ties with India, 159, 162; Caledonian orogenesis, **161**; subduction enigma, **174–76**, 178; Ordovician glaciation, **181**, *183*, **184**; collision with Europe denied, **182**; African polygon, 240; second order-polygons, *263*; radial movement, **266**; Mediterranean oroclines, **279**; offset against Europe, 302; mentioned, 103, 113, 271, 287, 289
Africanus, Julius, 70
Agassiz, Jean Louis Rodolfe, **66**, 67, 76
Ager, Derek, 60, 172, 307
Agricola, Georgius (Bauer), 48
Aguirre, L., 177
Ahmad, Fakhruddin, **98**, **160**, *161*, **167**, **179**

Ahmad al-Biruni, 11
Aiguilles Rouges (Switzerland), 248
Airy, George, **209**
Aitoff projection, *154*, 155, *165*
Akademie der Wissenschaften (Berlin), 36
Alabama: Carboniferous fan, 247
Alaska: orocline, *109ff*, 114, 288; Zodiac fan, **184–86**, *185f*; Malaspina glacier, *234*, 234–35; Gulf of, 288, 318; mentioned, 92, 156, 302
Alberta Association of Petroleum Geologists, ix, 219
Alberta University, 145
Albrecht (German physicist), 95
Aleutian: trench, *127*, **185**, 288, 292, 318; arc, 257; Islands, 292
Alexandria, Egypt, 10, 15, *15*, 20f
Algarve, Portugal, 282
Aligarh University (India), 160
All-American Observatory, 168, 171
Allegheny synclinorium, 245
Alluvium, 369
Alor, Indonesia, 298
Alpha of the Plough (essayist), 69
Alps: Ice Age, 65–66; orogenic pattern, 92, 118, 206ff, 245, **247–48**, **282**, 325; collision foreshortening myth, *108*, **240–41**, **285**; glacier tectonic analog, *234*, **234–35**; pre-Alpine basement, **248**; Tethys, 306

Amber, **34**
American Association of Petroleum Geologists: 1926 continental drift symposium, viii, 91–92; 1960 symposium, 119; *Bulletin*, 176
American Geological Institute, 131
American Geophysical Union, 104, 271, 292
American Journal of Science, 91
Amiens, France, 23
Ampferer, Otto, 100
Amphiboles, 198, 369
Amphibolite, 233, 369
Amru-Ibn-Al-Aas, 10
Anadyr, Gulf of (Siberia), 293
Anaximander, 63, 330f
Andalusite, 245, 369
Andes: roots, 25; alleged subduction, **176**; east-west tension, **176–77**, 187; mid-continental setting in Paleozoic Era, **315**
Andesitic volcanoes: relation to Benioff zone, *127*, 251; andesite line, *157*, 279, 302, 319; mentioned, 178, 187, 258, 369
Andhra, India, 296
Andromeda nebula (M31), 329, 337
Anelli, 207
Angaraland, 179
Anglo-Persian Oil Company, 104
Angular velocity, 356
Anhydrite, 227, 369
Annam (Indochina), 195
Antarctica: separation from India and Australia, 95, 156, 159, 171, **285**, *285*, 318; subduction enigma, *175*, **176**, 187; relation to South America, 315, 317; mentioned, 30, 92, 158, 209, 257, 271, 287, 289
Anticline, 369
Antilles, 279, **286**; paleomagnetic rotations, **297**
Antimatter, **351**, 353
Apennines, 207, 282
Appalachians: orogenesis, 144, **181**, 206, **245–47**, 250, 257; sinistral shear, *182f*, **183**; Aquarius fan, 184, *184*; diapiric, *246*, **246–47**; clastic fans, *246*, **246–47**, 256, 317; Tethyan analog, **307–9**, *308*; equatorial, 314, **320**
Arabia, 162, 285, 289
Arabian sphenochasm, *106*
Aragonite, 198, 369
Aral depression, *110*
Archaeocyathids, 158, 369

Archaeopteryx, **43**
Archbold, N. W., 158
Archean eon: missing crust, **162–64**, 172; greenstones, 164; evolution model, 312–13, *313*, 370
Archeofijia (Pacific land = South America), **315**
Archimedes, 21, 76
Arctic: Ocean, 92, 114, 151, 154, 288; orocline, *109–12*; paradox, **150–53**, 172
Arduino, Giovanni, 40, 48
Argand, Emile, 96, 118, 207
Argentina, 98
Arisili (New Guinea tribe), 4
Aristarchus, 10, **21**, 45
Aristotle, xii, 10f, 17, 20, 45, 58, 70, 330ff, 364f; spherical Earth, **14**, 17
Arizona: interfering torsions, 297; University of, 297; Meteor crater, 387
Armageddon, date of, 71
Armagh, Ireland, 70
Arunachalam, J., 338
Aryabhton (Hindu astronomer), 10
Ashmolean Museum (Oxford), 34
Asia: disjunctive seas of east Asia, 155, 236, 254, 256, 295, 317; relation to Gondwanaland, 162, 179, 279; Tethys, 240; paleogeographic ties with North America, **315**; mentioned, 273, 287, 293, 314
Assam, 279; orocline, 296, 300, 317f
Assumption of permanence of present states, **200**, 364
Asteroids, 215, 332; impact, **275**, 312, 326, 346; Titius–Bode law, 348–49
Asthenoliths, 189, 370
Asthenosphere, *127*, 262, 269, 370
Astrobleme, 215, 370, 387
Astromisches Jahrbuch, 348
Astronomical Society of South Africa, 140
Aswan, Egypt, 15
Asymmetry: orogenesis, xii, 245, 255f, 317; of the earth, opening-bud analogy, **146**, *146*; of expansion, **201**, 316f; of Mars, *314*
Atlantic City, New Jersey, 119
Atlantic fit, viii, ix, **89f**, **95**, *96*, 103–4, 158, 166, 297
Atlantic Ocean: origin through Moon ejection, 91f; relation to Alaskan orocline, *111f*; Wilson cycle, **144**; Mesozoic opening, 144, **181**, 257, 316–17; present widening, **169–70**;

Index 393

triple point, *175*, 176; Iapetus, **180–84**; midridge, *244*; globe model, *267*; computer model, *268*; mentioned, 155, 175, 178, 180, 288, 306, 312
Atlantis, 90
Atlas Mountains (Morocco), 279
Atoms, 356; synthesis, 344, 346
Attenuatella, 315
Attica, Greece, 187
Atwater, Tania, 131
Aufgeschwempte period, 49
Augen, 211
Augite, 262, 326, 370
Australia: fit against India, 95, **159–60**, *160*, 162, 250, 296; extension from China, 155 f, 285, **318**; NASA data, 170–71, 266; offset against Asia, 279, **301–2**; and New Zealand, 287, 302, 318; relation to South America, **315–17**; marsupial migration, 317; mentioned, 257, 289, 315, 326. *See also* Gaping gore
Australian and New Zealand Association for the Advancement of Science, x, 288, 305, 328
Australian Bureau of Mineral Resources, 162, 279
Australian Journal of Science, 126
Australian National University, 78, 106, 164
Austria, 333
Auvergne, France, **52–53**, 56
Avias, J., **315**
Avicenna (Ibn-Sina), 10
Axioms, cosmological, 330
Azimuthal projections, 166

Babar, Indonesia, 298
Back-arc-basins, 236, 254, 256
Background radiation, *see* Black body
Bacon, Francis, **89**
Bacon, Roger, 11
Bailey, E. B., 96
Baker, Howard, 90, **92**
Balearic islands, 282; sphenochasm, *108*
Bali, Indonesia, 298
Balleny fracture zone, 305, 317
Baluchistan orocline, *107*, 114, 279
Banda Sea, 279, 286, 295, 298, 302
Banded iron ore, **51**
Banks, Maxwell Robert, 106
Barnett, Cyril, 140, **146**
Barrell, Joseph, **60**
Basalt: Neptunist controversy, 49, **52–53**,

132; name coined by Pliny, 51; ocean floor, 123, 129; Jurassic tumor, 275; defined, 370; mentioned, 290 f, 326 f
Basin and Range province, **290**, *290*, 298
Batavia, Indonesia, 207
Batholiths, 245, 370
Bauer, Georg (Agricola), 48
Beagle, 120
Beaumont, Elie de, 59, 206
Becquerel, Antoine Henri, 76
Belemnites, 34, *34*
Belgrade, Yugoslavia, 347
Bell Laboratories (New Jersey), 332
Beloussov, Vladimir V., 128, 207, **208**, 269, 325
Beltian series, 311 f, *312*
Bengal, Bay of, 95, 159
Benioff, Hugo, **286**
Benioff zone, *127*, 197, 199, 236, **251–56**, *255*, 258, 369
Bentley, Richard, 341
Bentz, Alfred, *230*
Bergman, Torbern Olof, 48
Bering Sea, 286, 293; landbridge, 316
Berkeley, Bishop George, 357
Berkeley, California, 78
Berlin, Germany, 48, 92, 139, 327
Bernard Price Institute, 106
Bertrand, Elie, 37
Beryllium-ten anomaly, **197–98**
Betic cordillera, 282
Bex, Switzerland, 66
Big Bang, xi, 7, **332–38**, 343 f, 359, 364, 370
Big Crunch, 343
Big Pine fault (California), 276, 297
Biq Chingchang, 273
Birmingham, University of, 297
Biscay, Bay of, 282
Biscay sphenochasm, *108*
Bismarck Sea, 286
Black, Joseph, 47
Black body, 370; radiation, 332 ff
Blackett, Paul M. S., **106**, 121, 355
Black Forest (Germany), 247
Black hole, 329 f, **351–52**
Black Sea, 236
Blandford, Henry, 67
Blandford, William Thomas, 67, 98
Blastoid, 158, 370
Blinov, V. F., 137, 190, 327, 347
Blue Ridge (Tennessee), 246
Blueschists, 84, **198–99**, 370
Boccaccio, Giovanni, 36
Bode's law, 348

Bogolepow, Michael, **139**, 274
Bohemia, 46, 48, 50; massif, 247
Bohm, David, 356, 360
Bologna, Italy, 23
Boltwood, Bertram Borden, **74**
Bombicci (Italian geologist), 207
Bonarelli (Italian geologist), 207
Bondi, Hermann, 330, 340, 341 ff
Bonin-Marianas trench, 251, 305
Bonneville, Lake, 209
Borneo, 298, 302
Borough, William, **28**
Boudins, boudinage, **221**, 233, 370
Bouguer, Pierre, **24–25**, 26, 138; Bouguer anomalies, **25**
Boulder, Wyoming, 167
Bouvet Island, 288
Boyle, Robert, 37
Brachiopods, 150, 158, 315, 371
Bradbury, Harry J., **173**
Brahe, Tycho, 11, **21–22**, 59, 133, 335, 360
Brans, C., 141
Brazil, 106, 117, 121, 145
Breccia, 211
Breislak, Scipio, 46, **51**
Bremen, Germany, 332
Britain, 181, 183, 326 f
British Association for the Advancement of Science, 29, 75
British Columbia: University of, 118; mentioned, 292
British Institute of Geological Sciences, 148
British Museum, 148, 166
British Royal Artillery, 137
Brno, Moravia, 120
Brongniart, Alexandre, **40**, 46
Brontosaurus, 43
Brösske, Ludwig, 140, **145**
Brouwer, H. H., 298
Brown, B. W., **188–89**
Brownschweig Technical University (West Germany), **189**
Brown University, 118
Brunhes, Bernard, 122
Brunnschweiler, Rudolf O., 117
Bruno, Giordano, 11
Brussels, Belgium, 339
Bucher, Walter H., 118, 271
Buckland, William, 58, **64**, 66
Bud analogy of asymmetric expansion, **146**, *146*
Budapest, Hungary, 142
Buerlen, 97

Buffon, Comte de, 46, 48, 58, 72, 90
Bullard, Edward: 60, 121; Bullard fit, 104, 167, 278, 297, *299*
Burma, 159, 296
Burrett, Clive F., 85, 158
Buru, Indonesia, 298
Butler, Robert, 297
Butterfly families: trans-Pacific links, 316
Butung, Indonesia, 298

Caesar, Julius, 10
Caesarea (Palestine), 70
Calabria, Italy, 282
Calceolispongia, 160
Calcite, 198, 227
Calcutta, India, 159
Calderone, Gary, 297
Caledonian orogeny, 144, **181**, 257; sinistral shear, **183**, *183*, 307; diapiric, 208, 232; Tethyan analog, **307–12**, *308*; rotated, *308, 313*, 314, 371; equatorial, **320**
Calemenids, 158
Calgary, University of, 118
California, 140, 287, 292, 297; University of, 316
Callipus, 17
Cambrian Period, 74; faunas, 158, 160; sea levels, *192*; Himalayas, 249; fossils not enlarged by Earth expansion, 345; defined, 371; mentioned, 223
Cambridge University, 57, 64, 106, 123
Camper, Petrus, 39
Canada: Coast range, 292; Dominion Observatory, 292; mentioned, 237, 298, 326
Canberra, Australia, 169
Cape Town, South Africa, 26, 29
Carboniferous Period: Wegener, *94*; climate, 97, 192; faunas, 158, 303; Appalachian sinistral shear, 183; sea levels, *192*; Appalachian diapirs, 246, 246–47; Earth obliquity, 297; paleogeography, 99; defined, 371
Cardano, Geronimo, 36
Carey fracture zone, 305, 317
Cargo cults, 6
Caribbean, *111*, 113, 257, 282, **286**, 297, *299*
Carnegie Corporation, 106; grant to du Toit, 98
Caroline basin, *243*
Carpathian Mountains, 92
Carroll, Alexis, 12
Carter, W. D., 177

Carus, Titus Lucretius, 338
Cascade Range, **290**, *290*
Caspian Sea, 226
Cassini, Giovanni Domenico, **23**
Cassini, Jacques, **23**, 26, 138
Caster, Kenneth E., 117, 119
Catastrophism, 42, **57–60**, **371**
Cathaysian flora, 158, 162; trans-Pacific links, 315; defined, 371
Catherine I, Empress, 36
Cause of Earth expansion: cosmological, 139, 141, 147, 327; thermal expansion through radioactivity, 139, 147f; phase change, 141f, 326; declining gravity constant, 142, 144, 147, 327; pulsation, **145**, 325; meteoric accretion, 325–26
Cavendish, Henry, **83**
Cayenne, French Guiana, 23
Cayman: trench, 279; Sea, 297, *299*
Celebes Sea, 286, 298
Cenozoic Era, 79, 177, 371
Center of gravity, 226, 286f; sun, 62; defined, 371
Centimani (mythology), 306
Central America, 178–79, 316, 319; transcurrent faults, 297; Tethys, 307
Central Pacific rhombochasm, 318
Centre nationale de Recherche scientifique, 143
Cephalopods, 158f, 371
Cepheid variable stars, 329
Ceres, 275
Chalk Draw–Carta Valley fault zone (Arizona), 297, *297*
Challenger expedition, 74
Chamberlin, Thomas Crowther, **75**, 85
Chambers, Robert, 8
Chandler, Seth C., 73
Chaos (mythology), 306
Chao-Wen Chin, 345, 350
Charlottenburg, Berlin, 139
Charpentier, Jean de, **66**
Chert, 49, 371
Chéseaux, Philippe de, 332
Chicago, University of, 75
Chile, **176**, 177; trench, 178
China: ancient philosophy, 22, 27, 35; faunas, 102, 158f, 161, 316; extension from Australia, 155, 162, 285; offset against Australia, 279, **302**; dextral torsion, **293**; Borneo fit, 298
Ching Hai Lake, 293
Christianity: interfering god, 7; dogma straitjacket 9–10, *19*; medieval, 9–10, *20*, 21; Inquisition, 11; monotheism, 12; interpretation of fossils, **33–37**; censorship, 37; Noah's flood, 37f, 40, 64, 90; *Genesis*, 40, 53, 64, 91; Linnaeus evolution, **42**; Ice Age, **64**; date of creation, **70–71**; fundamentalism, 360
Christodoulidis, D. C., **171**
Chu-Hsi, 35
Cincinnati, Ohio, 117
Circadian growth lines of corals, **195–96**
Circum-continental rift system, *122*
Circum-Pacific torsion *277f*, 278, **286–97**, 302, 317–19, *302*; interaction with Tethyan torsion, 318–19
Ciric, Branislav, 347
Clairaut, Alexis-Claude, 24
Clarke Memorial Address, x
Cleavage, 83, 100, 371
Climate zones, 297
Clipperton fracture zone, 132, 279
Clube, S. V. M., 325
Cobbing, E. J., 176
Coesite, 262, 326, 371
Colberg, Loeffelholz von, 91
Colbert, E. H., 316
Collision tectonics, 217
Colloid, 55, 371, 386
Colombia, 176, 279, *280*, **282**, 297, 301
Colonna, Fabio, 37
Colorado, 143
Colorado Plateau, 139
Columbia University, 118, 147, 187, 271
Columnar jointing, *52*, 52f, 372
Comets, 32, 332
Concentric folding, **217–24**, *218*, *220*, 225, 240, 247, 249, 372
Condamine, Charles Marie de la, **24**
Conesauga Formation, 223
Conic projections, *165*, 166
Conjugate shears, **275–76**, *276*, 279, 372. *See also* Circum-Pacific torsion
Conodonts, 315, 372
Conservation laws, 82, 330, 343
Continental dispersion: maximum near Falkland Islands, x, **288**; minimum in Siberia, x, **289**
Continental drift: criticisms, viii, ix; 1926 symposium, viii, 91, 97; 1956 symposium, ix, xi, 106, 114, **117**, 143, 148, 278, 289, 295; evolution of concept, 91–99; advocates and opponents, **95–97**; decades in contempt, **101–4**, 117, 328; vindication, **105–19**; expansion alternative, **139–43**; 1981 symposium, 148–49; cartographic errors,

166–67; centers of maximum and minimum dispersion, **289**; mentioned, 118, 195
Continents: welded to mantle, **269–70**; move radially outward, 270; creep west, **273**, **285–86**
Contraction hypothesis, 126, 143f, 206, 261
Convection currents, 99, 147, 204; cause, **100**; history of concept, **100**; viscosity variation, **101**; effect of phase changes, **101**; salt dome analog, 227–28, *228*, 372
Conybeare (English geologist-clergyman), 58, 62, 64
Cooper's Chronicle, 70
Copernicus, Nicolaus, 11, **20–21**, 45, 120, 133, 173, 360, 363, 365
Coral Sea, 286
Cordilleran: Proterozoic, 257, *309*, 319; birth of Pacific, *310*, **311**, *311ff*, 313–14; equatorial, 310, **318**; Cambrian rotation, 314; Permian faunal links across Pacific, 315; subsequent evolution, 315–16; defined, 372
Core of the earth, 30–31, **201**, 326
Coriolis effect, **273**, 372
Cornell University, 195, 348
Corsica, 279, 282; rotation, 113
Cosmas, *9*, **9–10**
Cosmic constant, 337
Cosmic rays, 372
Cosmic repulsion, 336
Cosmological principle, **342–47**, 372; Perfect Cosmological Principle, 342, 350
Cosmos, 372; restricted definition, *337*, 342
Côte d'Azur, France, 282
Cotton, Leo A., **94**
Couple, 372
Courtenay, Bishop, *19*
Cox, Allan V., 292
Coxworthy, W. Franklin, 91
Craters: impact, 215, 312, *313*
Crawford, A. Ray, **162**, 301
Creation: date of, **70–71**; of time and space, 331; of God, 331. *See also* Matter, synthesis
Creer, Kenneth M., 140, **147**
Cremona, Italy, 36f
Cretaceous Period: chalk, 40; Moon origin, 92; paleomagnetic time scale, *124*; expansion since, 153, **170–71**; poles, 153, 288; faunas, 159–60; ocean-floor spreading, 174, 178, *180*, 317; Central America, 188, 316; sea levels, *192*; global torsions, 287; poles, 153, 288; paleoclimate, 297; Southeast Asia reconstruction, *300*; Pacific reconstruction, *303*; Tasman Sea reconstruction, *304*; marsupial migration, 316–17; asteroid impact, 326; defined, 372; mentioned, 151, 170, 174, 178, 188, 228, 288, 301, 316
Crinoid, 373
Critical temperature, 242, 373
Criticisms of Earth expansion: no major change in *g*, **190**; sea levels, 190, **191**, *192*; paleomagnetic radius, 190, **192–95**, *193*, *195*; moment of inertia, 190, **196**; radii of other planets, 190, **199–201**
Croatia, 241
Croll, James, **73**
Crook, Keith A. W., **164**, 172
Crookes's radiometer, 358
Crustal shortening, 205, 217
Crystal spheres (supporting celestial bodies), **17–18**, 360
Cuba, 282
Cuban Academy of Science, 282
Curie, Pierre, 76
Curie temperature, 31, 123, 126, 150, *242*, 373
Curitiba Museum (Brazil), 106, 117
Cuvier, Baron Georges Léopold, **39**, 45f, 133, 357; concept of evolution, **41–42**, 58f, 64
Cycadeoidea: trans-Pacific links, 316
Cycles, **62**; sea level, 191–92, *192*; fossil growth lines, **195–96**; orogenesis, 245f
Cyclopes (mythology), 306

Dachille, Frank, 325
Dal Piaz, G., 207
Dalrymple, G. Brent, 78
Daly, Reginald A., 96
Dalymayrac, B., 315
Damar, Indonesia, 298
Dana, James Dwight, 12, 206
Dark Ages, 9, 11, 45, 70
Darwin, Charles: evolution paradigm, **41–43**, 173, 365; uniformitarianism, 60f, 71, 75f; mentioned, xiii, 46, 120, 364
Darwin, George, 91, 95
Darwin plate, 261, 269–70
D'Aubuisson, Jean François, 46, **53**
David, Pierre, 122
David, T. W. Edgeworth, 30, 95

Index

Davie, John, 47
Dearnley, Raymond, **148**, 274
De Beaumont, Elie, 59, 206
De Charpentier, Jean, **66**
Declination, *see* Magnetic declination
Décollement, **222**, *223*, 249, 373
Deep-focus earthquakes: relation to Benioff zone, 251; mentioned, 189
De la Condamine, Charles, 24
De la Métherie, 46
Delesse, Achille, 105
De Maillet, Benoît, 46, 48
De' Medici, Cosimo, 36
De Mericourt, Peter, 27–28
Democritus, 6, 360
Dennis, John, 131
De Saussure, Horace, 65
Descartes, René, 7, 13, 29, 72
De Sitter, Willem, 341
Desmarest, Nicolas, 46, **53**
Devonian Period, 98; faunas, 158, 196; Caledonian–Appalachian orogeny, **181**, 307; Appalachian sinistral shear, *182f*, **183**, 307; sea levels, *192*; Appalachian diapir, **246–47**; trans-Pacific links, 315; mentioned, 196
De Witt, Bryce, 332
Dextral torsion, 186; dominates Pacific coast of Asia and North America, **293–96**; defined, 373. *See also* Circum-Pacific torsion
Diabase, 49, 373
Diamond, 262, 326f
Diapirs, **128**, **179**, 199, 207f, **225–36**, 237, **252**, 317, 325; imply tension, 254; east-west asymmetry, 256, 273; feedback foci, 317; defined, 373
Diastems, 60
Dibblee, Thomas W., 276
Dicke, Robert H., **141**, 144, 327
Dickins, James M., 117, 160
Didelphoid marsupials, 317
Dietz, Robert S., *154*, 155, 167
Diggers, Thomas, 332
Dike, 373
Diluvialism, **63**, 373
Dimensions: 336, 342, 373; equivalence of mass and energy, **353–55**; equivalence of T and L, **354**; equivalence of Hubble's constant and angular velocity, 356
Dinosaurs, 43, 316
Diplodocus, 43
Dirac, Paul A. M., **141**, 144f, 327, 329f, 345, 347; large dimensionless numbers, **304f**

Disjunctive seas, **256**
Dogma blindfold: Zittel, **3**; Comas, **9–10**; Caliph Omar, **13**; Roger Bacon, **11**; Copernicus and Luther, **11**; Lewis and Dana, **11–12**; Scheuchzer, **39**; Werner, **50**, 133; Richardson, **54**; Pleistocene glacial stages, **67–69**; Archbishop Ussher, **70–71**; Kelvin sycophants, **73–74**; retrospect, 76; Jeffreys, 81, **103–4**; Simpson, **102**, 133; Kuhn's *gestalt*, **120**; Planck's and Wegener's comments, **121–22**; Brahe, 133; von Buch, 133; Cuvier, 133; Kelvin, 133; Bailey Willis, 133; paleomagneticians, 151; Sutton, 172; *New Scientist*, **172**; Pope, **179**; Iapetus, **180–82**; Tanner, **187**; Brown, **188**; plate tectonicists, **197–98**; cosmology, 328
Dolerite, 106, 373; symposium, 126
Dong Ryong Choi, 314
Doppler, Christian Johann, 333
Drag, 373
Dravidians, 98
Drayson, Alfred Wilks, **138**
Duke University, 118
Dunglass, Baronet, 47
Dunkirk, France, 23f
Durban, South Africa, 117, 167
Durham University, 77f, 100
Düsseldorf, Prussia, 140, 145
Du Toit, Alexander L., **97–99**, 105, 121, 306
Dziewonski, Adam, **269**

e (mathematical constant), 373
Earth: flat, 9, **14**, 15, 132, 328, 363; Eratosthenes measures diameter, **15**, *16*; a sphere, **14**, 76, 132, 164; center of the universe, 16–17, 358; shape, **22**, 208; figure, **25–26**, 375; magnetism, **27–32**; core, 30–31, **201**, 326; primal heat, 62, 71–72; age, **70–79**, 132, 138; rotation, **72**, **186**, 245; axial wobble, **73**, **200**, 275; radioactivity, **76–78**; weight, **83**; mass increasing exponentially, 140, 321, 328; NASA radius measurement, **168–72**; moment of inertia, 200, 288; obliquity to ecliptic, **200–201**, 274–75, **297**; isostasy, 208–9; inhomogeneity, **261–62**; global differential motions, 272–73, *273*; polar wander, 274; geoid tumor, 275; asymmetry, **287**; paleoclimate zones, 297; rotation of equator, 309; evolution, *312*; Titius–Bode law, 348–49; Earth's crust defined, 374

Earth expansion, *see* Expanding Earth
Earthquakes: of Benioff zone, 253; polygon boundaries, 262–64; indicate mantle temperatures, **269–70**
East Asia orogenic arcs, **254–56**; dextral shear, 289
East China Sea, 286, 295
Easter Island, 168
East India Company, 29
East Indies, 156, 278, 287, *298*, **301**
East Pacific Rise, 171, 178
Eaton (American geologist), 46
Eauripik–New Guinea rise, *243*
Eclogite, 129, 242, 264, 326f, 374
Ecuador, 24
Eddington, Arthur Stanley, 330, 347
Edinburgh: 46; Royal Society of, 47; University of, 47, 57, 77, 100, 147, 196
Egyed, Lazlo, **142–43**, 190
Ehrensperger, Jacob, 347
Einstein, Albert: theism, 12, 331; action at a distance, 20; mass-energy equation, 331; cosmic constant, 337, 365; parity of gravitational and inertial mass, 339f, 364; field equations, 341, 353; no arbitrary constants, 355; imaginary solidity of matter, 356; mentioned, 363f
Elastic deformation, 334–35
Elements, genesis of, 332
Elliston, John, **55–56**, 238
Elton, Sam, **147–48**; **347–50**
Emba (near Caspian Sea), 226
Embleton, B. J. J., **266**, 270
Emiliani, Cesare, **68–69**
Emperor ridge, **319**, *319*
Enderby Land, 159
Entropy, 374
Eocene Epoch: Wegener, *94*; paleomagnetic time scale, *124*; Indian Ocean, *180*; Himalayas, *180*, 249; Zodiac fan, *185f*; Woodbine Formation, *229*; East Asia, 256; defined, 374; mentioned, 102
Eohippus: evolution paradigm, 41
Eo-Pacific, 261, 316f
Eötvos University (Budapest), 142
Epeirogenic jointing, **265**
Epicenter, 373
Epoch, 374
Equal-area projections, **165–66**
Equatorial orogenic girdle, **307–9**, *308*
Eratosthenes: measures Earth's diameter, 15, *16*, 22, 81, 83; mentioned, 10f, 21, 164

Erebus, Mount (Antarctica), 30
Erratics (glacial), 64, 305; defined, 374
Estonia, 117
Ether, 374
Eudoxus, 17f
Eugeosynclines, 147, 187, 198, **237–39**, 246; tectonic analogs of spreading ridges, **244–45**; facies, 247, 282; defined, 376 (*under* geosyncline)
Eukaryote, 374
Euler, Leonhard, 73, 95, 318; Euler pole, 374
Eurasia, 143, 179
Europe: Ice Age, *68*; oroclines, **282**; Africa collision false, **282**; offset against Africa, 302; mentioned, 192, 207, 240, 266, 287
Eustasy, 375
Evans, John W., 97, 117
Evaporite, 375
Everest, George, 26, **208–9**
Everest, Mount, 249
Evernden, Jack F., 78
Evolution of life, 8, 60; different models, **41–42**; "explosive," 43–44
Exeter Cathedral: Ptolemaic clock, *19*, 20
Exotic blocks, 250
Expanding Earth: history of concept, ix, x, 117, 128, 133, **137–49**; 1976 book, xi, xii; South Atlantic gap, *96*; Moscow conference, 137; development in German, **139**; 1960's revival, **145–49**; asymmetry, *146*, **200**; rate of expansion, 148, **170**, 191, 258, 316; evidence of expansion, **150–72**; NASA data, **168–72**; contrast with plate tectonics, 184, 187, **204**, 205, **269**, 307; hierarchy of extension, 194, **261–65**, *261*, *263ff*, **287–88**; universal dispersion, 261, **288–89**; adjustment of curvature, **264–65**; stages in expansion, 312, *313*; Sydney symposium (1981), 316; exponential increase, **319–20**, 325. *See also* Cause of Earth expansion
Exponential, 375

Fabric, 375
Facies, 375
Fairbridge, Rhodes, 143, **147**, 148, **187**
Falkland Islands, 98, 275
Fan, 106, 375
Far Eastern Geological Institute (Vladivostok), 316
Fault, 375
Faunal provinces and realms, **84–85**, 307

Favia speciosa, 195
Feistmantel, Otto, 98
Feldspar, 233, 262, 326, 375
Feldspathoid, 375
Ferromagnetism, 31
Fichtel (Hungarian geologist), 46
Figure of the earth, **25–26**, 375
Fiji, 302
Fisher, Osmund, 90f, 100, 102
Fitton, W. H., 54
Flandern, T. C. van, **142**
Flat Earth, 9, **14**, 15, 132, 328, 363
Flathead Lake (Montana), 291
Fleming, Reverend, 58
Flexure folding, **224**, 375
Floetz period, 49, 52
Florence, Italy, 36
Flores (Indonesia), 296, 298
Florida: State University, 141, 176, 188, 254, 285; shelf, 292, 297
Flow in solid state, **100–101**, 214, **233**, **271**
Flow lines, **217–19**, *218*, 223, 231, 240
Flysch, **164**, 172, 375
Folding, ix, xi, **206**, **217–24**, *218*, *220*, *222*, 376; flattening, **241**
Folding, experimental: Hall, 206; Reyer, 207; Ramberg, 208; Beloussov, 208
Folgerhaiter, Giuseppe, 106
Foliation, 211
Foraminifera, 376
Forbidden Sea, *9*
Förster (German geologist), 95
Fortis (Italian geologist), 46
Fossa magna, 305
Fossils, **33–40**, 80; defined, 33, 376; myths and superstitions, 33–35; spontaneous generation, 35, 37; Cuvier, **39–40**
Fracastoro, Girolamo, 18, 36
Fracture zones, **129**, *130*, **132**, 305
France, 22f, 59, 102, 151. *See also* French Academy of Science
Frauenburg, East Prussia, 21, 120
Freiberg, Saxony, **48**, 52; Bergakademie, 48, 56, 64
French Academy of Science, 23, 29, 106
Friedmann, Alexandre A., 330, 347
Füchsel, Georg Christian, 48f, 58
Fujiwara, Sakuhei, 273
Fulgurites, 34–35
Fundy, Bay of (Canada), 144
Fusulinid, 376

Gabbro, 242, 250, 262, 376
Gabon, 61

Gabrielse, Gerald, 291
Gaea (Earth goddess), 306
Galaxies, 210, 376; Gaussian distribution, 336; inter-galactic distance, 336, *337*; recession rate, 338; de Sitter recession, 341; statistical size, 344, 365; evolution, 347–48, 350; rotation, 355. *See also* Milky Way galaxy
Galileo Galilei, 11, 23
Gamow, George, 330, 332, 347
Ganges valley (India), 249
Gansser, Augusto, **179**, **248**, *248*
Gaping gore, 143, 156, **158**, 167, 172, 179, 301, 307
Garlock fault (California), 276, 297
Garnet, 129, 198, 264, 326, 376
Gastil, Gordon, 311, *311*
Geikie, Archibald, **50**, 55f, **73**, **75**
Geller, Margaret, 338
Gellibrand, Henry, **28**
Gemini, 342, 352, 376
General relativity, 337
Genesis, 40, 58, 91
Geneva, Lake, 66, 248
Geodetic measurements by NASA, **167–71**, *169*
Geoid, **26**, 273, 376
Geological Society: of Australia, ix, xi, 114, 126; of India, x, 126; of London, 55, 66, 104, 126, 172; of Scotland, 56; of Glasgow, 72, 75; of America, 78, 91, 191
Geological Survey: of South Africa, 98, 121; of Britain, 121, 148; of Canada, 123, 291, 311; of Chile, 176; of U.S., 177, 184, 245f, 292
Geological time scale, **79**
Geology: scorned by Kelvin, 80; numeracy, **80–85**; scorned by Jeffreys, 81; contrast with geophysics, **82**; comparison with other sciences, **82–83**
Geomagnetism, **27–32**
Geometrical progression, 376
Geophysical Surveys, 292
George V fracture zone, 305, 317
Georgia: Carboniferous diapir, 247
Geosynclines, 128, 214, 244, 376. *See also* Eugeosynclines; Miogeosyncline
Geotectonic revolution: three phases, 105
Gerard of Cremona, 36
German readers, 137, 207
Germany, 207, 226, 266, 326
Gesner, Conrad, 37
GIGO (Garbage In, Garbage Out), 74f, 81, 84f, *193*, **194**

Gilbert, William, **28–29**
Gilliland, W. N., 274
Glacier flow, 100
Glasgow, Scotland, 71
Glaucophane, 198, 376
Glen, William, 125
Glikson, Andrew Y., **162–64**, 172
Globular projection, *165*
Glossopteris flora, 99, 158, 376
Gneiss, 231, **232–33**, 245, 376; domes, 232–233, *232f*, 252, 376
Gnomonic projection, *165*, **166**
Goddard Space Center, 344
Gods: primitive concepts, **6–7**
Godsen (U.S. Geological Survey), 292
Goethe, Johann Wolfgang von, 46; *Faust*, vii
Gold, Tom, 342–43
Gonds (Dravidian race), **98**
Gondwana: flora, 98; stratigraphy, 121, 161
Gondwanaland: origin of name, **98**, 307; paleomagnetic confirmation, 106; gold medal, 114; defined, 376; mentioned, 99, 105, 162
Goodenough Island, 232 ff, *233*
Gorang, Indonesia, 298
Gore, 376. *See* Gaping gore
Gotland, 196
Göttingen, Hanover, 139
Gouge, 211, 377
Gough, D. Ian, 106
Graben: defined, 377. *See also* Rift valleys
Gracht, W. A. J. M. van der, 97
Gradualism, **61**
Graham, John W., 106
Grampian Mountains (Scotland), 54
Grand Saline salt dome (Texas), *231*, 232
Granite: basement, 49, 52f; controversy, **53–56**; colloid chemistry, **55**; orogenic cores, **245**, 258, 282, 321, 323; defined, 377
Granulite, 245, 377
Graphite, 262, 326f, 377
Gravitational constant, 141–47 *passim*, **327**, 336, 339f, 350, 377
Gravitational mass, 377
Gravity anomaly, **26**, 377; island arcs, 189
Gravity field, 242f, 273, 285, 292
Gravity tectonics, **204–9**, 325
Gravity waves, **252–53**
Graz University, 92
Great Australian Bight, 156
Great Barrier Reef (Queensland), 184

Great Basin (Nevada), 139
Great Britain, 181, 183, 326 f
Great circle, 377
Great Glen Fault (Scotland), 276
Great Lake (Tasmania), 106
Great Salt Lake (Utah), 209
Greek gods: genealogy, 306
Green, William Lowthian, **90**, 120, **138**
Greenland, 92, 97, 161, 209, *232*, 266, 287, 289
Greenough, George Ballas, 46, **66**
Greenstone, 377
Greenwich observatory, 209
Grenville front, 182, *182*, 377
Gresham College (London), 28
Gretener, P. E., 60
Griggs, David T., 100
Grindelwald glacier, 65
Groeber, P., 145
Groves, Ralph, 140
Growth lines of fossil corals, **195–96**
Guettard, Jean Etienne, 52–53
Guiana, French, 23
Gulf of Mexico, *111*, 226
Gunter, Edmund, **28**
Günz glaciation, 67
Gussow, William, 119
Gutenberg, Beno, **95–96**, 106
Guyot, 129, 377
Gypsum, 377

Haarmann, E., 207
Haeckel, Ernst Heinrich, 8
Haile, N. S., 301
Hall, James, **46–47**, 54, 58, **206**
Hallam, Anthony, 297
Haller, J., *232*
Halley, Edmund, 23, **28–31**
Halm, J. K. E., **140**, 190
Hamburg, Germany, 141, 327
Hartmann, Georg, 28
Harvard University, 85, 100, 118, 269, 338
Harz Mountains (Germany), 48
Hatherton, Trevor, **187**
Haughton, Samuel, **74**
Hawaii, 90, 138, 168f, 171, 289, 319; ridge, **318**, 319, *319*
Hawkins salt dome (Texas), *229*
Heat flux, terrestrial, 144, 269, 377; island arcs, 189, 235, 252; polygon boundaries, **262**; mentioned, 273, 318
Heezen, Bruce C., 121, *122*, **143–44**, 148; Heezen fracture zone, 302
Heide salt dome (Germany), 228, *228*, 236, 256

Heisenberg, Werner, 332
Heliocentric system, 20–21, 76, 173
Helium, 332, 346
Helmand depression (Afghanistan), *284*, 285
Helmert, F. R., 95
Helmholtz, Hermann von, **72, 77**
Helvetic nappes (Alps), **248**, 249
Heraclides, **20**
Heraclitus of Ephesus, 342
Herodotus, 35, 58, 70
Herschel, W., 207
Hesiod, 6, 306
Hess, Harry Hammond, ix, 100, 119f, 129, 141
Hierarchy of global polygons, 194
Hildenbrand (U.S. Geological Survey), 292
Hilgenberg, Otto C., **139–40**, 190, 327f
Hill, Mason L., 276
Hills, E. Sherbon, 224
Himalayas: orogenesis, xii, 92, 98, 113, 157, 206, 245, **248–50**, *249*, 325; not due to collision, **178–80**, *180*, 187, 307; paleogeography, **178–80**, 307; isostasy, **208–9**
Hindu Kush Mountains, 314
Hipkin, R. G., **196**
Hipparchus of Nicea, 17, **19**, 80–81
Hise, C. R. van, 224
Hispaniola: rotation, 113
Hixon, Hiram W., **138–39**
Hobart, Tasmania: continental drift symposium, **ix**, 106, 114, **117–18**, 126, 143, 148, 278, 289, 295; magnetic observatory, 29; dolerite symposium, 126
Holden, J. C., 154, *154*, 167
Holland Memorial Oration, x, xii
Holmes, Arthur, **77**, 78, 96, 98, 100, **147**, *263*
Homo diluvii testis, 39
Honduras, Gulf of, 140
Honolulu, Hawaii, 120, 171
Honshu, 265, 305; arc and trench, 251
Hooke, Robert, xiii, **37–38**, 40, 58, 133, 328, 357; Hooke's law, 335
Horizontal compression, 204
Hormuz Salt Series, 160
Horn, Cape, 289
Hornblende, 245
Hornfels, hornstone, 378
Horst, 256, 378
Hospers, Jan J., **123**
Hot spot, 320
Hoyle, Fred, 141, 342–43

Hubble, Edwin Powell, **329**
Hubble constant, 329, 336f, 340, 354ff
Hubble recession, 342f, 356; regression of, 347
Hubble's law, xi, 328–38 *passim*, 342, 349, 364, 378
Huchra, John, 338
Hudson River (New York), 106
Huelva, Spain, 282
Hugi, F. G., 65
Humboldt, Friedrich Heinrich Alexander, Baron von, 46, 50, 52, 59, 66, 90
Hungary, 326
Hutton, James, 38, **46–47**, 51, 57, 75, 120, 210; granite controversy, **53–56**; uniformitarianism, **58–61**, 71, 173; glacial transport, 65
Huxley, Thomas Henry, 24, **75**, 82, 84, 353
Huygens, Christian, 23
Hydrogen, 332, 345f, 355
Hypersthene, 378

Iapetus ocean myth, **180–84**, 187
Ibn-Sina (Avicenna), 10
Ice Age, **63–69**, *68*, 73, 76, 132, 378
Ichikawa, K., 314
Idaho: orocline, **290**, *290*, 296; University of, 315
Igneous, 378
Illinoian glaciation, 68
Imaginary numbers, 7
India: early philosophy, **16–17**, restored against Australia, 95, **159–60**, *160*; rotation, 113; paleogeographic enigma, **159–62**, 172; Indian polygon, 240; collision with Asia false, 282; offset against China, 302; mentioned, 121, 164, 170, 180, 208, 279, 289, 296, 316
Indian Science Academy, 167, 179
Indiana University, 90
Indian Ocean: beginning, 91, 155, 306, 316; spreading, 128, **178–80**, *180*; lineations, 159
Indicopleustrus (Cosmas), 9–10
Indochina, 296, 316
Indo-Gangetic trough, **249**
Indonesia, 113, 158, 240, 279, 282, 298, 316; Geological Research and Development Center, 279
Indus–Tsangpo suture, 161, 179, **180**, 249
Inertia, 378
Inertial mass, 340, 357, 378

Inquisition, 11, 37
International Geological Congress: of 1923, 118; of 1972, 244
Iran, 159f, 162, 226, 227, 279
Ireland, 181
Irian Jaya, 279
Iron, **346**
Irving, Edward M., 106, 117, 121, 292
Isaacson, Peter, **315**
Islam: god, 7; scholarship, 10; monotheism, 12
Island arcs, **187, 244–45, 254**
Isobath, 378
Isopors, 30
Isostasy, **25**; in rift valleys, **115–16**; discovery in Himalayas, **208–9**; time scale, *213*, 214, **215**; in orogenic cycle, 239; defined, 378; mentioned, 91, 247, 273, 288, 300
Isotherm, 379
Isotope, 379
Israel, 100
Isua Quartzite (Greenland), 8, 357
Italy, **36**, 279, 297; rotation, 113
Iturralde-Vinent, Manuel A., *281*, **282**
Ivanenko, D. D., 327
Ivrea zone (Italian Alps), 248

Jadeite, 129, 264, 326, 379
Jaeger, John Conrad, 106, 117
Jamaica: rotation, 113
Jameson, Robert, 46, 57
Jammu, India, 285
Japan: trench, 178; Sea, 251, 256, 286, 295; paleogeographic ties with North America, **314–16**; mentioned, 171, 265, 359
Jasper, 379
Java, 279, 296, 298
Jaz Murian (Iran), 285
Jean Charcot (research vessel), 256
Jeans, James, 344f
Jeffreys, Harold, viii, 76, 95, 101, 166, 270, 328, 365
Jena, Thuringia, 143
Jewish calendar, 70
Joachimsthal, Bohemia, 48
Johannesburg, South Africa, 106, 121
Johnston, C. R., 158
Johnston Memorial Address, xi, 335
Joints: columnar, 52, 52f; genesis, **265**; scale, 211; defined, 379
Joksch, H. C., **147**
Joly, John, **74**

Jordan, Pascaul, **141**, 145, 327
Journal of Geophysics, 266
Journal of Geophysical Research, ix, 171
Journal of Geology, 85
Jove–Titius law, **349**
Judaism, 7; calendar, 70
Jupiter: Great Red Spot, **200**; evolution, 346, 347–49; embryo star, **348**; Jove–Titius law, **348–49**; mentioned, 17f, 23, 138, 200, 274
Jura Mountains (Switzerland), 65, 223, 247
Jurassic Period, 43, 99, 151, 153, 379; Tasmanian dolerite, **106**; paleomagnetic time scale, *124*; sea levels, *192*; global tumor, 275; paleoclimate, 297; trans-Pacific faunal links, 316

Kai, Indonesia, 298
Kamchatka, 302
Kane (U.S. Geological Survey), 292
Kansan glaciation, 68
Kant, Immanuel, 338, 355
Karoo dolerites, 106
Kashmir, 279
Katz, H. R., 176
Kazakhstan, 159
Kazwini, Mohammed, 35
Kefauver, Estes, 274
Keindl, Josef, **140**, 190
Kelvin, Lord (William Thomson): age of Earth and Sun, **71–75**, 84f; scorns geologists, **80**, 328; Olbers' paradox, 332; mentioned, 59, 62, 76, 102, 119, 133, 196, 365
Kennedy, W. Q., 276
Kentucky, 246–47
Kepler, Johannes, **22**, 61, 335
Kermadec trench, *127*, 318; anomaly, **178**, 187
Khain, Ye. E., 145, 325
Khatanga rift valley, *110*
Kidd, Reverend, 64
Kiev, USSR, 347
Kinetic energy, 379
King, Clarence, 74
King, Lester C., 97, 99, 117, **162**
King, Philip B., **246**, *246*, **312**, *312*
Kingston, Ontario, 349
Kirillov, I. B., 137, 190, 327, 347
Kirwan (Irish geologist), 46
Kitikami Mountains (Japan), 314
Klepp, H. B., 349
Klippe, 249, 379
Knowable universe, 334–39 *passim*

Kolimskoye Mountains, 314
Köppen, W. P., 93, 122
Koran (al Quran), 10
Krassilov, V. A., **316**
Krebs, Wolfgang, **189**
Kreichgauer, K., 91
Krempf, Armand, **195**
Krikogen, 210, **235**, 247, 256, 317, 379
Krishnan, M. S., 121
Kropotkin, Peter, 62, **145**, 325
Krypton, 355
Kuh-i-Anguru salt dome (Iran), *227*
Kuhn, Bernard Friedrich, 65, 207
Kuhn, Thomas S., **120**
Kuhnian revolution, **120**, 128, 197
Kuroshio (Triassic Pacific land = North America?), **315**

Lactantius, Father Lucius, 9
Lactidurus, 14
LAGEOS (laser geodynamics satellite), *168*
Lamarck, Jean Baptiste de, 40, 60; evolution model, **41–42**
Laminar flow, **210**
Lamont Geophysical Observatory, 121, 143
Land hemisphere, 278, *278*, **287**
Language barriers of English readers, 126, 139, 143, **207**
Laplace, Pierre-Simon, 355
Lapland, 24
Lapparent, Valérie de, 338
Laramide orogeny, 292
Large numbers of Dirac, 141
Larson, R. L., 159
Laue, Max von, 356
Laurasia, 98 f, 105, 121, 162, 307, 379
Lau Ridge, 318
Lavoisier, Antoine, 133
Law of superposition, 379
Lawsonite, 198, 379
Lead isotope age measurement, **78**
Leclerc, Louis, 72
Lees, George M., viii, 104, 126, *227*
Lehigh University, 118
Lehmann, Johann Gottlieb, 48 f
Leibnitz, Gottfried Wilhelm von, 72
Lemaître, Abbé Georges, 329 f, 347
Lemurian compression, 155
Leonardo da Vinci, xiii, 36, 364
Lepëkhin, Ivan, 64
Lesser Himalaya, *248*, 249
Letavin, A. I., 137, 145
Lewis, Tayler, 11–12

Lhwyd, Edward, 34
Libavius, Andreas, 37
Ligurian orocline, *108*, 282; sphenochasm, *108*, 282
Lindemann, B., **139**, 140, 190
Lineation, 211, 240, 379
Linnaeus (Carl von Linné): evolution concept, **42**, 60
Lisbon scarp, *110*
Lister, Martin, 39
Lithosphere, *127*, 380; evolution of, 204, 265, 269, 273, **306–21**
Liverpool University, *152*
Llandovery Epoch, 158
Lodestone, **27**
Lomblem, Indonesia, 298
Lomboc, Indonesia, 298
Lomonosov, Mikhail Vasilievich, **45**, 58
London, 28, 140; University of, 356
Longwell, Chester R., ix, **118**
Lord Howe Ridge, 318
Los Angeles, California, 147
Louisiana State University, 118
Lowman, Paul D., **270**
Loxodrome, 166
Luc, Jean André de, 12, 65
Lucipara, Indonesia, 298
Luther, Martin, 11
Lyell, Charles, 45 f, 57 f, 66, 75; uniformitarianism, **58–62**, 71
Lynx-stones, 34
Lyons Conglomerate, **160**
Lystrosaurus, 161

Ma, Ting Ying, **195**
Maack, Reinhardt, 106, 117, 121
McDonough, T. R., 348
McDougall, Ian, 78
McDowall, A. N., *229*
McElhinny, Michael W., 84, *153*, **162**, **301**
McGill University, 76, 118
Mach, Ernst, 340
Mach's principle, 141, 340, 357, 380
Mackin, J. Hoover, **85**
Maclure (American geologist), 46
McRea, William H., 330, 344
Madagascar, 156, 159; dispersion from, 289
Madras, India, 159
Mafic rocks, 380
Magellanic Cloud, 337
Magma, 380
Magnes (Turkey), 27
Magnetic field: declination, 24, **27–29**,

30, 380; variation, 27f, 30; isopors, 30; storms, 30; poles, 30–31; westward drift, 30–31, **201**; cause of, 31; Curie point, 31; north-south reversals, **32**, *124*, 145; spreading-ridge profiles, *125*, *242*. See also Paleomagnetism
Maillet, Benoît de, 46, 48
Maine, 144, 247
Malaspina Glacier, *234*, **234–35**, 236, 239
Malaysia, 95, 102, 159, 279; tectonic rotation, 113, **301**
Malvoisine, France, 23
Manhattan Beach, California, 347
Manilya River (West Australia), 160
Manissa, Turkey, 27
Mantle, of Earth, 200, **269f**, 271, 274, 297, 380
Mantovani, R., 90, **92**, 138
Map projections, *165*, **165–66**
Marianas, 305
Mar-Istar, 17
Markl, R. G., 159
Marlow, Michael S., **188**
Mars: great rift zone, 199, 307; compared to Proterozoic Earth, 313, *314*; Titius–Bode law, 348–49; mentioned, 17f
Marsupials: trans-Pacific migration, 316–17
Marvin Glacier (Alaska), *234*
Mashkel depression (Pakistan), 285
Mason, Ronald G., 123
Massachusetts, 106, 144; Institute of Technology, 148; University of, 289
Massif, 380
Materialism, 7, 357
Matter: initial inheritance axiom, 330, 347, 364; synthesis, 344, 346; multiplicative creation, 345; additive creation, 345; imagined solidity, **356–57**
Matthews, Drummond H., 123, 128
Matthews, T. D., 359
Maupertuis, Louis Moreau de, **24**
Mauretania, 145
Maxwell, James Clark, 335; elasticoviscuous deformation, 335, 337
Medici, Cosimo de', 36
Mediterranean: oroclines, *108*, 113, 278, 282; widened, *108*, 140, 257, 285, 319; sinistral shear, *108*, 257; remnant of Tethys, *108*, 306; mentioned, *9*, 158, 287, 301
Medlicott, H. B., 98
Megashears, viii, *185f*, 214, **249**, 380

Melanesia: tectonic evolution, ix
Malanesian plateau, **318**
Melange, 211, **250**, 380
Melloni, Macedonio, 105–6
Menard fracture zone, 302
Mendel, Johann Gregor, 120
Mendeleev, Dmitri, 352
Mendocino fracture zone, 132
Mendocino orocline, **289**, *290*, 296; paleomagnetic confirmation, 113, 291
Mercanton, Paul, 106
Mercati, Michel, 37
Mercator projection, *112*, *165*, **165–66**, **288**
Mercury: phases, 21, 360; polygonal fractures, **199**, 312, *313*; Titius–Bode law, 348–49; mentioned, 18
Mericourt, Peter de, 27–28
Meridional extension, **282–85**, *283f*
Meservey, R., **148**
Meson, 380
Mesozoic Era: mass extinctions, 60; dispersion of Pangaea, 99, 148; accelerating expansion, 143, **153**, 258, 319f, 416; Tethys paleogeography, **178–79**, 307; first great ocean, 199, 315, **321**; Gondwana bulge, **200**; trans-Pacific faunal links, 316–17; defined, 380; mentioned, 164, 246f, 249, 255, 273
Metamorphism, 381
Metastable, 326, 381
Meteorites, 325f, 346
Meter: length standard, 22
Métherie, de la (French geologist), 46
Metonic cycle, 381
Mexico, 140, 156, 279, 297, 306, 316; Gulf of, 297, *299*
Meyerhoff, Arthur A., **160**, 161
Meyerhoff, Howard A., 161
Miami, University of, 68
Micro-continent "ferries," 315
Micrometeorites, 326
Midland Valley of Scotland, 54
Mid-ocean ridges, 91f, **116f**, 143, **167**, 172; genesis, **114–16**, 119, 121, 129, 144, 214; genetically similar to orogens, *242*, **243–45**, *244*
Migmatite, *232*, 245, 252, 381
Milan, Italy, 36
Milankovitch, M., 95
Milanovsky, E. E., 137, 145, 149, 325
Milky Way galaxy, 145, 329, 337, 342
Milne, Edward Arthur, 355
Mindel glaciation, 67
Minorca, 282

Miocene Epoch: age of, 74; paleomagnetic time scale, *124*; Indian Ocean, *180*; in Himalayas, *180*, 307; in Iran, 227; defined, 381; mentioned, 153, 256, 297
Miogeosyncline, **239**, 240f, 246ff, 249, 282, 312. *See also* Geosyncline
Missing links, 43
Mitchum, R. M., **191**
Miyabe, N., *265*
Moe, Indonesia, 298
Moho (Mohorovičić discontinuity), 238, **241**, *242f*, 255, 381
Mohorovičić, Andrija, 241
Molasse, 381
Møller, C., 339
Mollweide projection, 165, *165*
Molucca Sea, 295
Moment of inertia of Earth, **196**, **200**, 275, 381; continents exceed oceans, **274–75**, **286f**
Mongolia, 244, 316
Monotremes, 317
Montana, 291; Montana–Florida lineament, **292**, *293*, 299; University of, 311
Mont Blanc (France), 65, 248
Montreal, Canada, 76
Moon: ejected from Earth, 9, 92, 95; tidal drag, *18*, 208, 275; capture theory, 92; cosmic origin, 145; bulge, **201**; compared with primitive Earth, *313*; Titius–Bode law, 348
Moraine, 65, 381
Morley, Lawrence W., 123, 126
Moro, Lazzaro, 48
Morocco, 279, 282
Morris, W. A., **183**
Moscow, USSR: Society of Naturalists, 137; State University, 137, 145; conference on Earth expansion and pulsation, 137, 149; mentioned, 143, 347
Moses, 64, 70
Motagua fault (Guatemala), 279
Mountain building, 381. *See also* Orogenesis
Muehlberger, William R., *231*, 232
Murchison, Roderick Impey, 46, 66
Myers, L. S., 325
Mylonite, 211, 381

Nanking, China, 293
Nanyan, China, 293
Napier, W. M., 325
Naples Academy of Natural Sciences, 36
Nappe: distance of flow, **240–44**; rate of flow, **241**; defined, 382; mentioned, 214, 235, 240, 386
Narlikar, J. V., 141
NASA, 76, 149, **167–70**, *168f*, 171, 266, 348, 350
Nature, 349f
Naumann, C. F., 207
Nebraska, 102
Nebraskan glaciation, 68
Neiman, V. B., 137, 190, 327
Nelson, T. H., 274
Neogene, *153*, 382
Neolithic Papuans, **3**
Neptune, 199; Titius–Bode law, 348–49
Neptunism, **45–52**, 382
Netherlands, 128, 298, 301
Neuchâtel, Switzerland, 118, 207; Lake, 248
Neumayr, M., 306
Neutron star, 326, 382
Nevada, 297
Nevin, C. M., 224
New Caledonia: Permian link with Mexico, 315; Archeofijia (Pacific land = South America), **315**
Newcastle upon Tyne, University of, 100, 140, 147
New England University (Australia), 232
Newfoundland, 113, 279
New global tectonics, ix, x, 133
New Guinea, viii; tectonics, xi, 113, 178, 279, 282, 302, 318; natives, **3–5**; earthquake, 60; fossils, 158, 316; Tethys, 307; mentioned, 120, 155
Newton, Isaac: gravitation law, xi, 20–24 *passim*, 83, 209f, 327, 334, 337, 341ff, 364; oblate Earth, **23**, 83; feud with Hooke, 38, 328; viscous flow, 334; action at a distance, 341; corpuscular theory, 360; mentioned, 81, 120, 133, 364f
Newton–Hubble law, 336, 338, 341, 350
Newton–Hubble null, 336f, *337*, 343–50 *passim*
New York: city, 91, 97; state, 150, 247; City University of, 339
New Zealand: Tethys, 178; relation with Australia, 178, 302, 317–18; Alpine fault, 287, 295, 305; mentioned, 92, 102, 155, 187
Nice-Villefranche colloquium, 143
Nile delta, 70
Ninety-east Ridge, 287
Noah's flood, 37f, **64**, 90f, 188–89
Norfolk Ridge, 318

Norman, Robert, 28
North America, rotation; moved away from South America, 156, **285**; NASA measurements, **171–72**, **274**, *275*, 286, 309, **319**; polar wander, 266, 274, 286; Tethyan torsion, 278; circum-Pacific torsion, 287, 289–92; tectonic grain, *310*, 311; paleogeographic ties with Asia, **314–15**; mentioned, 246
North Carolina: University of, 118; paleogeography, 183; Ordovician diapir, 246
Northumberland, 54
Northwest Cape (Australia), 296
Norway, 232
Notation, fogs of, **357–60**
Noumea, New Caledonia, 315
Novaya Zemlya oroclines, *110*
Nuclear fission, 346; natural explosion, 61; perturbation of Earth axis, 274
Nucleon, 345f, 382
Nuées ardentes, 252, 382
Null universe, xi, 330, **338–42**
Nürnberg, Bavaria, 28
Nutation cycle, 62

Obliquity to the ecliptic, **201**, 382
Ob sphenochasm, 192
Ocean-floor spreading: history of concept, **114–17**, *115*, 138, 143–44; confirmation, **123–26**, *125*; plate tectonics model, 126–29, *127*; asymmetry, **132**, **256**; NASA data, 170; Africa enigma, **174–76**, *175*; circumcontinental, *175*, **257**; computer reconstruction, 267–69, *320*; global integration, 302; mentioned, ix, 181, 206, 318
Oceans: young age, 121, 147, **186**; sediments, 129, 177, 187; secular change in area, **143**; creep east, **273**, 286; oceanic hemisphere, *277f*, 278, **287**; no great oceans before Mesozoic, **321**
Oceanus (mythology), 306
Octantal rotations, **273**, *273*
Ohio, 102, 196; State University, 118
Okhotsk, Sea of, 286, 295, 302
Olbers, Matthäus Wilhelm, 332
Olbers' paradox, 332
Olbers' window, 333, *337*
Oligocene Epoch, *124*, *180*, 256, 382
Olivine, 382
Olivi of Cremona, 37
Ollier, C. D., 232–33, *233*
Oman Mountains, 285
Omar, Caliph, 10, 13
Ontogeny recapitulates phylogeny, 8

Ooze, 382
Ophiolites, **164**, 172, 197, 239, 246, 248f, 282, 382
Öpik, Armin A., 117
Orbigny, Alcide d', 59
Orbitolina, 160
Ordos basin (China), 285
Ordovician Period: faunas, 158, **181**, 183–84, 196; glaciation, **181**, *183*, 184; sea levels, *192*; diapirism *246*, 246–47; defined, 382
Oregon, 292
Orion constellation ("Year Man"), 4
Oriskany Sandstone, 223
Orocline: history of concept, viii, ix, xi, *107–12*, **113–14**, 121; Z- and S-torsions, **279**; defined, 382; mentioned, 126, 214
Orogenesis: crustal extension not contraction, **187–89**, **205–6**, 325; paradigm, **237–45**, *238*; involves crustal widening, 241; genetic identity with spreading ridges, 243–44, *244*, **257**; asymmetry, **245**; rate increasing, 245, **319**; cycles, 245, 325; axial zone, 246, 248, 253, 282; diapiric, **247**, *264*, 325; arcs convex eastward, 256, 273; cuts whole mantle, 257; Proterozoic pattern, *264*; equatorial orogenesis dominant, **320**; mentioned, 92, 113
Orotath, 298, 382
Orthographic projection, **165**, *165*
Orthomorphic projections, 166
Owen, Hugh G., 148, **166**, 172
Owen, Richard, **90**
Oxford University, 64, 73
Oyashio (Triassic Pacific land = North America?), **315**

Pacific Ocean: scar from Moon separation, 91f, 95; East Asian small seas, 91f, 236, **254–56**, 317; perimeter paradox, **148**, **156–58**, *157*, 172; alleged subduction, **178**; plate tectonic map, 178; asymmetry, 272–73, 317; Proterozoic birth, 307, **310–14**, **311–14**; lost continents, 315; trans-Pacific faunal links, **316–17**; Cretaceous growth, 317–19; Hawaii hotspot, **319**; Paleozoic, **345**
Padua Academy, 36
Pain, C. F., 232–33, *233*
Paired metamorphic belts, 197
Palawan, 298
Paleocene Epoch, *124*, 176, *180*, 382
Paleogene Epoch, *227*, 256, 287, 292, 382

Paleomagnetism: tectonic rotations, ix, 91, **113–14**, 301, **319**; development of concepts, 27, **105–6**; confirms Pangaea, 106, **121**; polarity reversals, **122–23**, *124–25*, 186; ocean-floor stripe pattern, **123–26**, *185*, *242*, 267; chronology, **125**; Arctic convergence, **150–51**; paleopole overshoot, **151–53**, *152–53*, 172; Archean shields not composite, **163**; Appalachian sinistral shear, **183**, *183*; Earth radius, **192–95**, *193*; polar wander, **266**, 274–75, *275*; Cretaceous poles, 288; tectonic terranes, 291–92; paleopole, 382; mentioned, 117f

Paleozoic Era: glaciation, 67, 98; age of, 74; paleogeographic enigma, **158–62**; Tethys, **178–79**, 307; Iapetus, **180–84**; Tethyan analog, **307**; Andes mid-continental, **315**; defined, 382; mentioned, 117, 164, 250, 273

Palissy, Bernard, 36f
Pallas, 275
Pamphili, Eusebius, 70
Panama, 302, 310
Pangaea, **93**; Mesozoic dispersion, 99; mantle bulge, **200**; dispersion, **288**; defined, 382; mentioned, 61, 90, 97f, 113, 117, 121, 140, 143, 153–62 *passim*, 167, 179, 257, 279, 285, 287, 306, 314
P'an Kiang, 159
Pannonian basin, 236
Pantar, Indonesia, 298
Papua, 232
Papuans, **3–7**
Parallax, 21, 360, 383
Parameter, 383
Parana Basin (Brazil), 106
Paris: observatory, 23; botanic gardens, 39; basin, 40; Polytéchnique, 76; Sorbonne, 76; mentioned, 23, 90
Parker, T. J., *229*
Parkinson, Wilfred Dudley, **169–70**, 266, 342, **353–55**, 364
Parmenides, 14
Pautot, Guy, 256
Pekeris, C. L., 100
Pelycosaurs, 43
Peneplain, 383
Pennsylvania, 144–45, 182, 247
Penzias, Arno A., 332
Perfect Cosmological Principle, 342, 350
Peridotite, 252, 383
Period, 383
Periodic table of elements, 352, 359

Permian Period: Indian Talchir beds, 67; faunas, 117, 158, **160**, 305, 315; equator, **150**, **315**; Earth radius, 153, **167**, 192–93; India-Australia paleogeography, *161*; Tethys, 178; sea levels, *192*; climate zones, 297; defined, 383; mentioned, 98, 192, 247
Perraudin, Jean-Pierre, **65**
Perry, Kenneth, 160, **167**, 172, 267–69, *268*, 270, 310, 319, *320*
Persian Gulf, 285
Peru, 24, 168, 171, 176
Peru–Chile trench; subduction anomaly, **176–77**, 187
Petrology, 383
Phase changes in crust and mantle, **101**, **200**, 242, 253, 318, 326f, 383
Philippines, 302, 316; Sea, 286, 295
Phillips, John, **73**
Picard, Jean, 23
Piedmont (U.S. Atlantic), 246, 383
Pickering, W. H., 90
Pigram, C. J., 279, 301
Placet, R. P. François, 90
Planck, Max, **121**
Planets: asymmetric expansion, 200; axial obliquity, 200; evolution, 346. *See also individual planets by name*
Plate tectonics: model, *127*, 153; Tethys contraction, 160, 178, 282, 307; Zodiac fan, **183–86**; blueschists, **198**; contrast with expansion model, **205**, 261, 307; mentioned, 144, 166
Platform, 383
Plato, 7, 11–23 *passim*, 27, 58, 342; Plato's problem, 17, 19
Platonic Academy, 36
Playfair, John, **47**, 58, 64, 71
Pleiades constellation ("Year Woman"), 4
Pleistocene Epoch: glaciation, 67; glacial stages, **67–69**; polar overshoot, 153, *153*; subduction, 177; defined, 383
Pliny the Elder, 15, 51
Pliocene Epoch, *124*, 153, *153*, *180*, 383; in Himalayas, *180*, 307
Pluto, 348–49
Pluton: mentioned, 214, 245, 252, 383
Plutonism, **51–57**, 383
Po basin, 236
Poise, 384
Polar wander path, **266**, 384
Polygon hierarchy of global expansion, **261–65**; movement radially outward, **266–69**, 273; movement factors, **305**; diapirism, 311
Pope, Alexander, **179**

Popes: Clement IV, 11; Sylvester II, 11; Paul III, 36
Porphyroid, 384
Porphyry, 49, 51f, 384
Port Moresby, Papua, x
Portsoy granite, 54
Potassium age measurement, **78**, 123
Potential energy, 384
Powell, John Wesley, 224
Precambrian Eon, 142, 148; natural nuclear explosion, 61
Precession of the equinoxes, 17, *18*, 19, **31**, 62, 201, 384
Price, Raymond A., 311
Primal heat of Earth, 62, 72
Primeval atom, 329
Primordial lead, 384
Primum mobile, 18f, 27f
Princeton University, ix, 100, 118ff, 327
Procrustes, 187
Productus, 160
Projections, *see* Map projections
Prokaryote, 384
Proterozoic Eon: no ophiolites or flysch, **164**; Grenville front, 182, *182*; Tethyan analog, 309; defined, 384; mentioned, 163, 199
Provinciality index, 85
Pterodactyls, 43
Pterosaurs, 43
Ptolemy (Claudius Ptolemaeus) 10, *19*, **20f**, 21, 27, 164, 358
Ptolemy Philometor, 21
Puerto Rico, 113, **282**
Pulsar, 384
Pulsation theory, 62, 137, **145**, 149, 384
Purcell series (Canada), 311–12, *312*
Pure numbers, 141
Pyrenees Mountains, 23f
Pyroxene, 198, 245, 384
Pythagoras: harmony of the spheres, 18, 21f, **80–81**, *81*, 83, 359; spherical earth, 9, 14, 173, 363ff

Quantum fluctuation, 331, 344, 353 *passim*, 385
Quark, 357, 384f
Quartz, 262, 326f, 356–57
Quasar, 332, 355
Quaternary Period, *125*, *152f*, 184, 192, 385
Quebec: Ordovician fan, 247
Queensland University, 305
Queen's University (Ontario), 349

Radioactivity: age of Earth, **75–78**, discovery, **76**; half lives, **77**; elements used for age determination, **78**; heat source for mantle convection, **100**
Radio galaxies, 332
Raff, Arthur D., 123
Raleigh, Lord, **76**
Ramberg, Hans, 128, **207**, 208
Ramirez, J. E., 176
Ramsay, Andrew, 67
Reade, T. Mellard, **73–74**
Reading University, 327
Redlichia fauna, 158
Red Sea, 114, 140, 257, 285
Red shift, 332f, 385
Relativity, 341, 353f, 385
Relaxation time, 215, 318, 326, 385
Renaissance: birth of, in Italy, 36
Reuss (Bohemia), 46
Revolution: in geotectonics, three phases, 105; planets, 385. *See also* Kuhnian revolution
Reyer, E., **207–8**
Reynolds, Osborne, **211**
Rhaetic Epoch, 106, 181, 385
Rheid, viii, ix, 385
Rheopexy, 386
Rhine graben, *110*
Rhombochasm, 132, 318, 386
Rhône valley (France), 65f
Rhumb line, 166
Rhynchocephalians, 43
Richer, Jean, 23
Riff Mountains (Morocco), 279; orocline, *108*, **282**
Rift valleys, genesis, **114–17**, *115*, 214, 239; mid-ocean, 121, 123; African, 128, 139, 175, 256f, 264; Appalachian, 144, 181; Andes, 176–77
Rigidity, 335, 387
Rig-Veda, 16
Riss glaciation, 67
Rocky Mountain Front, 172
Rocky Mountain Trench, 132, 186, 289, *290*, 291, *291*, 386
Rodgers, John, 118
Roman Island, Indonesia, 298
Rome, Italy, 36, 357; Roman empire, 45
Rome Formation, 223
Roots of mountains, 24–25, 235, 240, **241–45**, 386
Ross, Donald C., **245**
Ross Sea, 287, 295
Rotation, global, **309–14**; exponential increase, *275*, **319**; satellites, 386
Roti Island, Indonesia, 298
Royal Australasian College of Physicians, x

Index 409

Royal Society, 23, 29, 37, 104, 133, 209; founded, 36
Royal Society of Edinburgh, 47, 206
Royal Society of New South Wales, x
Royal Society of Sciences (Uppsala), 36
Royal Society of Tasmania, ix, xi, 114, 126, 144
Rubey, William, **191**
Rubidium age measurement, **77–78**
Ruiz, Carlos, 176
Rumania, 226
Runcorn, Keith, 100, 106, 119
Russia, 188, 207f, 325ff
Rutherford, Ernest, **76**
Ruwenzori horst, 256

Sagitov, R. M., 327
Sagittarius fan, 184–85, *185*
St. Amand, Pierre, 273
St. Helena, 29
St. Petersburg, Russia, 137; Academy, 36
Sakhalin, 305
Salajar Island, Indonesia, 380
Salisbury Crags, 54
Salt domes, 207, **225–31**, *226f*, *229f*, 387; overhang, 226; rim syncline, 228, 247; glacier analog, **234**, *234*
Salt "glacier," 227, *227*, 256, 387
Samoa, 279, 318
San Andreas fault, 132, 276, **289**, *290*, 292, 297
San Diego State University, *310*
Sandwich Islands, 90
San Jose State University, 315
Sanskrit philosophy, 15, 16–17
Santiago University, 176
Sardinia, 279, 282; rotation, 113
Saturn, 17f, 274; Titius–Bode law, 348–49
Saussure, Horace Venedicte de, 65
Savu Island, Indonesia, 298
Saxony, 46–53 *passim*
Scale of geotectonic phenomena, viii, x, xii, 82, 206, **209–16**, *212f*, 274, 328
Scandinavia, 181, 237
Schardt, H., 207
Scheibenberg Hill, 52f
Scheuchzer, Johan Jacob, **38–39**
Schimper, Karl, 66
Schist, 387
Schistes lustrés, 248
Schistophile minerals, **198**
Schistosity: indicates solid-state flow, **100**; mentioned, 211
Schlotheim (German geologist), 46
Schmidt, P. W., **266**, 270

Scholl, David W., **177**, **184–86**, *185f*, **188**
Schrödinger, Erwin, 331
Schuchert, Charles, 62, 117
Schwartz, E. H. L., 91
Schwarzschild, Karl, 352
Schwarzschild radius, 330, 386
Science, 271
Scientific American, 332
Scotia: arc, 286, 288; Sea, 289
Scotland, 181, 276
Scripps Institution of Oceanography, 123
Scrope, G. P., 207
Sea-floor spreading, *see* Ocean-floor spreading
Seamap fan, 184, *185*
Sears, James W., 311
Sedgwick, Adam, 58, 64
Sediment, 386
Seismic stratigraphy, **191**
Selenopetis fauna, 158
Senonian Epoch, 228, 386
Sensitivity, 386
Seram, 298; Tectonic rotation, 113, **301**
Serbian Royal Academy, 271
Serpentinites, 249, 252, 387
Seward, A. C., 97
Seward glacier (Alaska), *234*
Shakespeare, William, 81
Shan Mountains (Thailand), 159
Shantung Peninsula, 293
Shark teeth, *34*, 35, 38
Sharp, R. P., *234*
Shatsky rhombochasm, 318, *319*
Shear folding, 224, 387
Shears: sinistral, dextral, conjugate, **276**, *276*
Shen Kua, 35
Shenyang (Mukden), Manchuria, 293
Shields, Oakley, **316–17**
Shneiderov, A. J., 325
Shrinking Earth theory, 126
Sial, 387
Siberia, 151, 156, 158, 179, 192, 274; Cretaceous pole, 288; sphenochasm, 302, 321
Sicily, 279; orocline, *108*
Sierra Nevada, *244*, 245
Sigmoidal, 387
Signorini (Italian geologist), 207
Silesia, 50
Sill, 387
Sillimanite, 245, 387
Silurian Period, 158, 196, 387; sea levels, *192*
Sima, 387
Similar folding, **217–24**, *218*, *220*, 387

Simpson, George Gaylord, 84, **101–2**, 133, 365
Simpson, Robert W., 292
Singapore, 274
Sinistral, 387
Sinitsyn, V. M., 143
Sinusoidal projection, 165, *165*
Siva (Hindu god), 249
Siwalik hills (India), 249
Skerl, J. G. A., 89
Slate, 387
Smith, William, **40**, 67
Snider-Pellegrini, Antonio, **90**
Society of Economic Paleontologists and Mineralogists, 119
Soddy, Frederick, 76
Solar system, 347, 364; angular momentum enigma solved, 349
Solenhofen lithographic limestone, 43
Solid-state flow, **100–101**
Solomon Islands, 26, 282
Solomon Sea, 286
Solstice, 387
Solvay Conference (Brussels, 1958), 339
South Africa, Astronomical Society of, 140
South America: Africa match, **89–90**, *94*, *96*, **97–98**, **102–4**, *103*, 166, 268; Pacific rim, 156; NASA geodetic data, 170–71; subduction myth, 175–77, 226–27; Coriolis effect, 271; separation from North America, 285; general dispersion, 287; Pacific landmass source of sediments, **315**; trans-Pacific faunal links, **315**, **316–17**
South China Sea, 256, 286, 295, 298
Southeast Asia, 156, 158, **298–301**, *300*
Southern Ocean, 155 f, 170, 257
South Orkney Islands, 288
South Sandwich trench, *175*, 176
Spain, 164, 178, 307; rotation, *108*, 113, 279, 282
Sphenochasm, *107ff*, 132, 192, 282, 302, 387
Spreading ridges, *see* Ocean-floor spreading
Sri Lanka, 26, 159
Stadion: modern equivalent, 15
Stait, B. A., 158
Stanford University, 206
Stanley Memorial Lecture, x
Stanovoy Mountains, 314
Staub, Rudolf, 96, 100, 117
Steady state concept, 342 ff, 350
Steiner, J., 145, 325
Steinmann, G., 207

Steno (Niels Stenson), **38**, 48, 83, 357
Stereographic projection, *94*, 95, *96*, *107*, *154*, *165*, **165–66**, 319, *320*, 387
Stevens, Calvin, 315
Stevenson, Andrew J., *184*, **184–86**
Stewart, A. D., 327
Stishovite, 262, 326 f, 387
Stöklin, Jovan, **179**
Stone Age, vii, 3–5
Stothers, Richard, 344 f, 350
Strakhov, N. M., 142, 191
Stratum, 387
Stress-difference, 198, 208, 253, 387
Stromatoporoids, 158
Strutt, Robert John (Lord Raleigh), **76**
Subduction myth, x, 50, 126, 128, 155, 163, **170**, **174–89**, 197–99, 217, 388
Suess, Edouard, 99, 121, 179, 306
Sula Island, Indonesia, **279**
Sulawesi (Celebes), 279, 298, 302; Sea, 295
Sulu Sea, 286, 295, 298
Sumatra, 158, 279, 296, 298
Sumba Island, Indonesia, 298
Sumbawa Island, Indonesia, 298
Sun, 145, 200, 274 f; sunspots, 30; source of heat, **71–72**, **75**; luminosity change, 349 f
Sunda: arc, 257, 302; orocline, 296, 318
Supandjono, J. B., 279
Superposition, law of, 61, 83, 388
Sussex University, 344
Sutton, John, 172
Suture, 388
Swansea, University College of, 60, 307
Sweden, 128, 207, 349
Switzerland, 100, 332, 347; Swiss Plain, 65, **248**
Sychev, P. M., **188**
Symposia: on continental drift (New York, 1926), viii; Hobart (1956), ix, xi, *115*, 126, 143, 289; on Earth expansion (Sydney, 1981), xi, 316; on syntaphral tectonics (Hobart, 1963), 238; on Southwest Pacific geodynamics (Noumea, 1976), 315
Syncline, 388
Synclinorium, 249, 388
Syneresis, 388
Syntaphral tectonics, 238
Syrene, 15, *16*
Syzygy, 388

Tadzhikistan, 10
Talchir boulder bed, 67, *161*
Tallahassee, Florida, 188

Index

T'ang dynasty in China, 22, 27
Tanimbar Island, 298
Tanner, William F., **176–77**, **188**, **254**, 274, 285
Tao-Yuan, 35
Tasman fracture zone, 305, 317
Tasmania: Hydroelectric Commission, 106; University of, 117; glaciation, 184; paleoclimate, 297; mentioned, 120
Tasmanide orogen: Tethyan analog, **309**; equatorial, **320**; mentioned, 316
Tasman Sea, 155, 178, 286, 302, 317
Taurus fan, 184, *185*
Taylor, Frank Bursley, 90, **91–92**
Tectonic diapirs, **385**
Teichert, Curt, 159f
Telliamed, *see* Maillet
Temple, P. G., 274
Tennessee, 247
Termier, Henri and Geneviève, 142, 191
Terranes, tectonic, 291–92
Tertiary Period: Paris basin, 40; orogenic belts, 92; ocean-floor spreading, 174–75, 201; Zodiac fan, 184–86; sea levels, *192*; Himalayas, 249f; East Asia, 254–55, 273; Cordilleran, 273, 292; defined, 388
Tethyan torsion, *111*, *157*, 180, 183, **250**, *277f*, **278–82**, 288, 293, 301, 319, *320*; history of concept, viii, 90, 138; equatorial, 156, 287–88; integration with conjugate torsion and spreading zones, 171, **297–305**, 320
Tethys: mythology, 306; forebears of, xii, **307–12**, *308*; paleogeography, **178–79**, 257, *308*, 314–15; transverse extension, **282–85**, *283f*; offset by circum-Pacific torsion, 295, 317; defined, **306–7**, 388; equatorial, **320**
Texas: University of, 85; Permian equator, **150**; Hawkins salt dome, *229*; Grand Saline salt dome, *231*; mentioned, 196, 297
Teyler Museum (Haarlem), 39
Thailand, 158f, 279, 296
Thales of Miletus, 27
Tharp fracture zone, 302
Thecodonts, 43
Theobald, W., 67
Therapsids, 43
Thermodynamics, 388
Thixotropy, 388
Tholeiite, 388
Thomas, G. E., 274
Thomas, George A., 117, 160

Thomson, William, *see* Kelvin
Thorium age measurement, 77
Three Kings Ridge, 318
Thresholds of physical behavior, **199**, **210–11**
Thrust, 388
Thunderstones, 34, *34*
Thuringia, 48, 53
Tibet, 159, 161f, *248*, 250, 316
Tidal friction of Moon, **196**, 388
Tien Shan Mountains, 314
Till (glacial), 63, 389; tillite, 305, 389
Time scale: paleomagnetic, **123–26**, *124*
Timor, 158, 298
Tintina Trench, 292
Titans (mythology), 306
Titius, Johann Daniell, 348; Titius–Bode law, 348; Jove–Titius law, 349
Tokyo, Japan, 171
Toledo, Spain, 36
Tonalite, 245, 250, 389
Tonga, 178, 251, 318; Tonga–Kermadec trench, *127*
Tonguestones, *34*, 35, 38
Tonkin, Gulf of, 293
Toronto, University of, 29, 118, 129, 144, 148
Torricelli Mountains, 60
Torsions: global, **297**, 301, 310; sinistral and dextral, **279**; defined, 389. *See also* Circum-Pacific torsion; Tethyan torsion
Transcurrent faults, 131f, 282, 301, 389. *See also* Megashears; Torsions
Transform faults, **129–31**, *130*, 302, 389
Transverse Ranges (California), 292, 297
Trapeznikov, Yu. A., 145
Trenches: relation to transforms, 31; defined, 126, *127*, 389; plate-tectonics model, *127*; extensional, 147, **187–88**; Peru–Chile, 176–77, *177*; no subduction, 177f, **186–87**
Triassic Period: faunas, 103, 158; paleoclimate, 151, 197; length of degree, 153; Tethys, 178; sea levels, *192*; Asia and North America linked, **314–15**; defined, 389; mentioned, 228
Trichinopoly, India, 159
Trilobites: Appalachian mismatch, **181–82**; mentioned, 117, 158, 389
Trinidad, 279
Trinity College (Dublin), 74
Triple points: South Atlantic, *175*, 176
Tryon, Edward P., 330, **339**, 343, 351, 364
Tuff, 389
Tukangbesi, Indonesia, 340

Tunnel diode, 331
Turbidity currents, 184, 238, 389
Turbulent flow, **210–11**
Turin Academy of Science, 207
Turkey, 158
Turkmen, 285
Turtle Island, Indonesia, 298
Tuscany, 36
Tyrrhenian Sea, 236; orocline and sphenochasm, *108*

Udintsov fracture zone, 302
Uebergangsgebirge period, 49
Ultradense matter, 326
Ultramafic rocks, 189, 389
Umaria, India, **160**, *161*
Umbgrove, J. F. H., 301
Unanfänglich period, 49
Uncaused cause, 330 f
Uncertainty principle, 331 f
Unconformity, 389
Undation theory, **207–8**
Underplating alleged, 177 f
Underthrust, 389
Uniformitarianism, **57–62**, 71 ff, 389
Unique events, **46**
Unique initial term, 330
Universe, 389; mass of, 141, 143, 343; restricted definition, *337*, 342; open or closed, 343. *See also* Null universe
Upper mantle; phase changes, **200**
Uppsala, Sweden, 208
Uranium, 346; age measurement, **77**
Uranus: Greek mythology, 306; Titius–Bode law, 348–49
Urbino, Italy, 235, 247, 256
U.S. Naval Observatory, 142
Ussher, Archbishop James, 64, **70–71**, 81, 85

Vablonovoy Mountains, 314
Vacquier, Victor, 123, 125
Vacuum fluctuation, *see* Quantum fluctuation
Vail, P. R., **191**
Valley and Ridge country, 222, 245, 247
Vallier, Tracy L., **177**, **184–86**, *185*
Van Bemmelen, R. W., 128, 208
Vancouver Island, 292
Van der Gracht, W. A. J. M., 97
Van Flandern, T. C., **142**
Van Hise, 224
Variation, *see under* Magnetic field
Vaselov, B. I., 327
Vatican fossil collection, 37
Vedas, 16–17

Velocity of light, 334–40 *passim*, 350–55 *passim*; pure number, 355, 364
Venetz, Ignace, **65**
Venezuela, *111*, 156, 279, *280*, **282**, 289, 301
Venice, 48
Vening Meinesz, Felix A., 100
Venus, 18, 20; phases, 21, 360; Titius–Bode law, 348–49
Verbeekinidae, 307
Vermont: Ordovician fan, 247
Verona, Italy, 36, 40
Vertical tectonics, 189, **205–8**
Very long base interferometry (VLBI), 168, *168*
Vesta, 275
Vesuvius, Mount (Italy), 51, 106
Vicente, J. C., 176
Victoria, Lake (Africa), 264
Vienna, Austria, 140, 306; Academy of Science, 271
Vine, Frederick J., **123–26**, *125*, 128
Viscosity, 209, **211**, **215**, 227, 239, 252, 335
Vital force, 363
Vladivostok, Siberia, 251
Vogel, Klaus, 140, 167, 172, **266**, *267f*, 270
Voigt, Johann Karl Wilhelm, 52, **53**
Voisey, Alan Heywood, 117
Voltaire, 24
Von Buch, Leopold, 46, 50, **52–53**, **56**, 59, 64, 66, 76, 133
Von Colberg, Loeffelholz, 91
Von Goethe, Johann, vii, 46
Von Helmholtz, Hermann, **72**, **77**
Von Humboldt, Baron, *see* Humboldt
Von Leibnitz, Gottfried, 72
Von Linné, Carl, **42**, 60
Von Zittel, Karl, 3, 5
Vosges Mountains (France), 247

Wadia, D. N., 97
Wales, 196
Walker, R. T., 143
Walker, W. J., 143
Wallace, Alfred Russel, 46; Wallace's Line, 301, 389
Walzer, Uwe, **143**, 269
Warring, C. B., 90
Washington (state), 292
Waterhouse, Bruce, **148**, 305
Wall, James, 47, 57
Wegener, Alfred Lothar: life and work, **92–95**, *94*; dogma conservatism, **121–22**; mentioned, viii, ix, 75,

Index 413

98–105 *passim*, 113, 121–22, 139ff, 153, *154*, 155, 206, 261, 306, 325, 363ff
Wegmann, Eugene, 118, 207
Weights and Measures, International Conference on, 355
Weizmann Institute (Israel), 100
Wells, John, **195–96**
Werdau, Saxony, 140
Werner, Abraham Gottlieb, **46–57**, 64, 133, 365
Wesleyan University, 118
Western Ontario University, 118
West Indies, 156, 278, 287, **297**
West Virginia, 223, 247
Wetar Island, Indonesia, 298
Wettstein, Henry, 90f
Wezel, Forese C., **235**, 247, **256**, 317
Wheeler, John Archibald, 331
Whewell, William, 57, 67
Whinstone, 54, 390
White dwarf, 326, 390
White Sea depression, *110*
Williams, George, **297**
Willis, Bailey, 133, 206, 224, 365
Wilson, Hugh, **188**
Wilson, J. Tuzo, **129–31**, *130*, **144**, 145, 365
Wilson, Robert W., 332
Wilson, Roderic L., *152f*
Wilson cycle, **144–45**, 390
Windermere Group, *312f*

Winterthur, Switzerland, 347
Wisconsinan glaciation, 68
Wise, Donald U., 289
Wobble of Earth axis, **201**, 390
Wöhler, Friedrich, 363
Wolf, Rudolf, 31
Wood, Robert Muir, **172–73**
Woodhouse, John, **269**
Woodward, John, 39
Würm glaciation, 67
Wyoming, 267

Xenophanes, 35, 63
X-ray diffraction, 356

Yale University, ix, 12, 62, **117**, 118f, 141, 144, 173, 206
Yarkovski, I. O., **137**, 190, 327
Yellow Sea, 286, 295
Yen Chen-Chang, 35
Ylem, 329
Young (writing in 1810), 90

Zagros Mountains, 92, 279
Zell, W., 177
Zeno, 15, 76
Zittel, Karl von, 3, 5
Zodiac fan anomaly, **184–86**, *185*, 288, 293
Zonenshayn, L. P., **244–45**
Zoroastrian god, 7
Zürich, Switzerland, 31, 38f, 91, 117

Library of Congress Cataloging-in-Publication Data

Carey, S. Warren (Samuel Warren)
Theories of the earth and universe.

Includes index.
1. Geology—Philosophy. 2. Expanding earth.
I. Title.
QE6.C37 1988   551.1   87-6433
ISBN 0-8047-1364-2 (alk. paper)